# CLIMATE CHANGE MITIGATION AND AGRICULTURE

This book reviews the state of agricultural climate change mitigation globally, with a focus on identifying the feasibility, opportunities and challenges for achieving mitigation among smallholder farmers. The purpose is ultimately to accelerate efforts towards mitigating land-based climate change. While much attention has been focused on forestry for its reputed cost-effectiveness, the agricultural sector contributes about 10–12% of emissions and has a large technical and economic potential for reducing greenhouse gases. The book does not dwell on the science of emissions reduction, as this is well covered elsewhere—rather, it focuses on the design and practical implementation of mitigation activities through changing farming systems.

The book includes chapters about experiences in developed countries, such as Canada and Australia, where these efforts also have lessons for mitigation options for smallholders in poorer nations, as well as industrialising countries such as Brazil and China. A wide range of agro-ecological zones and of aspects or types of farming, including livestock, crops, fertilizer use and agroforestry, as well as economics and finance, is included. The volume presents a synthesis of current knowledge and research activities on this emerging subject. Together the chapters capture an exciting period in the development of land-based climate change mitigation as attention is increasingly focused on agriculture's role in contributing to climate change.

**Eva 'Lini' Wollenberg** is a social scientist and faculty member at the University of Vermont, USA, coordinating mitigation research for the Climate Change, Agriculture and Food Security (CCAFS) Research Program of the CGIAR.

**Alison Nihart** is a masters student in Natural Resources at the University of Vermont, USA, and a research assistant for the Climate Change, Agriculture and Food Security (CCAFS) Research Program of the CGIAR.

**Marja-Liisa Tapio-Biström** is an agricultural economist and coordinator of the FAO Mitigation of Climate Change in Agriculture (MICCA) program.

**Maryanne Grieg-Gran** is Principal Researcher in the Sustainable Markets Group at the International Institute for Environment and Development (IIED), London, UK.

**About CTA**

The Technical Centre for Agricultural and Rural Cooperation (CTA) is a joint international institution of the African, Caribbean and Pacific (ACP) Group of States and the European Union (EU). CTA operates under the framework of the Cotonou Agreement and is funded by the EU. Its mission is to advance food and nutritional security, increase prosperity and encourage sound natural resource management. It does this by providing access to information and knowledge, facilitating policy dialogue and strengthening the capacity of agricultural and rural development institutions and communities in ACP countries.

For more information on CTA visit, www.cta.int

# CLIMATE CHANGE MITIGATION AND AGRICULTURE

*Edited by Eva Wollenberg,*
*Alison Nihart,*
*Marja-Liisa Tapio-Biström and*
*Maryanne Grieg-Gran*

publishing for a sustainable future
London • New York

First published 2012
by Earthscan
2 Park Square, Milton Park, Abingdon, Oxon OX14 4RN

Simultaneously published in the USA and Canada
by Routledge
711 Third Avenue, New York, NY 10017

*Routledge is an imprint of the Taylor & Francis Group, an informa business*

The CGIAR Research Program on Climate Change, Agriculture and
Food Security (CCAFS) is a strategic partnership of the CGIAR and
the Earth System Science Partnership (ESSP). CCAFS brings
together the world's best researchers in agricultural science,
development research, climate science and Earth System science, to
identify and address the most important interactions, synergies and
tradeoffs between climate change, agriculture and food security. The
CGIAR Lead Center of the program is the International Center for
Tropical Agriculture (CIAT) in Cali, Colombia. For more
information, visit www.ccafs.cgiar.org.

*British Library Cataloguing in Publication Data*
A catalogue record for this book is available from the British Library

*Library of Congress Cataloging in Publication Data*
A catalog record for this book has been requested

ISBN: 978-1-84971-392-4 (hbk)
ISBN: 978-1-84971-393-1 (pbk)
ISBN: 978-0-203-14451-0 (ebk)

Typeset in Baskerville
by Wearset Ltd, Boldon, Tyne and Wear

MIX
Paper from
responsible sources
FSC
www.fsc.org    FSC® C004839

Printed and bound in Great Britain by the MPG Books Group

# CONTENTS

CONTENTS

CONTENTS

CONTENTS

CONTENTS

# LIST OF ILLUSTRATIONS

## Figures

## Tables

# CONTRIBUTORS

**Ane Alencar** is a geographer, holds a Ph.D. in Forest Resources and Conservation from the University of Florida, and is currently leading projects related to mapping the dynamics of deforestation and forest fires in the Amazon and the analysis of impacts of infrastructure works on the region.

**Candra Wirawan Arief** is the Field Coordinator for the Northern Sumatra Program at Conservation International Indonesia, based in Medan.

**Oliver Bach**, Senior Manager for Standards and Policy of the Sustainable Agriculture Network (SAN) Secretariat, has worked with the Rainforest Alliance and the SAN since 1997, and since 2006 has overseen the SAN standards development and review process, developing criteria for oil palm and sugarcane plantations, a standard for sustainable cattle ranching, and policies on ecosystem conservation and prohibited pesticides.

**Christopher M. Bacon** is a political ecologist and an Assistant Professor in Environmental Politics and Policy at the Environmental Studies Institute, Santa Clara University, where his research explores how environmental governance relates to sustainable livelihoods, biodiversity conservation, and environmental justice in the Americas.

**Justin Baker** is a Research Associate in Economic Analysis at the Nicholas Institute for Environmental Policy Solutions at Duke University, with broad research interests in natural resource management under reduced-carbon policies.

**Tobias Bandel** is managing partner at Soil and More International and is mainly responsible for project development and implementation, carbon credits (Clean Development Mechanism), certification systems, production protocols, and commercial activities.

**Ellysar Baroudy** manages the World Bank's BioCarbon Fund, has an environment background and has worked on sustainable development issues for 15 years.

**David Neil Bird (Neil)** joined Joanneum Research in August 2005, where his main areas of interest and work are: evaluation of emissions and emission reductions in land-based projects (AFOLU) and bioenergy systems including research on the influence of changes in surface albedo on environmental benefits; and development of unique CDM A/R projects including methodological development where necessary.

**Frederick Boltz**, Senior Vice President for Global Initiatives and Climate Change Lead for Conservation International, is a natural resource economist with over 20 years' experience in tropical conservation and development.

**Jan Börner** is an agricultural and natural resources economist with research and field experience in South and Central America, Africa, and Europe currently working with the Center for International Forestry Research (CIFOR) in Brazil.

**Giacomo Branca** is an economist in the Agricultural Development Economics Division of the Food and Agriculture Organization (FAO), where his research interests include Payments for Environmental Services (PES), with a specific focus on sustainable agriculture, climate change mitigation, and food security.

**Louise Buck** is a faculty member in the Department of Natural Resources at Cornell University and Director of the Landscapes and Leaders Program of EcoAgriculture Partners in USA.

**Jeronim Capaldo** is an economist in the United Nations' Global Economic Monitoring Unit, and formerly worked at the Food and Agriculture Organization (FAO) focusing on models of adoption of sustainable agricultural practices, climate change risk, and food security.

**Isabel Castro** is a forest engineer and works as a research assistant at the Amazon Environmental Research Institute (IPAM).

**Sebastián Castro** is a Ph.D. student at the Plant & Soil Science Department of the University of Vermont, and a Principal Investigator at the Earthwatch Institute, Costa Rica, and his interests include agroecology, nutrient cycles in agro-ecosystems, and global change ecology.

**David Chadwick** is an agricultural and environmental scientist who has worked at Rothamsted Research, North Wyke for over 17 years, where his research focuses on improving the understanding of nutrient utilization of organic amendments in agricultural systems and the processes and pathways resulting in diffuse pollution to both air and water.

**Declan Conway** is a professor at the School of International Development/ Water Security Centre, University of East Anglia, where his research is on climate and society, with a focus on water resources and development.

**Wilson Lopes da Silva** works at the Instrumentation Center of the Brazilian Agricultural Research Corporation (EMBRAPA) and is specialized in environmental chemistry applied in agriculture, mainly spectroscopic studies of organic matter from soil, water, and rural effluents.

**Allan Dale**, Adjunct Associate Professor at the Cairns Institute, James Cook University, is a researcher and policy advocate focused on improved regional development and natural resource governance.

**Stephanie Daniels** brings a background in smallholder sourcing and international development to the management of value chain partnerships at the Sustainable Food Laboratory, focused on large-scale impact on sustainable livelihoods and climate change in the food sector.

**Jan de Leeuw** is a systems ecologist and team leader of the Drylands Program at the International Livestock Research Institute (ILRI) in Nairobi, Kenya.

**Alessandro De Pinto** is a Research Fellow in the Environment and Production Technology Division of the International Food Policy Research Institute (IFRPI), where his work focuses on natural resource management and adaptation and mitigation to climate change.

**Luc Dendooven** is a research professor at the Research and Advanced Studies Center of the National Polytechnic Institute of Mexico (CINVESTAV) and specializes in soil-plant research with focus on C and N cycling.

**Brian Ford-Lloyd** is Professor of Plant Conservation Genetics at the School of Biosciences, University of Birmingham, UK.

**Ursula Georgeoglou** is a tropical forest ecologist and a research assistant at the Agroecology and Rural Livelihoods Group at the University of Vermont, where she is collaborating with research and projects that envision a sustainable approach among conservation, production, and livelihoods.

**James Gockowski** is Technical Program Leader for the Sustainable Tree Crops Program at the International Institute of Tropical Agriculture (IITA) and has been working on rural development and environmental issues along the forest–agriculture interface of West and Central Africa since 1980.

**Gianluca Gondolini** is the Sustainable Agriculture Projects Manager – Latin America at the Rainforest Alliance, where he manages the implementation of agricultural mitigation projects and adoption of sustainable agriculture practices amongst Latin American producers, as well as promotion of certification according the Sustainable Agriculture Network Standard.

**Katie Goodall** is a Ph.D. student at the University of Vermont researching coffee cooperatives and their influence on bird distribution and conservation in the north central mountains of Nicaragua.

**Bram Govaerts**, a cropping systems agronomist and soil scientist, is head of the Mexico-based Conservation Agriculture Program at the International Maize and Wheat Improvement Center (CIMMYT) and leader of the Take it to the Farmer project within MasAgro.

**Maryanne Grieg-Gran** is Principal Researcher in the Sustainable Markets Group at the International Institute for Environment and Development (IIED), London.

**Hal Hamilton** is founder and co-director of the Sustainable Food Laboratory, a consortium of businesses, not-for-profits, and universities collaborating on sustainability innovations in mainstream food supply.

**Celia Harvey** is an ecologist with expertise in Conservation Biology, Agroforestry and Climate Change Mitigation, and currently serves as Vice President of Global Change and Ecosystem Services at Conservation International.

**Tanja Havemann**, Director of BeyondCarbon GmbH, has been involved with investments, project management and policy development related to carbon, with an emphasis on sustainable land management, since 2003.

**Petr Havlík** is an agricultural economist and global change modeler, and has a joint appointment between the International Institute for Advanced Systems Analysis in Laxenburg, Austria, and the International Livestock Research Institute (ILRI) in Nairobi, Kenya.

**Jeff Hayward**, Director of the Climate Program at the Rainforest Alliance, has 20 years' experience in natural resource management and currently works to address the climate mitigation and adaptation potential in agriculture and forestry through policy, finance, and capacity building.

**Mario Herrero** is a senior scientist and team leader at the International Livestock Research Institute (ILRI) in Nairobi, Kenya, where he specializes in agro-ecological systems analysis.

**Jonathan Hillier** holds a Ph.D. in mathematics from the University of York, UK, has worked in several areas of biological modeling from plant–pest–predator interactions, plant architectural modeling and remote sensing, and currently works on the modeling of greenhouse gas emissions of food and energy crop production.

**Terry Hills** is an Advisor on Climate Change Adaptation for Conservation International, based in Brisbane, Australia.

**Neeta Hooda** has been engaged in climate change and land-use issues for over 15 years and is now involved in REDD+ and in piloting agriculture mitigation projects that use sustainable land management practices through engagement of local communities.

**Fredrich Kahrl** is a Ph.D. candidate in the Energy and Resources Group at the University of California, Berkeley, where his research focuses on the engineering, economic, and environmental science dimensions of energy and agricultural systems.

**Peter Läderach** is the leader of the Decision and Policy (DAPA) program for Central America and the Caribbean at the International Center for Tropical Agriculture (CIAT), Managua, Nicaragua, where he is focusing on the development of methodologies to quantify the impact of climate change on agriculture, livelihoods, and economies.

**Bo Lager** holds an M.Sc. in Agroforestry and is Program Director for Vi Agroforestry, East Africa, where the focus of his work is on agroforestry, sustainable agriculture, carbon finance, and climate change.

**Jean Lee** is currently pursuing her Ph.D. at the University of Vermont and focusing on pro-poor strategies of carbon mitigation programs and their impacts on livelihoods in sub-Saharan Africa.

**Leslie Lipper** is a senior environmental economist in the Agricultural Development Economics Division of the Food and Agriculture Organization (FAO), where she leads a research program on the economics of sustainable agricultural systems that includes analysis of climate-smart agricultural systems and the institutions needed to support them.

**Yuelai Lu** is specialized in the socio-economic aspects of sustainable soil management and is currently coordinating the UK–China Sustainable Agriculture Innovation Network (SAIN), based at University of East Anglia.

**Saodah Lubis** is the team leader for the Coffee and Carbon Program in Aceh for Conservation International Indonesia, based in Banda Aceh.

**Marilia Magalhaes** is a consultant at the Global Facility for Disaster Reduction and Recovery (GFDRR/World Bank) where she works with disaster operations portfolio management and conducts research on Climate Change Adaptation.

**Daniella Malin** is a project leader with the Sustainable Food Laboratory, where she directs the Food Lab's climate initiative, is a project director with the Market Mechanisms for Agricultural Greenhouse Gases (M-AGG) and serves as facilitator for the climate metric of the Stewardship Index for Specialty Crops.

**Wendy Mann** is working with the Agricultural Development Economics Division of the Food and Agriculture Organization (FAO) on international and national policy to enable national action on climate-smart agriculture.

**Victor Mares** is a systems agronomist working for the Crop Management and Production Systems Division at the International Potato Center (CIP).

**Ladislau Martin Neto** is Coordinator of EMBRAPA Virtual Laboratory Abroad (Labex) USA, where his research focus is on soil carbon dynamics and reactivity using different analytical techniques.

**Simone Mazer** is a forest engineer and works as a research assistant at the Amazon Environmental Research Institute (IPAM).

**Nancy McCarthy** is an economist and lawyer based at LEAD Analytics; her research interests include economics of climate change in agricultural, natural resource management and risk, and legal and regulatory frameworks for sustainable development.

**V. Ernesto Méndez** is an Associate Professor of Agroecology and Environmental Studies at the University of Vermont, where his research and teaching efforts focus on developing and applying interdisciplinary approaches that analyze interactions between agriculture, livelihoods, and environmental conservation in tropical and temperate rural landscapes.

**Monica Mezzalama** is a plant pathologist at the International Maize and Wheat Improvement Center (CIMMYT), Mexico.

**Debora M.B.P. Milori** works for the Brazilian Agricultural Research Corporation (EMBRAPA) and is an expert in the use of photonics as applied in agriculture and environment sciences.

**Peter A. Minang** is the Global Coordinator of the Alternatives to Slash-and-Burn (ASB) Partnership for the Tropical Forest Margins at the World Agroforestry Centre (ICRAF) in Nairobi, Kenya.

**Mark Moroge** is an Associate of the Rainforest Alliance's Climate Program, which develops and implements cross-cutting systems, projects and tools to build upon the Rainforest Alliance's existing sustainable forestry, agriculture and tourism activities by giving them an explicit focus on climate change adaptation and mitigation.

**Katlyn S. Morris** is a Ph.D. candidate in the Department of Plant & Soil Science at the University of Vermont, where she researches and teaches the environmental, social, and economic aspects of food crop production in rural areas and seasonal food insecurity among coffee-producing households.

**William (Bill) B. Morris** is a specialist in land change science and remote sensing of agriculture, working with the Agroecology and Rural Livelihoods Group at the University of Vermont.

**Paulo Moutinho**, Executive Director of the Amazon Environmental Research Institute (IPAM), is a biologist, holds a Ph.D. in Ecology, and is one of the authors of the proposal for compensating reduced deforestation that provides a basis for Reducing Emissions from Deforestation and forest Degradation (REDD).

**Brian Murray** is Director for Economic Analysis at the Nicholas Institute for Environmental Policy Solutions, and Research Professor of Environmental Economics at Duke University, and is widely recognized for his work on the economics of climate change policy, including the design of cap-and-trade policy elements to address cost containment and inclusion of offsets from traditionally uncapped sectors such as forestry and agriculture.

**Constance L. Neely** is an agro-ecologist and senior consultant on climate change, land, livestock and livelihoods with the Food and Agriculture Organization of the United Nations in Rome, Italy, and the World Agroforestry Centre (ICRAF) in Nairobi, Kenya.

**Christine Negra**, Program Director at the H. John Heinz III Center for Science, Economics and the Environment (USA), works with partners in government, academia, industry, and NGOs to bring technical analyses and consensus-based solutions into global change policy dialogues.

**Ken Newcombe** is the CEO of C-Quest Capital, a carbon finance business dedicated to originating and developing high-quality emissions reduction around the world.

**Jess Newman** is a Transatlantic Fellow at the Ecologic Institute in Berlin, Germany, and graduated from Harvard College with a B.A. in Environmental Science and Public Policy in 2010, where her senior thesis analyzed design elements correlated with participant livelihood benefits in pro-poor carbon payment projects in East Africa.

**Alison Nihart** is a masters student in Natural Resources at the University of Vermont, USA, and a Research Assistant for the Climate Change, Agriculture and Food Security (CCAFS) Research Program of the CGIAR.

**David Norse** is Emeritus Professor of Environmental Management, UCL Environment Institute, and is an active researcher on policy solutions to food security and agro-environmental problems in China and globally.

**Ylva Nyberg** is a Ph.D. candidate in Crop Production Ecology, with a specialization in agroforestry, at the Swedish University of Agricultural Sciences (SLU), and also serves as a Technical Advisor to Vi Agroforestry, East Africa.

**Lydia Olander** is the Director of the Ecosystem Services Program at the Nicholas Institute for Environmental Policy Solutions at Duke University and in this role is directing the Technical Working Group on Agricultural Greenhouse Gases and the National Ecosystem Services Partnership.

**Erika de Paula Pedro Pinto** is an ecologist and works as a project manager at the Amazon Environmental Research Institute (IPAM).

**Adolfo Posadas** works for the International Potato Center (CIP), where his focus is applied physics, mathematical modeling and non-destructive techniques applied to agriculture and soils studies.

**Professor David Powlson** works on the cycling of carbon and nitrogen in agricultural soils and their interactions with the wider environment, is a Lawes Trust Senior Fellow at Rothamsted Research, UK, and a Visiting Professor in Soil Science in the School of Human and Environmental Science, University of Reading, UK.

**Noel Preece** is a terrestrial ecologist, has worked in many of Australia's main biomes, and has a three-decade long career as a land manager, environmental consultant, ecological and carbon researcher, and entrepreneur.

**Roberto Quiroz** is a biophysical scientist at the International Potato Center (CIP) based in Lima, Peru, where he conducts research on methods to assess the impact of climate variability and change on agriculture and the feedback of agricultural practices to the environment.

**Ricardo Rettmann** is an environmental manager and works as a research assistant at the Amazon Environmental Research Institute (IPAM).

**Claudia Ringler** is a Senior Research Fellow in the Environment and Production Technology Division of the International Food Policy Research Institute (IFPRI), where she works on natural resource management, in particular, water resources management and adaptation to and mitigation of climate change.

**Ricardo Romero-Perezgrovas** is a Ph.D. student at K.U.Leuven, Belgium, where his research focuses on the impact of Conservation Agriculture on socio-economic system parameters with the Mexico-based Conservation Agriculture Program at the International Maize and Wheat Improvement Center (CIMMYT), Mexico.

**Mariana Rufino** is a livestock systems scientist at the International Livestock Research Institute (ILRI) in Nairobi, Kenya.

**Ken D. Sayre** is a systems agronomist specialized in conservation agriculture-based cropping systems at the International Maize and Wheat Improvement Center (CIMMYT), Mexico.

**Sara J. Scherr**, the Founder, President and CEO of EcoAgriculture Partners, is an agricultural and natural resource economist specializing in land and forest management policy in tropical developing countries.

**Götz Schroth** is Global Director Agroforestry of Mars Incorporated and Collaborating Professor of Agroecology at the Federal University of Western Pará, Santarém, Brazil, where his interests are in sustainable land use and biodiversity of agricultural landscapes.

**Matthias Seebauer** joined UNIQUE forestry consultants in 2008, where he has been involved in the development of a number of agricultural carbon finance and other land-based carbon finance projects.

**Christina Seeberg-Elverfeldt** is the Climate Change and Environment Officer for the Mitigation of Climate Change in Agriculture (MICCA) Programme at FAO, where she focuses particularly on the pilot projects for incorporating mitigation into smallholder farming systems.

**Aline Segnini** is doing post-doctoral work at the Brazilian Agricultural Research Corporation (EMBRAPA) and specializes in soil carbon sequestration and spectroscopy applied in soils.

**Bambi Semroc** works with the private sector on improved business practices and currently serves as Senior Director of the Conservation Tools for Business program within the Center for Environmental Leadership in Business at Conservation International.

**Seth Shames** is the Coordinator of EcoAgriculture Partners' Payments for Ecosystem Services program, which supports emerging markets for ecosystem services from carbon, watersheds and biodiversity in agricultural landscapes through research, policy development and project design.

**Pete Smith** is the Royal Society-Wolfson Professor of Soils & Global Change at the University of Aberdeen, UK, where his research focuses on modeling interactions between soils, agriculture, and climate change.

**Kai Sonder** is an Agronomist and Head of the GIS Unit at the International Maize and Wheat Improvement Center (CIMMYT), Mexico.

**Lucimar Souza** is a psychologist and works as Altamira's local coordinator at the Amazon Environmental Research Institute (IPAM).

**Osvaldo Stella Martins** is a Mechanical Engineer, holds a Ph.D. in Ecology, and is the Project Coordinator for the Climate Change Program at the Amazon Environmental Research Institute (IPAM).

**Naomi Swickard**, AFOLU Manager, oversees the AFOLU program for the Verified Carbon Standard (VCS) Association and is based in Bangkok, Thailand.

**Marja-Liisa Tapio-Biström** is an agricultural economist and coordinator of the FAO Mitigation of Climate Change in Agriculture (MICCA) program.

**Timm Tennigkeit** is managing partner at UNIQUE forestry consultants, Germany, and works on natural resource management and climate finance-related topics in Africa and Asia.

**Philip K. Thornton**, Ph.D., is a research team leader of the Climate Change, Agriculture and Food Security (CCAFS) program at the International Livestock Research Institute (ILRI) in Nairobi, Kenya, where he specializes in agricultural systems analysis.

**Piet van Asten** is a systems agronomist at International Institute of Tropical Agriculture (IITA) in Uganda, where he has been working on sustainable intensification of perennial-based cropping systems in Africa's humid areas for the past eight years.

**Meine van Noordwijk** is the Chief Science Advisor of the World Agroforestry Center (ICRAF) and is based in Bogor, Indonesia.

**Penny van Oosterzee** is an environmental scientist, science writer, and adjunct senior research fellow at James Cook University currently working on a model for aggregating terrestrial carbon across landscapes based on landholder land-use activities.

**Nele Verhulst** is a Ph.D. student at K.U.Leuven, Belgium, and doing her research with the Mexico-based Conservation Agriculture Program at the International Maize and Wheat Improvement Center (CIMMYT), Mexico.

**Christof Walter**, Sustainable Agriculture Manager at Unilever, has over 10 years' experience in sustainable agriculture, corporate sustainability and sustainable sourcing and works on the relationship between farming and climate change, including small-scale farmers in supply chains, and agri-sourcing opportunities.

**Fred Werneck** is manager of responsAbility Ventures I, the social venture capital fund of Swiss-based responsAbility Social Investments.

**Tim Wheeler** is a Professor at the Walker Institute for Climate System Research, University of Reading, UK, with a specialty in crops and climate change.

**Johannes Woelcke** is a Senior Economist in the Africa Region, Agriculture and Rural Development, at the World Bank.

**Eva "Lini" Wollenberg** is a social scientist and faculty member at the University of Vermont, USA, coordinating mitigation research for the Climate Change, Agriculture and Food Security (CCAFS) Research Program of the CGIAR.

**Sven Wunder** is the Principal Economist and Head of the Brazil office of the Center for International Forestry Research (CIFOR), where his main work areas are payments for environmental services, deforestation and forest–poverty linkages.

**Galdino Xavier** is an agronomy engineer and works as a research assistant at the Amazon Environmental Research Institute (IPAM).

**Giuliana Zanchi** currently works at Joanneum Research, in Graz, Austria, where her main areas of expertise are: assessment and modeling of carbon emissions and removals from biomass and soil in forest and agricultural ecosystems; and assessment of GHG impacts of bioenergy use, specifically of forest-based bioenergy.

**Yatziri Zepeda**, formerly Ecosystem Services Coordinator at Conservation International in Chiapas, Mexico, is an environmental economist working on payment for ecosystem services and sustainable land-use initiatives.

**Olaf Zerbock**, formerly Advisor for Climate Change Initiatives at Conservation International, is currently a Forestry and Climate Change Specialist with the US Agency for International Development.

# ACKNOWLEDGEMENTS

The editors would like to thank Sara Scherr, Jerry Nelson, Pramod Aggarwal, Sonja Vermeulen, Tim Hardwick, Paige Gunning, and the authors who contributed to the book for their guidance and assistance in preparing the volume.

# ABBREVIATIONS

A/R            Afforestation/Reforestation, usually referred to as A/R in
               the context of the Clean Development Mechanism
ABMS           Activity Baseline and Monitoring Survey or System
ABO            Area-based offset crediting scenario
AFOLU          Agriculture, Forestry and Other Land Use
AGB            Above-ground biomass
ALM            Agricultural Land Management (see also SALM)
APEX           Agricultural Policy Extender model
ArcGIS         Geographic Information System tool
ASB            Alternatives to Slash-and-Burn Partnership, a global
               partnership devoted to research on the tropical forest
               margins
BAU            Business as usual (scenario). Also referred to as
               "baseline," this is the land use and greenhouse gas
               emissions profile for a mitigation project area prior to
               intervention, which serves as a benchmark to measure the
               impact of a mitigation project.
BNDES          Brazilian Development Bank
C              Carbon
CA             Conservation agriculture
CAADP          Comprehensive African Agricultural Development
CAR            Climate Action Reserve
CARE           A humanitarian organization focused on fighting global
               poverty.
CC             Climate change
CCAFS          The CGIAR Research Program on Climate Change,
               Agriculture and Food Security, a strategic partnership
               between the Consultative Group on International
               Agricultural Research (CGIAR) and the Earth System
               Science Partnership (ESSP).
CCBA           Climate, Community and Biodiversity Alliance, a
               partnership of international NGOs and research

institutes seeking to promote integrated solutions to land management around the world.

CCX        Chicago Climate Exchange, a voluntary, legally binding greenhouse gas reduction and trading system for emission sources and offset projects in North America and Brazil until mid-2010.

CDM        Clean Development Mechanism of the Kyoto Protocol. Mechanism for "a country with an emission-reduction or emission-limitation commitment under the Kyoto Protocol (Annex B Party) to implement an emission-reduction project in developing countries." These projects can earn the country certified emission reduction (CER) credits, which can be used to meet their Kyoto targets. CDM is designed to motivate emissions reductions through sustainable development while providing multiple options for how developed countries may meet their targets.

CENTURY    Biogeochemical model of plant–soil nutrient cycling used to simulate carbon and nutrient dynamics.

CEPF       Critical Ecosystem Partnership Fund

CES        Compensation for ecosystem services

CFI        Carbon Financial Instrument. Unit of trade on the Chicago Climate Exchange

CGIAR      Consultative Group on International Agricultural Research

$CH_4$        Methane

CHNS       Analysis to determine levels of carbon, hydrogen, nitrogen and sulfur in soil

CI         Conservation International

CIAT       International Center for Tropical Agriculture (member of CGIAR)

CIFOR      Center for International Forestry Research (member of CGIAR)

CIMMYT     International Maize and Wheat Improvement Centre (member of CGIAR)

CIP        International Potato Center (member of CGIAR)

$CO_2$        Carbon dioxide

$CO_2e$       Carbon dioxide equivalent. A measure of greenhouse gases expressed as the equivalent concentration as $CO_2$.

COMACO     Community Markets for Conservation

COMET-FARM  Whole farm accounting version of COMET-VR

COMET-VR   Voluntary Reporting of Greenhouse Gases-Carbon Management Evaluation Tool

COP        Conference of the Parties, governing body of the UNFCCC. A number after the acronym indicates the "nth" conference for that convention.

| | |
|---|---|
| CS | Carbon stock |
| CT | Conservation tillage or conventional tillage |
| CTB | Conventionally tilled raised beds |
| CWR | Conservation of crop wild relative |
| DAYCENT | Daily Century model. Daily time-step version of the CENTURY biogeochemical model |
| Defra | UK Department of Environment, Food and Rural Affairs |
| DfID | Department for International Development (United Kingdom) |
| DNDC | DeNitrification-DeComposition model. Computer simulation model of carbon and nitrogen biogeochemistry in agro-ecosystems, used to predict crop growth, soil temperature and moisture regimes, soil carbon dynamics, nitrogen leaching, and emissions of trace gases. |
| DOE | United States Department of Energy |
| DPM | Decomposable plant material |
| DRIFT | Diffuse Reflectance Infrared Fourier Transform |
| EMBRAPA | Instrumentation Center of the Brazilian Agricultural Research Corporation |
| EPA | United States Environmental Protection Agency |
| EPIC | Erosion Productivity Impact Calculator model |
| EPR | Electron Paramagnetic Resonance |
| EQIP | Environmental Quality Incentives Program of the U.S. Natural Resources Defense Council |
| ERPA | Emission Reduction Purchase Agreement |
| ES | Ecosystem service |
| EU-ETS | European Union Emission Trading System |
| FAO | Food and Agriculture Organization of the United Nations |
| FASOMGHG | Forest and Agricultural Sector Optimization Model— Green House Gas version |
| FSA | Farmer Self Assessment |
| FTIR | Fourier Transform Infrared |
| FUNBIO | Brazilian National Biodiversity Fund |
| GACA | Global Agriculture Climate Assessment project |
| GHG | Greenhouse gas(es) |
| GLOBIOM | A global recursively dynamic partial equilibrium model integrating the agricultural, bioenergy and forestry sectors with the aim to give policy advice on global issues concerning land-use competition between the major land-based production sectors. |
| GM | Genetically modified |
| GPS | Global positioning system |
| GWP | Global warming potential |
| ha | hectare |

| | |
|---|---|
| HA | Humic acids |
| $H_{LIF}$ | Soil organic matter humification index |
| HM | Humic substances |
| ICRAF | The World Agroforestry Centre (member of CGIAR) |
| IFPRI | International Food Policy Research Institute (member of CGIAR) |
| IITA | International Institute for Tropical Agriculture (member of CGIAR) |
| INM | Improved Nitrogen Management |
| IPAM | Amazon Environmental Research Institute |
| IPCC | Intergovernmental Panel on Climate Change |
| IPCC default factors | Average rates of greenhouse gas flux |
| IPM | Integrated Pest Management |
| IT | Information technology |
| JVs | Joint ventures |
| K | Potassium |
| KACP | Kenya Agricultural Carbon Project (also sometimes called the Western Kenya Smallholder Agricultural Carbon Project, or Vi Agroforestry Project) |
| LIBS | Laser Induced Breakdown Spectroscopy |
| LIFS | Laser Induced Fluorescence |
| LULUCF | Land Use, Land-Use Change and Forestry |
| MFIs | Micro Finance Institutions |
| MICCA | Mitigation of Climate Change in Agriculture (MICCA) Project, a five-year project by FAO. |
| MIS | Management information system |
| MLRA | Major land resource area |
| Mn | Manganese |
| M-PESA | The Kenyan mobile phone-based money transfer system. M is for mobile and *pesa* is Swahili for money. |
| MRV | Monitoring, reporting and verification |
| N | Nitrogen |
| $N_2O$ | Nitrous oxide |
| Na | sodium |
| NOx | nitric oxide and nitrogen dioxide |
| NAMAs | Nationally Appropriate Mitigation Actions |
| NARES | National Agricultural Research and Extension Systems |
| NASA-CASA | United States National Aeronautics and Space Administration Carnegie-Ames-Stanford Approach |
| NCAS | National Carbon Accounting System (Australia) |
| NCAT | National Carbon Accounting Toolbox (Australia) |
| NDVI | Normalized difference vegetation index |
| NEPAD | The New Partnership for Africa's Development, a program of the African Union |

| | |
|---|---|
| NGGI | National Greenhouse Gas Inventory |
| NGO | Non-governmental organization |
| NIRS | Near Infrared Spectroscopy |
| NMR | Nuclear Magnetic Resonance |
| NRM | Natural resource management |
| OBO | Output-based offset approach to crediting greenhouse gas mitigation from agricultural and land-use activities through reduced emissions intensity |
| ODA | Overseas Development Assistance |
| OM | Organic matter |
| P | Phosphorus |
| PB | Permanent raised beds |
| PC, PCA | Principal component, Principal component analysis |
| PES | Payments for ecosystem services |
| PFTs | Plant Functional Types |
| PNNL | Pacific Northwest National Laboratory |
| QA/QC | Quality Assurance/Quality Control |
| RE | Regional ecosystem |
| REDD | Reducing Emissions from Deforestation and forest Degradation |
| REDD+ | REDD with addition of conservation, sustainable management of forests and enhancement of forest carbon stocks |
| REITs | Real estate investment trusts |
| RothC | Rothamsted C soil decomposition model, a biogeochemical soil carbon turnover model |
| RPM | Resistant plant material |
| SAIN | UK–China Sustainable Agriculture Innovation Network |
| SALM | Sustainable Agricultural Land Management (see also ALM, SLM) |
| SAN | Sustainable Agriculture Network |
| SBI | Subsidiary Body for Implementation under the UNFCCC |
| SBSTA | Subsidiary Body for Scientific and Technological Advice under the UNFCCC |
| SHIFT | Studies on Human Impact on Forests and Floodplains in the Tropics Program |
| Sida | Swedish International Development Cooperation Agency |
| SLATS | Statewide Landcover and Trees Study |
| SLM | Sustainable land management |
| SMEs | Small and medium-sized enterprises |
| SNR | Signal-to-noise ratio |
| SOC | Soil organic carbon |
| SOM | Soil organic matter |
| SRES | Special Report on Emissions Scenarios |
| SSA | Sub-Saharan Africa |

| | |
|---|---|
| STCP | The Sustainable Tree Crop Program |
| SWAT | Soil & Water Assessment Tool |
| t | tonnes |
| TIST | International Small Group and Treeplanting Program |
| TOC | Total organic carbon |
| UNFCCC | United Nations Framework Convention on Climate Change. The UNFCCC is an international treaty created in 1992 regarding how to decrease and cope with global warming and its effects. The 1997 Kyoto Protocol is an addition to the treaty with more legally binding components. |
| USD | United States dollar |
| USDA | United States Department of Agriculture |
| UV-vis | Ultraviolet-visible |
| VCS | Verified Carbon Standard, global standard for voluntary offset projects |
| VCU | Verified Carbon Unit of the VCS |
| VVB | Validation/verification body of the VCS |
| WCA | West and Central Africa |
| WKSACP | Western Kenyan Smallholder Agricultural Carbon Project (see also KACP) |
| WOCAT | World Overview of Conservation Technologies and Approaches |
| Zn | Zinc |
| ZT | Zero tillage |

# Part I

# INTRODUCTION

# 1

# CLIMATE CHANGE MITIGATION AND AGRICULTURE

## Designing projects and policies for smallholder farmers

*Eva Wollenberg, Marja-Liisa Tapio-Biström and
Maryanne Grieg-Gran*

## Introduction

Tackling climate change requires attention to agriculture. Raising crops and livestock directly contributes an estimated 10–12% of anthropogenic greenhouse gas emissions globally[1] (Smith *et al.* 2008), or about one-third of emissions if indirect impacts on land-use change and land degradation are considered (Smith *et al.* 2007a: 499). Agricultural emissions are also expected to increase in the next 30 years as population, income, agricultural intensification and diet preferences for meat and dairy increase, particularly in low- and middle-income countries.

In light of these trends and the need to feed a world of an estimated 9 billion people in 2050, some have called for the "redesign of the whole food system" to achieve sustainability and climate change mitigation (Foresight 2011). Fortunately, even modest shifts in agricultural practices can reduce net emissions.[2] The biophysical potential for mitigation in agriculture is comparable to that of the energy and industrial sectors and exceeds that of the transport sector (Smith *et al.* 2007a). As agriculture is a major driver of deforestation and other land-use change, measures that halt expansion of agriculture and retain carbon-rich forests or grasslands would reduce future emissions further. Such reductions require, however, meeting the demand for agricultural products from less land.

Agriculture thus holds enormous potential for mitigating climate change. Yet, how to achieve agricultural climate change mitigation in practice still raises major challenges, particularly for resource-poor and smallholder farmers in developing countries. The purpose of this book is to

3

review the state of knowledge of the practice of climate change mitigation in agriculture and provide guidance for its further development.

Chapters from more than 100 authors, involving many of the leading actors in agricultural climate change, examine what we know and what we still need to learn about agricultural mitigation. The authors review technical options, incentives, project design, measurement and monitoring systems, supporting policy measures and impacts. Together they demonstrate the scope for mitigation in different agricultural sectors, major mitigation initiatives and approaches for pursuing mitigation. The chapters indicate the urgency of identifying institutional mechanisms, incentives and policies that work in particular places for particular people. The chapters should provide a foundation for further action by identifying the experiences, methods and principles that can inform the design of agricultural mitigation.

We focus on the estimated 500–800 million resource-poor and smallholder farmers in developing countries for their important role in climate change mitigation in agriculture and the special challenges they face in pursuing their livelihoods and food security. The majority (74%) of agricultural emissions originate in low- and middle-income countries, where smallholders predominate. The emissions in these countries are expected to increase rapidly as a proportion of the global total—especially in Latin America and Africa where expanded fertilizer use, cattle raising and agriculture generally, are expected (Smith *et al.* 2007a). To have a major impact on global emissions it therefore will be necessary to reach smallholder farmers and not only target developed countries or large farmers and agri-business.

Yet smallholder farmers have particular needs for food and economic security, a low capacity to absorb risk, and poor access to finance and information. Mitigation measures need to give attention to the characteristics of these farmers to be relevant to their needs. Technical options backed by appropriate incentives that achieve mitigation, as well as food security and improved livelihoods will be necessary. Clear benefits for farmers need to be available from mitigation, whether in the form of improved productivity, positive social impacts such as empowerment, or payments. Also, mitigation measures must support farmers' adaptation to climate change if mitigation is to be achieved in the long term. Linking mitigation to rural development policies will be necessary to coordinate interventions and achieve the large scales necessary for impact on the climate.

Policies and projects therefore need to be designed with the aspirations and constraints of smallholders in mind. Given this goal, the book addresses three challenges. First, how can institutions and incentives best support smallholder farmers to participate in and benefit from agricultural mitigation? Second, what kinds of low-cost, rapid measurement and accounting approaches are needed to assess how smallholders' practices affect greenhouse gas (GHG) emissions? Third, what are the impacts and trade-offs of mitigation on smallholders' livelihoods and food security?

## Scope

Agriculture refers here to the cultivation of land for crops, the tending of livestock and agroforestry. Mitigation includes direct emissions from agriculture, as well as indirect emissions from land use change.

We intentionally focus on agriculture or the production of food at the farm level, rather than mitigation across the food value chain or food system. Life cycle analysis shows that for most food products, emissions from production at the field level greatly exceed those generated from processing, transportation and distribution or storage (Garnett 2008, Weber and Matthews 2008), especially in developing countries where the majority of production is consumed domestically.[3] We do not examine energy or biofuel aspects of mitigation, as these deserve extensive treatment that is beyond the scope of the current volume.

Although the book is focused on developing countries, we have included examples from developed countries where they provided lessons or models of potential interest to smallholder-based mitigation.

We refer henceforth to agricultural mitigation or mitigation for brevity. We use the term smallholder to describe smallholder farmers in developing countries unless noted otherwise.

## Organization of the book

The book is organized in four parts. Part I (Introduction) establishes the context for agricultural mitigation. The authors review available technical options and current projects, and discuss where opportunities exist to jointly pursue food security, livelihoods and adaptation to climate change.

Part II then explores the institutional arrangements and incentives to support mitigation systems. As we discuss below, the design of institutions and incentives is critical to make mitigation economically viable for farmers and encourage shifts in farming practices. This section includes chapters on smallholders' access to the carbon market and experiences to date. Shames *et al.* proposes design principles for improving smallholders' participation and benefits. Haywood *et al.* examine the role of certification as a driver for mitigation and identifies emerging best practices. The policies supporting agricultural mitigation and the lessons from experience in developing Reduced Emissions from Deforestation and Forest Degradation (REDD) are covered by Negra and Wollenberg.

The section also covers the financial arrangements and markets that support mitigation. East Africa is a leader in agricultural carbon market projects, hence the book draws heavily from examples in this region. The

Kenya Agricultural Carbon Project (KACP), in particular, has pioneered the sale of carbon credits from soil management in smallholder farming and is reviewed in the chapters by Lager and Nyaberg, Shames *et al.*, and Lee and Newman.

Due to the importance of land-use change and the conversion of forests for agriculture, Part II includes several chapters about institutional arrangements for managing across agriculture and forests. The chapters provide examples of landscape approaches from Brazil and Australia that capture land-use change and the interactions between agriculture and forests. The section concludes with the experiences of a network in China seeking to promote sustainable agriculture that also reduces impacts on the climate.

In Part III we highlight key developments and issues concerning the methods for quantifying and accounting for greenhouse gas emissions. The chapter by Hillier *et al.* introduces the Cool Farm Tool, an emissions calculator designed for farmers to help them understand how and where to best reduce emissions. Other chapters review methods for rapid and low-cost measurement, including soil carbon measurement, modeling and activity-based monitoring. 'Emissions intensity', or output-based measures for carbon crediting, are discussed in the chapter by Baker and Murray as a means for better capturing mitigation gains relative to agricultural productivity. The section concludes with an overview by Swickard and Nihart of the Verified Carbon Standard (VCS) as the current predominant platform for agricultural carbon credits, and the lessons VCS has learned.

Part IV gives examples of mitigation in different agricultural sectors. The chapters in this section reflect the state-of-the-art of knowledge in each sector and identify gaps in knowledge. They give special attention to trade-offs among outcomes.

In the remainder of this chapter we synthesize findings from the chapters related to: (1) technical options for mitigation; (2) institutional and economic options; and (3) measurement and accountability. We also introduce basic concepts and discuss the status of projects, policies and major initiatives to show the state of play for agricultural mitigation. We conclude with principles and next steps for the design of agricultural mitigation policies and projects for smallholders.

## Technical options

The potential for agricultural mitigation has been well summarized by the Intergovernmental Panel on Climate Change (Smith *et al.* 2007a). All three major greenhouse gases—carbon dioxide ($CO_2$), nitrous oxide ($N_2O$) and methane ($CH_4$)—play a role in agriculture. Nitrous oxide and methane contribute nearly all of agriculture's direct emissions (Figure 1.1 and Table 1.1). Indirectly, land-use conversion due to agriculture is a major source of carbon emissions. Eighty percent of direct and indirect emissions from agriculture are estimated to be related to livestock (Neely and Leeuw, this volume).

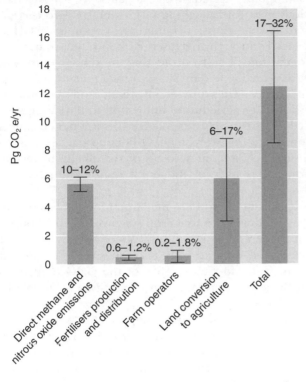

Total global contribution of agriculture to greenhouse gas emissions, including emissions derived from land-use changes. The overall contribution includes direct (methane and nitrous oxide gases from agriculture practices) and indirect (carbon dioxide from fossil fuel use and land conversion to agriculture). Percentages are relative to global greenhouse gas emissions

*Figure 1.1* Global contribution of agriculture to greenhouse gas emissions (source: Bellarby *et al.* 2008: 5).

*Table 1.1* Sources of direct and indirect agriculture greenhouse gases

| Sources of agriculture GHG | Million tonnes $CO_2e$ |
| --- | --- |
| Nitrous oxide from soils | 2,128 |
| Methane from cattle enteric fermentation | 1,792 |
| Biomass burning | 672 |
| Rice production | 616 |
| Manure | 413 |
| Fertilizer production | 616 |
| Irrigation | 369 |
| Farm machinery (seeding, tilling, spraying, harvest) | 158 |
| Pesticide production | 72 |
| Land conversion to agriculture | 5,900 |

*Source:* Bellarby (2008: 6).

From a technical standpoint, agricultural mitigation is feasible and significant. Three options exist for managing GHGs: (1) reduce current methane and nitrous oxide emissions; (2) increase removals of GHGs from the atmosphere, mostly through increased carbon sequestration; or (3) avoid creating new emissions, e.g., by protecting the storage of carbon in existing biomass or soil carbon, or by substituting renewable energy sources for petroleum energy. The sum of emissions and removals is referred to as net emissions. For agricultural mitigation to have a significant impact on GHGs, large areas need to be covered, as the net emissions reduced per unit of land area is relatively small. Other approaches to managing climate change, such as changing aerosols or the albedo of vegetative cover warrant further exploration (Ridgewell *et al.* 2009), but are beyond the scope of this book.

Reducing $CO_2$ emissions from the soil has the largest biophysical potential for mitigation among options in annual crop systems (i.e., not considering agroforestry or other perennial agricultural systems). Reducing net $CO_2$ emissions from cultivated organic soils or sequestering carbon in mineral soils is estimated to provide about 89% of the GHG reductions possible by 2030 (Smith *et al.* 2007a).[4] This is equivalent to about 4,895–5,380 Mt $CO_2e$/yr (Smith *et al.* 2008).

Maintaining and increasing existing stocks of sequestered carbon in the soil and perennial vegetation are therefore a high priority. The potential impact of carbon sequestration is almost 10 times more than the potential from reducing both $N_2O$ and $CH_4$ emissions, based on IPCC figures (Smith *et al.* 2007a). Carbon sequestration becomes limited over time, however, as soils reach carbon saturation points. Preventing loss of any newly sequestered carbon from fire, tillage and land conversion also will be necessary[5] (Neely and de Leeuw, this volume). Biochar may prove to be a more stable carbon sink that is less vulnerable to these perturbations and could be an alternative to burning plant material to maintain soil fertility (Lehmann *et al.* 2006).

Agroforestry and perennial systems sequester significant amounts of carbon in above-ground biomass. Estimates for the mitigation potential of agroforestry by 2040 are about 2,200 Mt $CO_2e$/yr (Verchot *et al.* 2007, based on IPCC 2000). Carbon storage in non-perennial systems is relatively insignificant. One concern is that agroforestry systems tend not to be included in accounts of agriculture, as they have been considered part of the forest sector in past IPCC reports. Comparisons of mitigation potential within the sector have been therefore difficult.

Similarly, an even larger, more immediate mitigation impact is possible by preventing land-use change whereby agriculture replaces more carbon-rich forests and grasslands. This will often have the added benefits of enhancing environmental services and biodiversity. The high potential impact and low cost of reducing forest conversion will provide the rationale for promotion of Reduced Emissions from Deforestation and Forest

Degradation (REDD) (Stern 2006). But the effectiveness of REDD will rely heavily on farmers' ability to increase productivity on existing agricultural land to meet the rising demand for food expected over the next 40 years (Grieg-Gran 2010), otherwise food security could be compromised. Managing emissions from land-use change therefore requires working with both the agriculture and forestry sectors. Sustainable intensification in agriculture can contribute to mitigation where it results in the sparing of forest or other carbon-rich lands.

The chapters in this book further investigate and summarize technical options for mitigation. Seeberg-Elverfeldt and Tapio-Biström provide a comprehensive review of agricultural mitigation interventions and what is known about their potential impacts. Other chapters review interventions specific to conservation agriculture (Verhulst *et al.*), sustainable land management (McCarthy *et al.*), rangeland management (Neely and de Leeuw), nitrogen fertilizers (Norse *et al.*), livestock (Herrero *et al.*), coffee systems (Mendez *et al.*), and other agroforestry systems (Semroc *et al.*).[6] Additional chapters review the possibilities for mitigation by reducing agricultural expansion to maintain forests in the Amazon (Stella *et al.*, Börner and Wunder), Ghana (Gockowski and Van Asten) and Cameroon (Minang and Van Noordwijk).

The reviews of technical options demonstrate the richness of mitigation options that are available and already used by many farmers. *The chapters suggest that future mitigation will depend more on whether practices are sufficiently attractive to large numbers of farmers rather than whether the mitigation is technically feasible.* Biophysical mitigation potentials are strongly mediated by economic factors and implementation conditions. There is a large gap between what is already technically feasible and what farmers are willing or able to do. Effective project and policy design is necessary to overcome many of the adoption barriers and opportunity costs of new practices to narrow this gap (see Wunder and Börner).

Yet the chapters also show the gap in knowledge around evidence for agricultural mitigation impacts and data to support mitigation decisions. Impacts of management practices on emissions levels need to be more widely and rigorously tested in practice. The emissions factors[7] associated with agricultural practices and resulting mitigation potentials are characterized by high variability among agro-ecosystems, measurement uncertainties and lack of data, especially from developing countries.

Much of the currently available data and experiences with mitigation are from temperate, developed regions or represent conditions in only a few tropical countries (e.g., Brazil). For example, the IPCC default value for indirect $N_2O$ emissions from N fertilizer production and use is probably far too low and should be replaced with improved national-level values (Norse *et al.*, this volume). Despite the wide agreement on the enormous mitigation potential of soil carbon sequestration, more evidence is needed to support its use in developing countries. The variability of soils and practices, high risk of shifts in practices and difficulty of attributing soil carbon changes to management changes when measurements need to be made over large areas and long time periods, pose significant challenges.

One additional issue, which is not yet well understood, is the likely impact of climate change on the potential for mitigation in agriculture. Increased drought, erratic rainfall and higher incidence of wild fires will negatively affect biomass production and subsequently levels of carbon sequestration. The uncertainty and risk associated with climate change should be built into the design of carbon projects and policy, and farming systems should be supported to increase their adaptive capacities.

There is considerable variation between and within countries in emissions sources and potentials. In Brazil, much attention has been given to integrated models for avoided deforestation and sustainable intensification of agriculture, as land-use change accounts for the majority of Brazil's emissions (Stella and Paula, this volume). In contrast, 83% of emissions in East and West Africa are from enteric fermentation in livestock (Brown *et al.* 2011).

Strategies for managing GHGs therefore need to take account of this variation and be designed to local agroclimatic and socio-economic conditions and farm practices. Understanding local social contexts helps projects better address issues related to land tenure, institutional arrangements, incentives and technical capacities (Semroc *et al.*, this volume). Reduced agricultural expansion is difficult to achieve where high population density, high demand for agricultural production and low biophysical, technical or social capacity to increase production exist (Palm *et al.* 2010). An adaptable menu of policy and project approaches will be necessary to enable flexibility in developing suitable interventions (Tschakert 2007).

Measures to avoid perverse outcomes based on subsequent increased pressure for use or emissions intensity also need to be considered. Increasing pasture productivity to sequester more carbon can increase stocking rates of livestock, resulting in overall increases in emissions, unless compensating measures are taken to reduce livestock-related emissions (Robbins 2011). While the increased efficiency of production per unit of land, or of emissions per unit of output is desirable, this should not occur at the expense of an increase in total emissions. Similarly, increased productivity on forest edges can encourage agricultural expansion (Ewers *et al.* 2009).

## Institutional and economic options

Institutional arrangements and economic incentives will be needed to increase mitigation in agriculture. The chapters in this book document different initiatives and potential drivers of mitigation. Here we review the policy context for mitigation in agriculture, possible incentives and the financing of mitigation. We also discuss institutional arrangements for effectively working with local communities, and expanding mitigation to large areas and numbers of farmers.

## Policy context

International negotiations on climate change related to land use and land-use change have drawn attention to the mitigation potential in agriculture and the possibility for carbon markets and finance to provide incentives. But the international policy framework has not prioritized agriculture thus far. Other initiatives in agriculture, led by non-governmental organizations (NGOs), the private sector, subnational entities, countries, donors and the World Bank, meanwhile have established precedents and expanded the base of knowledge that should pave the way for a future agricultural mitigation program under the United Nations Framework Convention on Climate Change (UNFCCC).

Scope for agricultural mitigation exists under the UNFCCC and Kyoto Protocol and agricultural targets are included for Annex I countries (see Box 1), but comprehensive management of agricultural emissions is not required and no mechanism for offsets exists in international policy. Since 2009, stimulated in part by progress on REDD+, some country negotiators have promoted more attention to agriculture in the UNFCCC. As with REDD, progress will depend on establishing a clear link between mitigation and economic development opportunities, clear technical foundations, demonstrable implementation mechanisms, finance and governance (Negra and Wollenberg, this volume).

---

**Box 1 The UNFCCC and agricultural mitigation**

The 1992 United Nations Framework Convention on Climate Change (UNFCCC) created the scope to include agriculture as an eligible mitigation action. The 1997 Kyoto Protocol for implementing the Convention then created targets for agricultural methane and nitrous oxide emissions in Annex I (developed) countries. The Protocol also made agricultural soil carbon a voluntary mitigation action for Annex I countries under the Land Use, Land Use Change and Forestry (LULUCF) category. In 2006 the IPCC established guidelines for GHG accounting in Agriculture, Forestry and Land Use (AFOLU), replacing former LULUCF guidelines.

The Kyoto Protocol also established the Clean Development Mechanism (CDM), a mechanism enabling developed countries to meet their emissions targets by buying offset credits from developing countries, and thereby contribute to sustainable development, which started registering projects in 2001.

The CDM gave little space to land-based mitigation activities because of concerns about the uncertainties in emission estimates and lack of permanence.

---

11

CDM was designed instead primarily with energy projects in mind. Forest-based mitigation was restricted to afforestation and reforestation and excluded the conservation of standing forest.

As a result, CDM accounting requirements are onerous to meet in agricultural systems and slow approvals have resulted in limited development of CDM methodologies in agriculture. As of mid-2011, 17 approved methodologies and 28 registered afforestation and reforestation (A/R) projects related to smallholder agriculture had been approved.[8] The CDM methodologies for agriculture have limited application, as the methods often require time-consuming field measurement and access to analytical laboratories. Lessons from CDM experiences nonetheless are relevant to mitigation in agriculture in terms of general project accounting (see Baroudy and Hooda, this volume), as well as afforestation and reforestation projects that involved farmers (see chapter by Lee and Newman for project lessons from a CDM afforestation and reforestation project).

Since 2007, agriculture has been a component of negotiations in two tracks of the UNFCCC: (1) the Kyoto Protocol Ad Hoc Working Group, where agriculture has featured in debates about emissions targets and the scope of CDM, and (2) the Ad Hoc Working Group on Long-Term Cooperation (AWG-LCA) created through the 2007 Bali Action Plan, which established reporting of nationally appropriate mitigation actions (NAMAs) and classified agriculture as a sectoral approach to mitigation.

The AWG-LCA has also been the track where negotiations on REDD have taken place. REDD has created policy attention to agriculture as a driver of deforestation, although few countries have identified clear strategies for addressing agriculture under REDD (Kissinger 2011). The AWG-LCA also catalyzed a technical work program, a period of intensive review and development of technical options and finance for REDD that provides a model for what could be done for agriculture.

In 2009, UNFCCC discussions began to include Low-Emissions Development Strategies as longer-term frameworks for managing emissions (Clapp *et al.* 2010).

The dichotomy between mitigation and adaptation that is a feature of the UNFCCC framework is not optimal for agriculture. Decisions about farming practices and farming systems affect adaptation and mitigation in interrelated ways. As mentioned above, mitigation cannot succeed without also being adaptive to climate change. Approaches that integrate adaptation and mitigation will be needed for agriculture. As the chapters in this volume make evident, linking mitigation to food security and livelihoods supports a more integrated approach with climate change adaptation. A sectoral and landscape-based approach that applies this type of multi-objective framework would support a more integrated approach to policy.

In addition to efforts within the UNFCCC, agricultural ministers started joining forces in 2010 to support other forms of intergovernmental collaboration for agricultural mitigation. Multilateral initiatives by the World Bank, the Food and Agriculture Organization of the UN, and others have

12

been assisting to convene people, establish databases and methods, support capacity building, and establish mitigation pilot projects. The Global Research Alliance for Agricultural Greenhouse Gases, and the Consultative Group on International Agricultural Research, established major research initiatives to support agricultural mitigation in 2009.

At the national and subnational level, the Alberta Offset System in Canada and New Zealand's Emissions Trading Scheme provide examples of subnational and national offset systems that include agricultural mitigation. New Zealand's system may be the first national effort planning to enforce compliance with emissions targets in agriculture. Compliance is scheduled to begin in 2015. Australia has been seeking to establish an emissions trading scheme, and initiated a National Carbon Offset Standard in 2010. Some other ways in which countries are establishing mitigation policies related to agriculture are through national climate change action frameworks (e.g., Kenya), low emissions development strategies (e.g., China, Brazil, Mexico) and NAMAs (see Box 1). The development of meaningful national level policy is essential to achieving the large scale of mitigation effort required for real impacts.

### Benefits associated with mitigation

Regardless of policy developments at the international level, mitigation practices need to benefit smallholders to be practicable. Incentives that could interest farmers in agricultural mitigation are (1) improved farm production or efficiency, (2) income and other benefits from selling offsets in the carbon market or payments for ecosystem services (PES) schemes, and (3) premiums on commodities or other benefits for sustainable or 'low-carbon' (more properly termed low net emissions or low climate impact)[9] products, available through supply chains from international corporations and certification. Other benefits such as increased biodiversity or watershed services may provide incentives as well, but are not discussed here. A common theme in the chapters is that carbon market benefits are likely to be minor in relation to yield gains without major increases in benefits or reductions in costs associated with market exchanges.

The book reviews the evidence for these different incentives for mitigation for smallholders and explores the conditions that would improve the economic feasibility of mitigation and smallholders' access to the benefits. In doing so, the chapters provide examples of a range of institutional arrangements associated with these incentives, including carbon market projects (Shames *et al.*); payments for ecological services (Wunder and Börner); certification (Hayward *et al.*); federal-state negotiated frameworks for integrated; landscape-scale mitigation in Australia (Van Oosterzee *et al.*); linking sustainable agricultural development to REDD+ in Brazil (Stella *et al.*); and bilateral intergovernmental cooperation for assistance in sustainable agriculture in China (Lu *et al.*).

Although we focus on the incentives for smallholders, other particip-ants in mitigation schemes (e.g., government entities, project developers, corporations) will also require appropriate incentives and institutional arrangements for mitigation to be successful. Analysis of incentives across all participants can help show which partners are likely already to have strong motivation to act and where extra support is needed.

### Benefits from farm production and efficiency

Economic benefits from gains in production or efficiency are inherent to many mitigation practices. Smith (2007b) reviewed evidence for economic impacts among 18 mitigation measures and found most practices yielded clear economic benefits (e.g., nutrient management, rice management, grazing land management, land restoration and water management), while others resulted in potential trade-offs (e.g., converted croplands) and/or uncertain impacts (e.g., agroforestry, reduced tillage). Shames *et al.* (this volume) observed that most interventions resulted in yield increases of about 100%, according to a study of 41 sustainable land man-agement interventions globally (citing Pender 2008) and a review of 45 sustainable land management interventions in sub-Saharan Africa (citing Pretty *et al.* 2006). Verhulst *et al.* similarly found that conservation agricul-ture yielded net returns that were twice as high as for conventional agri-culture for six sites in Mexico.

From an analysis of the production of food, yield variability and expo-sure to extreme weather events associated with different mitigation activ-ities, McCarthy *et al.* found that many mitigation practices and technologies can increase food production and the adaptive capacity of the food production system. The largest scope for achieving synergies between food security and mitigation appears to be in sustainable land management (i.e., conservation tillage, integrated nutrient management, agronomic management to increase organic residues and residue manage-ment) and agroforestry or restoration of degraded lands that also increase food or income (McCarthy *et al.*, see also Smith and Wollenberg). Tennig-keit *et al.*[10] propose that carbon management in sub-Saharan Africa could lead to transformative effects on the capacity of the region to meet future food needs.

Capturing these synergies may involve upfront investment and oppor-tunity costs, particularly for smallholders in the short term. McCarthy *et al.*, summarizing existing project-level estimates of these costs, indicate that potential disincentives to adoption are considerable. Overcoming these barriers to adoption should be a priority and may require more attention to upfront investments in technological practices. Providing funds, for example for tree planting, check dams, or small-scale biochar processing equipment would increase the speed of implementing mitiga-tion and compensate for the delayed flows of benefits.

Trade-offs also can occur, especially where trees compete with food crops in agroforestry systems, or residues are limited and needed for feeding livestock. Assessing impacts across a range of ecosystem services is necessary to ensure that food provisioning is maintained (Mendez *et al.*, Minang and Van Noordwijk). Several chapters (McCarthy *et al.*, Herrero *et al.*, Minang and Van Noordwijk, and Smith and Wollenberg) present frameworks for assessing trade-offs and synergies. Herrero *et al.* suggest that a livelihood-based rather than a commodity-based approach to mitigation is necessary to ensure that smallholders' multiple needs are met. They observe that mitigation decisions rest on the objectives of the system, the GHG efficiency gap in terms of GHG emissions per unit of product and the cost per GHG emissions reduction for the mitigation strategies selected. While improving GHG efficiencies in specific systems may result in large gains per unit of product, these have to be weighed against whether the system produces low or high absolute amounts of GHG, whether the systems' objectives include improved efficiency or productivity, and the impacts on total emissions.

If the need to increase food production requires greater intensification of agriculture, care will be needed to consider the impact on absolute emissions in the long run. More attention is needed to understand the short- and long-term benefits and trade-offs of increased GHG emissions from intensified agriculture and indirect effects on the area of forest elsewhere. Reduction in GHG emission intensity (see Baker and Murray) will be needed through more efficient use of inputs such as fertilizers, water and fossil fuel energy, reduced waste and shifts to foods that yield lower emissions. To better link mitigation and food production at landscape scales, REDD policies will need to address strategies for sustainable intensification of food production and institutional mechanisms for maintaining the farm–forest boundary. The experience of the ProAmbiente project in Brazil described by Stella *et al.* should provide relevant lessons.

A final issue that requires attention is that gains in productivity or efficiency alone will provide only limited opportunities for improved livelihoods. Improved access and control over assets such as land and capital will be necessary for long-term livelihood security and significant improvements in well-being. Possibilities for enhancing livelihood assets in the natural resource base should be considered in concert with mitigation actions (Smith and Wollenberg).

### Benefits from carbon markets and payments

Payments for agricultural carbon credits may provide additional incentives to farmers. We distinguish between carbon markets and payments. Farmers may receive mitigation payments by selling carbon credits or through other mechanisms, such as government subsidies, development projects or certification programs. Trade in agricultural carbon offsets is

15

still limited. Only a few agricultural carbon market projects or payment schemes have been tried in developing countries. Most experience has been in North America.

A number of factors have constrained the development of carbon markets and payments for smallholders, including low returns to agricultural carbon per hectare; poor competitiveness of agriculture vis-à-vis other sectors; high transactions costs of achieving scale across many smallholders; high risk and lack of finance opportunities inherent to smallholder agriculture; and the normal barriers to adoption of new technologies. De Pinto *et al.* conclude that smallholders may not be the best candidates for carbon market programs. In addition, many farmers manage common-pool resources (rangelands, community forests, coastal zones) where boundaries, rights to benefits and collaborative management are unclear or contested.

The nascent nature of the carbon market means that evidence on the benefits to smallholders is scanty, however, feasibility studies provide an indication of potential returns where major constraints are overcome. Tennigkeit *et al.*[11] model the possibilities for profitability from sequestered carbon in agriculture for sub-Saharan Africa based on ideal conditions. They show that revenues increase significantly with participation in carbon projects, but are primarily due to increases in maize yields rather than carbon payments. Estimated annual carbon payments to farmers from a 200,000 hectare project range from US$1.15/ha for a 'no external inputs package' to US$27.40/ha for agroforestry. Estimated annual net revenues to farmers, including maize yields and accounting for costs, range from US$10/ha for no external inputs to US$309/ha for a package with increased seed and fertilizer inputs, and US$177 for agroforestry.

Tennigkeit *et al.* also speculate on the larger potential outcomes of agricultural carbon for the region, although in practice adoption rates, the viability of sequestering soil carbon smallholder systems, the need for upfront finance and limits to access of fertilizer and seed would severely constrain these estimates. Assuming a sequestration rate of $40\,MtC/yr$ and a discounted carbon price of $US\$10/tCO_2$, Tennigkeit *et al.*[12] estimate that mitigation projects in sub-Saharan Africa could generate annual revenues of US$1.47 billion, or almost twice the overseas development assistance for agriculture in the region.

Several chapters in the book examine the evidence from ongoing carbon projects involving smallholders. Lager and Nyberg reflect on the lessons learned from the Kenya Agricultural Carbon Project (KACP), which in 2010 became the first project to sell credits from soil carbon. The authors conclude the real benefits of the project arise from improved yields and food security, and other benefits related to increased market access, advisory services and land-use practices. They also urge that public sector or donor funds are needed to cover establishment costs. Lee and Newman review findings from KACP as well as from six farm forestry carbon projects in East Africa, which has become one of the hubs of innovation for farmer involvement in carbon markets. They found that farmers who did participate in the carbon market complained that payments were too low, although they valued the reliability of payments.

A number of chapters also provide ideas about how economic efficiency can be improved in carbon projects by reducing costs or increasing benefits. These lessons should be relevant to both market-based or payment schemes. Shames *et al.* offer design principles for reducing costs and risk, while increasing benefits for farmers in carbon projects. Havemann identifies institutional models for aggregation of farmers, including coops, contract farming, sharecropping, or third-party aggregators. In addition, the level of payments to farmers can be designed to better target the needs of different farmers, thereby avoiding overpaying some farmers and wasting financial resources or underpaying others and getting no results (de Pinto *et al.*). Benefits for farmers can be improved through supporting their capacity for negotiation, securing rights to land and carbon, and accessing higher levels of project finance (Shames *et al.*). Project-level benefits can be increased by bundling mitigation with ecological services such as biodiversity and water management (de Pinto *et al.*). Compliance policies requiring limits to emissions would improve carbon prices (Negra and Wollenberg).

We conclude from the analyses of carbon projects in the book that revenues are possible from the carbon market, but more information is needed about whether smallholder farmers' participation in the carbon market makes economic sense, improving farm production and efficiency to yield livelihood benefits and mitigation is likely to lead to high impact, at least until carbon prices shift significantly. In particular, we need to better understand how to improve carbon benefits, if carbon benefits can be realized widely and whether they lead to positive long-term impacts for different kinds of farmers, including men and women smallholders.

Given the importance of outcomes in yields relative to carbon payments, care will be needed to ensure that the novelty of carbon trading and enthusiasm for its potential profits do not draw excessive attention and resources at the expense of less novel, but equally as important opportunities to scale up existing technologies such as conservation agriculture or agroforestry that provide benefits directly from increased productivity or efficiency.

### Benefits from corporate supply chains

In addition to incentives from improved livelihoods and trading carbon offsets, the private corporate sector appears well positioned to provide technical support and premiums for mitigation to their farmer suppliers. Increasing consumer demand for low-carbon products in developed countries and the growth of corporate social responsibility programs have supported expansion of these programs. Companies like Unilever, Campbell's soup and Danone have been supporting agricultural mitigation among their farmer suppliers to offset corporate emissions. Havemann notes that domestic supply chains in developing countries are unlikely to be able to command premiums on climate-friendly products, which implies that farmers with access to international markets will benefit most from these arrangements. Certification schemes can help farmers access premiums in the markets (see chapter by Hayward *et al.*).

## Institutional arrangements for scaling up smallholder mitigation

For agricultural mitigation to be operational and rapidly reach large scales, supporting institutional arrangements and policies are necessary. Carbon market projects, sustainable land management projects and corporate supply chains have received the most attention as institutional arrangements for implementing mitigation, but institutional arrangements that play auxiliary, supporting functions in providing finance, aggregating farmers, solving tenure issues, enabling trading platforms or supporting farmer learning are also critical. National policies will be essential for coordination and impact.

Aligning with existing organizations, finance channels and development policies will be important for rapid results and efficient use of resources. Strengthening and expanding existing programs for sustainable land management, conservation agriculture, agroforestry, land restoration or livestock management will enable the most rapid progress. Significant resources already have been invested and capacity built worldwide for these practices.

Infrastructure already exists for many aspects of mitigation. For example, the Community Markets for Conservation (COMACO) program in Zambia has established extension and payment mechanisms necessary to reach 50,000 farmers (Shames *et al.*). Networks of rural and community banks may provide a foundation for carbon finance (Baroudy and Hooda). Werneck explores the possibilities of using microfinance channels to support carbon payments. He cautions that agricultural carbon may not yet be sufficiently profitable to be competitive with other microfinance investments. Development policy could build in mitigation and adaptation finance and sustainable land management to achieve larger-scale impacts. The New Partnership for Africa's Development (NEPAD) Planning and Coordinating Agency, for example, would be a strong platform for taking on mitigation initiatives (Baroudy and Hooda).

Linking with existing strong organizations at the local level is also essential. Experiences in Chiapas, Mexico and Western Kenya suggest that communities with pre-existing strong local organizations will be better prepared to link to new initiatives and reap their benefits (Shames *et al.*, Semroc *et al.*). Partnership with a strong implementing partner that understands local conditions has enabled Plan Vivo to work successfully with smallholders (Swickard and Nihart).

Finance will be necessary to support mitigation, although precisely how much is not clear. Estimates are needed to guide countries' budget allocations. Agriculture is largely missing from current climate finance. At the 2009 UN climate talks, countries pledged to secure US$100 billion annually from public and private funds for adaptation and mitigation by 2020 in developing nations, but these funds will be shared among diverse sectors. According to UNFCCC agreements, developed countries are to finance NAMAS but the modalities have not been specified. Private investors have begun to mobilize finance for climate-related ventures in agriculture based on speculated increases in carbon prices as regulations tighten and societal demands for products with lower climate impacts increase, however investments have been mostly experimental in nature.

Baroudy and Houda draw on lessons from the experience of the World Bank's BioCarbon Fund, one of the earliest efforts to organize investment in agricultural carbon, to note the need for initial project development finance and a stronger market for agricultural offset credits if carbon projects are to increase in number and thrive. They urge that projects for agricultural carbon finance should be more holistic in addressing not only farmers' needs for funding, but also their needs for crop or livestock yields, or added value. Non-carbon-related investments in agriculture through government programs and private sector initiatives will also be important if they are linked to low net emissions farming practices and systems.

Australia's experience from the Degree Celsius Wet Tropics Biocarbon Sequestration and Abatement Project may provide lessons about using decentralized natural resource management government bodies to coordinate landscape-scale mitigation (van Oosterzee et al.). Shames et al. state that working at larger scales also provides opportunities for linking with landscape-scale planning processes. They make the case that using a landscape design approach enables optimal location of mitigation activities, linking of projects to reduce costs and stakeholder coordination. Carbon credits in one part of the landscape also may be used to compensate for credits in another part (Semroc et al.).

Institutional arrangements should include monitoring of on- and off-site impacts and safeguards if we are to understand reductions of net emissions and how mitigation can best serve poor farmers. Lee and Newman found that most projects want to track their impacts on farmers' well-being, but did not have enough resources to do so. Public sector support is likely to be necessary. Safeguards can be developed for food security, livelihoods, economic development, pro-poor outcomes, and environmental impacts and increased climate change-related risks.

Networks will be necessary for capacity building and sharing of technical knowledge (see Lu et al. for example of Sustainable Agriculture Innovation Network in China). Awareness of new practices and support for their adoption will need to be fostered. Both market and non-market based on-the-ground projects will be an important source of innovation and learning, and their lessons should be linked to the policy making process (Negra and Wollenberg). As Verhulst *et al.* discuss, traditional extension services are not likely to be successful, and instead more dynamic, decentralized networks of farmer groups, technology developers, extension workers, local entrepreneurs and researchers will be needed.

## Measurement and GHG accounting

Measurement and accounting of emissions can be complex. Agricultural mitigation involves not only three gases, but also the interactions among them. Measurement of $N_2O$ and $CH_4$ can be expensive and sensitive to diurnal and seasonal fluctuation. High spatial variability and patchiness make sampling and modeling more difficult. Many activities cannot be monitored with remote sensing. Data about the nature and extent of farming activities are also poor in most developing countries, so even where reliable emissions factors are available, applying them is problematic. Data and methods for carbon are generally better developed than for methane and nitrous oxide in developing countries. A priority for advancing mitigation is therefore the development of low-cost, consistent and streamlined methods for measuring GHG emissions and mitigation impacts.

Monitoring and accountability will need to address the different purposes of the related schemes such as national reporting and planning, carbon accounting for investors, sustainable agriculture certification, low-carbon labeling, scientific research or helping farmers to lower their emissions independent of third-party observation or compensation schemes. For each of these purposes, a balance between cost and rigor of information collection will be needed. The IPCC Tier I, II and III methods provide initial guidance on levels of rigor appropriate to roughly the national, project and farm scales.[13] Where high accuracy is not required, activity- or practice-based estimates based on regional or country-specific emissions factors may be sufficient (see Seebauer and Tennigkeit). In addition, user-friendly methods that are accessible and transparent to a wide range of stakeholders, including farmers, will be needed for some purposes.

The book presents some emerging tools and techniques to address these challenges. Reducing costs is a common theme. Milori *et al.* provide an example of innovation, describing a low-cost, but sophisticated spectroscopic technique that the Brazilian Agricultural Research Corporation (EMBRAPA) has developed for sampling soil carbon on farmers' fields. The mobile device reduces the need for transport to laboratories and sample preparation.

Olander provides a comprehensive review of process-based models for GHG emissions relevant to agriculture. She finds that the models cover most possible management practices, but that modeling of practices related to nitrous oxide, methane management and biochar need further testing. As the available models were developed mostly for commodity crops, they are less appropriate for the range of crops and complex farming systems of smallholder agriculture, although research is currently ongoing to fill this gap. Further testing and calibration in developing countries may help to reduce the data needs for models.

Seebauer and Tennigkeit describe the application of the 'activity baseline and monitoring survey' methodology (ABMS),[14] an approach to soil carbon accounting, in the Kenya Agricultural Carbon Project (KACP). ABMS combines monitoring of farm practices with soil carbon modeling to derive locally applicable default values for soil carbon sequestration. This provides a cost-effective compromise between direct measurement, which can be very expensive, and estimates based solely on activities. In the Kenya case, ABMS involved administering a questionnaire about current practices and intentions to adopt new management practices on an annual basis for the first 10 years and at 3–5 year intervals thereafter. ABMS can also be used to monitor adoption rates. The authors observed that the principle of conservativeness in estimating carbon sequestration can be used to compensate for uncertainty where achieving accuracy is costly.

Hillier et al. have been developing a single tool that farmers can use to estimate GHG emissions at the farm level.[15] As farmers often have the best knowledge of their own farms, they are in a better position to characterize emissions-related management practices. Using the tool also educates farmers about mitigation options. The Sustainable Food Lab is currently testing the "Cool Farm Tool" among 17 entities in 14 countries. Hiller et al. observe that residue management appears to be an important option in many systems.

The availability of standards[16] offering protocols and trading platforms is still limited for developing countries. Seebauer and Tennigkeit note that carbon offset standards have been developed primarily "for and by experts in data-rich countries," which have made them difficult to apply in developing countries. Swickard and Nihart provide an account of the lessons learned from the Verified[17] Carbon Standard, which is the only agricultural standard currently with global application. Established in 2006, VCS introduced agriculture, forestry and land-use (AFOLU) protocols in 2008 and has led the way for agricultural mitigation in the voluntary market. The chapter by Lager and Nyberg provides an example of a project registered under VCS. An important innovation of the VCS, which helped to overcome some of the risk associated with agricultural credits, was the requirement to establish reserves or buffers of emissions credits as insurance to compensate for impermanence (see also Baroudy and Hooda).

The Plan Vivo standard is also notable for its principles related to community livelihoods, poverty reduction and environmental health. The standard has registered five projects with smallholders in developing countries, all of which include some element of agroforestry. The standard applies only to the above- and below-ground biomass of trees rather than soil carbon or agricultural crops. The oldest project, Scolel'Te in Mexico, was first verified in 2001. See chapters by Semroc and Lee and Newman for examples of projects using the Plan Vivo standard.

Most efforts to measure emissions have focused on area-based measures, i.e., emissions per hectare. An area-based measure is appropriate for inventories and other efforts to sum total emissions across the landscape. Baker and Murray discuss an alternative measure, 'emissions intensity,' that describes emissions per unit of agricultural output. Emissions intensity reflects the efficiency of agricultural production relative to GHG emissions and, like life cycle analysis, is useful for developing improved efficiencies.

An output-based measure can indicate the direct effects of mitigation interventions on site, as well as the indirect effects of on-site changes in yield on production and emissions elsewhere. If on-site yields increase, for the same level of emissions, less area or less intense forms of production are needed elsewhere to produce the same aggregate level of output. This constitutes a positive form of leakage (i.e., mitigation reductions) that could be credited using the output-based approach.

While these authors highlight some potential limitations to use of this approach in carbon crediting systems, such as perverse incentives to increase the intensity of inputs and coping with yield volatility, output-based measures of emissions have an immediate role as a tool for improving efficiency. The potential for sustainable agriculture to deliver production at relatively low GHG emission intensities should be promoted as an important efficiency advantage over conventional agriculture. Identifying opportunities for such efficiency gains should be a priority, while recognizing that gains in efficiency should not be traded for increases in total emissions.

## Conclusions

How we use land, cultivate plants and raise animals has significant implications for greenhouse gas (GHG) emissions, but shifts in agricultural practices can help to reduce gases in the atmosphere. Smallholders in developing countries can play a critical role. Dealing with climate change through agriculture though will not be a matter of solely promoting new practices. It also will require addressing larger issues of food security, poverty alleviation, social justice and the assumptions underlying sustainable agricultural development.

As the chapters in this book demonstrate, many technical options are already economically feasible and likely to be compatible with the food security and livelihood needs of smallholders. New means are rapidly emerging to improve and expand adoption of these options. The use of carbon markets for smallholder-based mitigation requires further refinement to improve benefits and testing before it can be widely applicable. Benefits from improved farming practices where mitigation is a co-benefit are likely to have the largest impacts on both mitigation and food security. Action is needed now to further improve the viability, scope and accessibility of mitigation options for smallholders and rapidly scale up initiatives. Effective incentives and support for adoption of new practices will be necessary across the full range of players involved in mitigation.

Three principles should guide the future development of mitigation for smallholder farmers to ensure positive impacts:

1   **Clear benefits:** Practices must make economic sense to farmers by increasing productivity, reducing risk, providing additional sources of revenue such as payments or certification premiums, or providing other benefits;
2   **Balancing multiple objectives and pursuing synergies:** Policies and programs should prioritize mitigation options that achieve the multiple objectives of poverty reduction, food security, climate change adaptation and environmental sustainability, recognizing that trade-offs in favor of poverty reduction and at the cost of mitigation may be necessary in some places; and
3   **Self-determination:** Smallholders, including men and women from different positions of power and influence, should be aware of, participate in, and make decisions about mitigation projects or policies that affect them.

As demands for agricultural production increase in the future, novel approaches will be needed to intensify agriculture while also reducing agriculture's impacts on the climate and securing ecological services (see also Hayward *et al.*, and Lu *et al.*). In promoting mitigation measures, policy makers and practitioners therefore should give immediate priority to actions that:

• take stock of existing good practices, demonstrate the economic feasibility of these practices and identify what is needed to maintain and expand them;
• give special attention to sustainable land management that maintains and increases carbon sequestration in the soil and perennial vegetation;
• support early adoption and action among relevant entities to expand the evidence base for relevant practices and supporting incentives and institutional arrangements;
• provide upfront finance to initiatives through credit mechanisms or investment support;

23

- build capacity rapidly through the development of curricula, training and exchange programs;
- build learning hubs and other awareness building and technical support mechanisms to increase innovation and adoption of practices;
- develop and apply decision support tools to prioritize mitigation actions and investments; and
- use analysis at different scales, including food product-, farm- and landscape-level to identify impacts on multiple objectives and prioritize interventions.

In the medium term, additional priorities will be to:

- expand adoption of mitigation practices that also result in increased food yields and food security;
- increase options for reducing the GHG emissions intensity of food (GHG emissions per unit of output) *and* achieve reductions in emissions on an area basis;
- reduce waste and increase efficiency of inputs;
- exploit opportunities to bundle the climate regulation services of mitigation with other ecological services such as water quality and biodiversity; this will increase the efficiency of incentives and infrastructure;
- develop institutions and incentives for landscape-based mitigation to optimize interactions between agriculture and other land uses such as forestry or sustainable biofuels, while also enhancing flexibility to choose among diverse mitigation options and avoid leakage; and
- develop practices that are adaptive to future climate change.

Agricultural mitigation will need to advance at the national, regional and international levels to achieve impact. A strong vision for agriculture that contributes to economic development and mitigation in both the short- and long term will be essential. Climate change policies that incorporate agriculture-related objectives, and conversely, agricultural development policies that encourage lower climate footprints will facilitate larger impacts.

Frameworks for evaluating efficiencies and multiple objectives can help identify where trade-offs and synergies occur. Analysis according to different scales, including for a given farming practice or food, as well as across farming systems and landscapes, will help show who is likely to benefit and how to improve efficiencies of specific practices or systems. Low-cost measurement and monitoring should help provide data to support decisions. Accounting systems for GHG emissions will need to be developed at the national and international levels to coordinate results and provide transparency.

Public financing for mitigation will be helpful to support early action for more pilot projects and finance, as well as capacity building. Governments will be responsible for defining and enforcing clear carbon and land rights in ways that can benefit both men and women.

The ultimate aim of all these efforts is that we learn how to better manage greenhouse gases in agriculture and agricultural landscapes. It is an urgent task.

## Notes

1 Direct emissions from agriculture are estimated to be 5.1–6.1 Gt $CO_2e$/yr (Smith *et al.* 2008).
2 Mitigation in agriculture is estimated to be able to reduce emissions by 1.5–4.3 Gt $CO_2e$/yr by 2030 at future carbon prices of US$20–100 (Smith *et al.* 2008), potentially offsetting 24–84% of current agricultural emissions.
3 Even in the UK and US, agricultural production at the field level accounts for 50–60% of emissions (Garnett 2011, Weber and Matthews 2008).
4 The IPCC (2007) recognizes improved cropland and grazing land management, the rehabilitation of degraded land, set-asides, land-use change to grassland, agroforestry and improved livestock and manure management as the most promising technical options for mitigation.
5 Preventing such losses will also be important where land-use change threatens existing storage of carbon.
6 These chapters represent a sample of the many interventions and agricultural systems for which analysis of mitigation is possible. Some interventions have been comprehensively treated in the literature, e.g., irrigation in wet rice systems (Wassman 2007), agroforestry and shifting cultivation (Palm *et al.* 2005). Others have been generally reviewed, e.g., maize, wheat and rice systems (Reynolds 2010), or reviewed for other, non-mitigation purposes, e.g., an extensive literature exists on soil organic matter management, peatlands, and the rehabilitation of degraded lands.
7 Emissions factors measure the amount of GHG released into the atmosphere for a given practice or land use. The emissions factor times the level of activity indicates the amount of GHG released.
8 Approved methodologies directly related to agricultural production include, for example, methane emissions reduction by adjusted water management practice in rice cultivation and offsetting synthetic nitrogen with inoculants in legume-grass rotations, methane recovery from animal manure management systems. A couple of other methodologies address methane recovery from manure, e.g., for biogas. See http://cdm.unfccc.int/methodologies/SSCmethodologies/approved.
9 While the term 'low carbon' is appropriate to mitigation activities that reduce $CO_2$ emissions, these emissions are relatively low in agriculture. In agriculture, products that sequester carbon or reduce other greenhouse gases will result in mitigation. We suggest that an alternative term that reflects these other pathways to mitigation is more appropriate.
10 See chapter titled "Potential of agricultural carbon sequestration in sub-Saharan Africa."
11 See chapter titled "Economics of agricultural carbon sequestration in sub-Saharan Africa."
12 See chapter titled "Potential of agricultural carbon sequestration in sub-Saharan Africa."
13 Hillier *et al.* summarize the features of the tier requirements.
14 Part of approved Verified Carbon Standard protocol.
15 The Food and Agriculture Organization has developed the FAO Ex-Act tool for generating data at the larger, project scale.
16 The International Organization for Standardization (ISO) 14064–2 "Specification with Guidance at the Organization Level for Quantification and Reporting

of Greenhouse Gas Emissions" provides a foundation for other standards and protocols. ISO requirements are (1) a streamlined life cycle assessment for project and baseline conditions; (2) evaluation of possible baseline scenarios; (3) identification of relevant GHG emissions controlled by the project, and upstream and downstream impacts of the project; and (4) identification of sources and sinks.

17 Formerly known as the Voluntary Carbon Standard.

# References

Bellarby, J., Foereid, B., Hastings, A. and Smith, P. (2008) Cool Farming: Climate impacts of agriculture and mitigation potential. Report produced by the University of Aberdeen for Greenpeace, Greenpeace.

Brown, S., Grais, A., Ambagis, S. and Pearson, T.R.H. (2011) Baseline GHG Emissions from the Agricultural Sector and Mitigation Potential in Countries of East and West Africa. Final Report to ILRI and ICRISAT.

Clapp, C., Briner, G. and Karousakis, K. (2010) *Low-Emission Development Strategies (LEDS): Technical, Institutional and Policy Lessons* [Online]. Available at: www.oecd. org/dataoecd/32/58/46553489.pdf (Accessed: 29 April 2011). Organisation for Economic Co-operation and Development (OECD).

Ewers, R.M., Scharlemann, J.P.W., Balmford, A. and Green, R.E. (2009) Do increases in agricultural yield spare land for nature? *Global Change Biology*, vol. 15, no. 7, pp. 1716–1726.

Foresight (2011) The Future of Food and Farming Final Project Report. The Government Office for Science, London.

Garnett, T. (2008) Cooking up a storm: Food, greenhouse gas emissions and our changing climate. Food Climate Research Network Centre for Environmental Strategy, University of Surrey, Guildford, UK.

Grieg-Gran, M. (2010) Beyond forestry: Why agriculture is key to the success of REDD+. IIED briefing, International Institute for Environment and Development, London, UK.

IPCC (2000) Land-use, land-use change and forestry, Special report of the Intergovernmental Panel on Climate Change, Cambridge University Press, UK.

Kissinger, G. (2011) *Linking forests and food production in the REDD+ context*, Working Paper 1 [Online]. Available at: ccafs.cgiar.org/resources/working-papers (Accessed 12 October 2011) CGIAR Research Program on Climate Change, Agriculture and Food Security (CCAFS), Copenhagen, Denmark.

Lehmann, J., Gaunt, J. and Rondon, M. (2006) Bio-char sequestration in terrestrial ecosystems – a review, *Mitigation and Adaptation Strategies for Global Change*, vol. 11, pp. 403–427.

Palm, C.A., Smukler, S.M., Sullivan, C.C., Mutuo, P.K., Nyadzi, G.I. and Walsh, M.G. (2010) Identifying potential synergies and trade-offs for meeting food security and climate change objectives in sub-Saharan Africa, *PNAS*, vol. 107, no. 46, pp. 19661–19666.

Palm, C.A., Vosti, S.A., Sanchez, P.A. and Ericksen, P.E. (2005) *Alternatives to Slash and Burn: The Search for an alternative*, Columbia University Press, New York.

Pender, J. (2008) 'The world food crisis, land degradation and sustainable land management: linkages, opportunities and constraints', International Food Policy Research Institute, Washington, DC.

Pretty, J.N., Noble, A.D., Bossio, D., Dixon, J., Hine, R.E., Penning de Vries, F.W.T. and Morison, J.I.L. (2006) Resource-conserving agriculture increases yields in developing countries, *Environmental Science and Technology*, vol. 40, no. 4, pp. 1114–1119.

Reynolds, M. (2010) Climate Change and Crop Production, CABI, Wallingford, UK.

Ridgwell, A., Singarayer, J.S., Hetherington, A.M. and Valdes, P.J. (2009) Tackling regional climate change by Leaf Albedo Biogeoengineering, *Current Biology*, vol. 19, pp. 146–150.

Robbins, M. (2011) *Crops and Carbon: Paying Farmers to Combat Climate Change*, Earthscan, London, UK.

Smith, P., Martino, D., Cai, Z., Gwary, D., Janzen, H.H., Kumar, P., McCarl, B., Ogle, S., O'Mara, F., Rice, C., Scholes, R.J., Sirotenko, O., Howden, M., McAllister, T., Pan, G., Romanenkov, V., Rose, S., Schneider, U. and Towprayoon, S. (2007a) 'Agriculture', in Metz, B., Davidson, O.R., Bosch, P.R., Dave, R. and Meyer, L.A. (eds) Climate Change 2007: Mitigation, Contribution of Working Group III to the Fourth Assessment Report of the Intergovernmental Panel on Climate Change. Cambridge University Press, Cambridge and New York, NY.

Smith, P., Martino, D., Cai, Z., Gwary, D., Janzen, H.H., Kumar, P., McCarl, B., Ogle, S., O'Mara, F., Rice, C., Scholes, R.J., Sirotenko, O., Howden, M., McAllister, T., Pan, G., Romanenkov, V., Schneider, U. and Towprayoon, S. (2007b) Policy and technological constraints to implementation of greenhouse gas mitigation options in agriculture, *Agriculture, Ecosystems & Environment*, vol. 118, pp. 6–28.

Smith, P., Martino, D., Cai, Z., Gwary, D., Janzen, H.H., Kumar, P., McCarl, B., Ogle, S., O'Mara, F., Rice, C., Scholes, R.J., Sirotenko, O., Howden, M., McAllister, T., Pan, G., Romanenkov, V., Schneider, U., Towprayoon, S., Wattenbach, M. and Smith, J.U. (2008) Greenhouse gas mitigation in agriculture, *Philosophical Transactions of the Royal Society B*, vol. 363, pp. 789–813.

Stern, N. (2006) *Stern Review: the economics of climate change*, Cambridge University Press, Cambridge, UK.

Tschakert, P. (2007) Environmental services and poverty reduction: Options for smallholders in the Sahel, *Agricultural Systems*, vol. 94, no. 1, pp. 75–86.

Verchot, L.V., Van Noordwijk, M., Kandji, S., Tomich, T., Ong, C., Albrecht, A., Mackensen, J., Bantilan, C., Anupama, K.V. and Palm, C. (2007) Climate Change: Linking adaptation and mitigation through agroforestry, *Mitigation and Adaptation Strategies for Global Change*, vol. 12, pp. 901–918.

Weber, C. L. and Matthews, H. S. (2008) Food-Miles and the Relative Climate Impacts of Food Choices in the United States, *Environmental Science & Technology* vol. 42, no. 10, pp. 3508–3513.

Wassman, R., Lantin, R.S. and Neue, H. (eds) (2007) *Methane Emissions from Major Rice Ecosystems in Asia*, Springer-Verlag, New York.

# 2

# AGRICULTURAL MITIGATION APPROACHES FOR SMALLHOLDERS

## An overview of farming systems and mitigation options

*Christina Seeberg-Elverfeldt and*
*Marja-Liisa Tapio-Biström*

## Introduction

In this chapter we provide an overview of the mitigation activities carried out worldwide through agriculture, forestry and other land-use (AFOLU) activities, and review the potential for mitigation from different farming systems and practices. The examples are mainly located in developing countries; however, sometimes data was only available from developed countries. We conclude with an outlook for the necessary actions for integrating small-holder mitigation approaches into the current climate change framework.

## Global overview of AFOLU mitigation activities

Agriculture is part of a larger AFOLU or terrestrial carbon landscape[1] that can contribute to climate change mitigation. To build a foundation for furthering AFOLU-based mitigation, in 2010 the Food and Agriculture Organization (FAO) conducted a comprehensive search of AFOLU mitigation projects. Eleven registries were searched, including crediting schemes (compliance and voluntary markets) and third-party databases for registered and certified projects for forestry (afforestation/reforestation (A/R) and Reduced Emissions from Deforestation and Forest Degradation (REDD)), manure treatment from livestock, agricultural and rangeland soil carbon activities. The projects were registered by the crediting scheme for credits or listed in a third-party database as being implemented. The results were used to establish the AFOLU mitigation project database with 497 projects (FAO 2010a). It has been created to obtain a better overview of the developments in the sectors. The database is accessible through the internet (www.fao.org/climatechange/67148/en/) and updated on an annual basis.

The database shows that the majority of AFOLU projects were in Latin and North America (61%), followed by Africa (20%) and Asia (12%); very few projects (2%) were in Europe or Central Asia (Figure 2.1). Manure projects were prominent in Latin America (80% of projects in the region), as well as in Asia (66%), while Africa hosted mainly forestry (64% of projects in the region) and some agricultural soil carbon projects (10%). North America had the highest number of terrestrial carbon projects (39%). Among the developing countries, Africa had the highest share (38%) of terrestrial carbon projects, followed by Latin America (14%). Asia and the Pacific had very few projects (7%), similarly to Europe and Central Asia (2%).

Asia had the majority of Clean Development Mechanism (CDM) projects, which in the land-based sectors only permits credits for manure, agricultural residues and A/R. Sixty percent of all registered CDM projects were in China and India, and of these only four were terrestrial carbon projects. In the AFOLU mitigation project database these two countries implemented only 11 AFOLU projects, which can be explained by an emphasis on projects in the energy and industrial sectors rather than AFOLU.

Generally, not many terrestrial carbon projects were prepared for CDM (Figure 2.2), due to the demanding data requirements, as well as the complexity of the auditing process and methodologies. Only 15 out of 2,262 registered CDM projects were forestry related (UNEP Risoe Centre 2010).

Instead, most AFOLU projects were registered with crediting schemes in the voluntary market (40%) or operated independently (20%) not registered under a crediting scheme. Several crediting schemes for voluntary offsets have been created for forestry mitigation activities representing about a quarter of the projects in the database. Fifty percent of these projects were implemented outside crediting schemes. Soil carbon projects represented

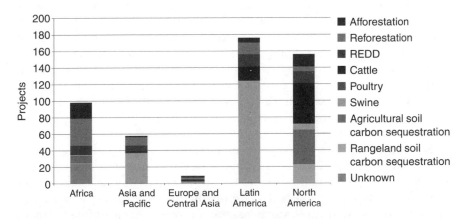

*Figure 2.1* Project breakdown by region and project type (source: FAO 2010a).

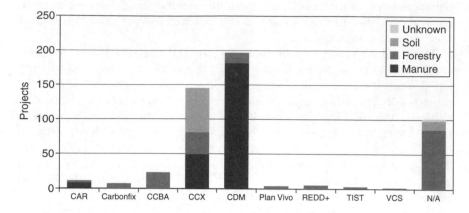

*Figure 2.2* Project breakdown by scheme and project type (source: FAO 2010a).

about 15% of projects in the database. Eighty-three percent of these were registered under the Chicago Climate Exchange (CCX) scheme in the USA, while 17% operated outside credit schemes and the market. These numbers suggest that there is significant untapped potential in the agriculture and forest sectors for more participation in the carbon market.

## Review of mitigation potentials for different farming systems and practices

The greenhouse gas (GHG) mitigation potential in AFOLU is related to the potential for carbon sequestration, and the reduction of methane ($CH_4$) and nitrous oxide ($N_2O$) emissions. Carbon sequestration refers to removing carbon from the atmosphere and depositing it in a sink, either in soil or biomass (UNFCCC 2010). Ideally, the carbon sequestration potential should be reported as a rate (mass per units of area and time) as it is a dynamic process over a certain time period (Nair *et al.* 2009). However, often carbon sequestration is reported as stocks, as time studies are rare. For vegetation carbon pools, the sequestration potential is projected according to biomass growth and productivity. Table 2.1 gives an overview of the GHG mitigation potentials of practices relevant to smallholders in different climate zones. These numbers represent global averages with a wide variation for the technical mitigation potential. The actual realized mitigation is potentially much less due to adoption rates, carbon prices, etc.

Significant in this table is the large difference of mitigation potentials between humid and dry areas and between warm and cool areas. Literature shows a huge diversity of sequestration potentials among management practices. Mitigation impacts depend on the area's soil type, topography, land-use category, climatic variables, vegetative cover, farm management practices, and the interactions among practices. We reviewed a variety of

*Table 2.1* Annual mitigation potential for different climate regions for agricultural mitigation practices

| *Improved land management practice* | *all GHG (t $CO_2e$/ha/yr)* | | | |
|---|---|---|---|---|
| | *Cool–dry* | *Cool–moist* | *Warm–dry* | *Warm–moist* |
| Improved agronomic practices | 0.29 | 0.88 | 0.29 | 0.88 |
| Soil nutrient management | 0.33 | 0.62 | 0.33 | 0.62 |
| Tillage and residue management | 0.17 | 0.53 | 0.35 | 0.72 |
| Water management | 1.14 | 1.14 | 1.14 | 1.14 |
| Set-aside and land cover (use) change | 3.93 | 5.36 | 3.93 | 5.36 |
| Agroforestry | 0.17 | 0.53 | 0.35 | 0.72 |
| Pasture management | 0.13 | 0.80 | 0.11 | 0.81 |
| Restoration of organic soils | 33.51 | 33.51 | 70.18 | 70.18 |
| Restoration of degraded soils | 3.53 | 4.45 | 3.45 | 3.45 |
| Application of manure/bio-solids | 1.54 | 2.79 | 1.54 | 2.79 |
| Bioenergy | 0.17 | 0.53 | 0.35 | 0.72 |

*Source:* IPCC (2007).

studies to obtain data on the carbon sequestration and emission reduction potential of different practices. The review is not aimed to be complete, but covers a range of practices which are relevant for mitigation, and are applicable to smallholder farmers in developing countries. As the research methods vary across studies, comparisons are difficult. However, the data can be used to obtain an iteration overview of the available potentials of different agricultural practices.

### Agroforestry

Agroforestry is widely practiced among smallholder farmers in developing countries and supports adaptation through diversified incomes from tree and field crops, increased resilience to climate extremes, and enhanced productivity. Trees provide a higher carbon sequestration potential than pastures or field crops (Nair *et al.* 2009), however, a huge range of potentials is reported across different agroforestry systems (Table 2.2). In general, the carbon sequestration potentials in arid, semi-arid and degraded sites are lower than in humid and fertile areas; and tropical agroforestry systems have higher potentials than temperate ones (Nair *et al.* 2009).

### Grassland management

Grasslands are estimated to store up to 30% of the world's soil carbon, to which the carbon stored in grasses, shrubs, trees and bushes can be added (Grace *et al.* 2006). Declining plant productivity and soil organic matter can be reversed through grazing management, fallow systems and sowing native or improved seeds, which also sequester atmospheric carbon and

*Table 2.2* Carbon sequestration potential in different agroforestry systems

| Farming practice | Region | t $CO_2$/ha/yr | Source |
|---|---|---|---|
| Fodder bank | Ségou, Mali, W. African Sahel | 1.06 | Takimoto et al. (2008)* |
| Coffee-based system | Central Kenya | 1.80 | Forest Trends (2010) |
| Live fence | Ségou, Mali, W. African Sahel | 2.17 | Takimoto et al. (2008)* |
| Tree-based intercropping | Canada | 3.05 | Peichl et al. (2006)* |
| Traditional agroforestry parklands | Ségou, Mali, W. African Sahel | 4.00 | Takimoto et al. (2008)* |
| Silvopasture | W. Oregon, USA | 4.07 | Sharrow and Ismail (2004)* |
| Agrisilviculture | Chattisgarh, Central India | 4.62 | Swamy and Puri (2005)* |
| Silvopastoralism | Kurukshetra, India | 5.03 | Kaur et al. (2002)* |
| Full-sun cocoa | Southern Cameroon | 7.19 | Gockowski and Sonwa (2011) |
| Shaded cocoa | Southern Cameroon | 13.03 | Gockowski and Sonwa (2011) |
| Home and outfield gardens | Panama | 15.74 | Kirby and Potvin (2007)* |
| Agroforestry woodlots | Kerala, India | 23.97 | Kumar et al. (1998)* |
| Indonesian homegardens | Sumatra | 29.36 | Roshetko et al. (2002)* |
| Cacao agroforests | Turrialba, Costa Rica | 40.66 | Beer et al. (1990)* |
| Agroforestry woodlots | Puerto Rico | 44.19 | Parrotta (1999)* |
| Mixed species stands | Puerto Rico | 55.82 | Parrotta (1999)* |

Sources with * reviewed in Nair et al. (2009).

Table 2.3 Carbon sequestration potential of different grassland management practices

| Farming practice | Region | $t\,CO_2/ha/yr$ | Source |
|---|---|---|---|
| Transition from heavy to moderate grazing | Eurasia | 0.18 | Conant and Paustian (2002) |
| Transition from heavy to moderate grazing | Australia/Pacific | 0.33 | Conant and Paustian (2002) |
| Avoided land cover/land-use change | Global | 0.40 | Tennigkeit and Wilkes (2008) |
| Transition from heavy to moderate grazing | North America | 0.59 | Conant and Paustian (2002) |
| Transition from heavy to moderate grazing | Africa | 0.77 | Conant and Paustian (2002) |
| Improved grazing management, rangeland | USA | 1.26 | Eagle et al. (2010) |
| Fertilization on grasslands | Global | 1.76 | Tennigkeit and Wilkes (2008) |
| Improved grazing on rangelands | Global | 1.98 | Conant et al. (2001) |
| Grazing management | Global | 2.16 | Tennigkeit and Wilkes (2008) |
| New grasslands | USA | 2.20 | FAO (2010b) |
| Transition from heavy to moderate grazing | South America | 2.53 | Conant and Paustian (2002) |
| Fire control on grasslands | Global | 2.68 | Tennigkeit and Wilkes (2008) |
| Improved grazing management, pasture | USA | 4.26 | Eagle et al. (2010) |
| Vegetation cultivation | Global | 9.39 | Tennigkeit and Wilkes (2008) |

increase soil carbon (Conant and Paustian, 2002). Table 2.3 shows management practices that can improve the carbon sequestration potential of grasslands, the sequestration potential being overall low per hectare, apart from humid and/or intensively managed pasture systems where the sequestration potential is considerably higher.

## Crop farming systems

In crop farming systems, residue management, manuring, cover crops, nitrogen-fixing crop rotations, composting and the application of organic soil amendments, as well as organic farming practices can improve soil carbon sequestration. Due to increased input from crop residue retention, increased crop intensification and rotation, as well as reduced carbon decomposition, conservation agriculture has a positive impact on carbon sequestration (Ortiz-Monasterio *et al.* 2010). Some examples of these impacts are presented in Table 2.4, showing that the overall potential per hectare is low. However, not all studies are conclusive (Govaerts *et al.* 2009) and results point in different directions with respect to the impact of zero tillage on soil carbon. Thus, more research is needed, especially in tropical areas, where still much information is missing (Ortiz-Monasterio *et al.* 2010).

Integrated Food Energy Systems can allow simultaneous food and energy production through multiple-cropping systems, or systems mixing annual and perennial crop species, i.e., maximizing synergies between food crops, livestock, fish production and energy. Bio-slurry as a by-product from biogas installations can be used as a fertilizer to improve soil quality or as fish feed, replacing inorganic fertilizer or commercial feed.

Emissions can also be reduced through an efficient and integrated nutrient management. Examples are farm nutrient management plans or nutrient budgets which allow for planning the nutrient input. This implies to assess the agro-ecological conditions, as well as cropping patterns on an

*Table 2.4* Carbon sequestration potential in crop farming management practices

| Farming practice/system | Region | t $CO_2$/ha/yr | Source |
|---|---|---|---|
| Diversify annual crop rotations | USA | 0.66 | Eagle *et al.* (2010) |
| Organic agriculture | Global | 0.73–1.46 | Niggli *et al.* (2009) |
| Conventional to no-tillage | USA | 1.12 | Eagle *et al.* (2010) |
| Conventional to conservation tillage | USA | 1.23 | Eagle *et al.* (2010) |
| Conventional to no-tillage | Global | 2.09 | West and Post (2002) |
| Maize-based farming system (increasing residue production, tree plantations) | Western Kenya | 2.10 | Forest Trends (2010) |
| Application of organic matter (manure) | USA | 2.63 | Eagle *et al.* (2010) |

individual basis. Another practice which improves nitrogen use efficiency is precision farming.

### Rice systems

Rice systems require special attention with respect to their mitigation potential, as the flooding regime determines the management pattern and represents the main source of $CH_4$ emissions. Some practices, such as draining the wetland rice, can reduce $CH_4$ emissions, but at the same time $N_2O$ emissions can increase by creating nearly saturated soil conditions (IPCC 2007).

Management practices have the potential to reduce emissions (Table 2.5), especially in irrigated rice systems, of which four broad categories exist: water management, organic matter, soil amendments and others (Yagi et al. 1997). According to Wassmann et al. (2000) optimizing irrigation patterns with additional drainage periods in the field or early timing of mid-season drainage can reduce $CH_4$ emissions by 7–80% with minimal impact on yields. Methane emissions can be reduced through the application of organic amendments such as compost by 58–63%, through biogas residues by 10–16%, and direct wet seeding by 16–22%. Additionally, positive impacts can be attained through the applications of incorporating fallow residues (11%), mulching of rice straw (11%), as well as sulfate (9–73%). Other options are different tillage practices, such as no tillage. Upland crop rotations, as well as selecting and breeding rice cultivars which emit less $CH_4$ are further alternatives (Yagi et al. 1997). Research indicates the magnitudes of $CH_4$ and $N_2O$ mitigation options, however outcomes are extremely site- and practice specific.

Rainfed and deepwater rice have limited mitigation options and potential gains are low. Any changes in water regime can increase $N_2O$ emissions, thus irrigation interventions are only recommended for rice systems with high $CH_4$ baseline emissions that would outweigh any flushes of $N_2O$. Yan et al. (2009) estimated that by applying rice straw off-season and draining all continuously flooded rice fields at least once during the growing season, the net reduction in $CH_4$ emissions would be 30%. They conclude that although draining continuously flooded rice fields may lead to an increase in $N_2O$ emissions, the global warming potential resulting from this increase is outweighed by the reduction from the $CH_4$ emissions.

Table 2.5 Mitigation potential of rice management practices

| Farming practice/system | Region | t $CO_2e$/ha/yr | Source |
|---|---|---|---|
| Rice variety development for $CH_4$ | various sites | 1.34 | Eagle et al. (2010) |
| Rice water management for $CH_4$ | various sites | 1.94 | Eagle et al. (2010) |

## Conclusions and outlook

The database on mitigation projects reveals the scarcity of mitigation projects in land-based sectors both in terms of practices and geographic coverage. The technical potential for mitigation from AFOLU sectors is considerable, but largely untapped at the moment. We need to rethink how agriculture and forestry can play a bigger role in current market-based mechanisms, but also how these mechanisms, as well as the international climate change mechanisms can be reformed and accommodated to include terrestrial carbon activities. Thus, different actions need to be taken at different levels.

Agriculture must be part of the legal framework for climate change under the United Nations Framework Convention on Climate Change (UNFCCC), to open an avenue for specific climate financing through mitigation funding. A concrete step will be to establish a work program under the UNFCCC Subsidiary Body for Scientific and Technological Advice. This can systematically support the debate on outstanding scientific, methodological and technical issues to maintain the policy work which is necessary to incorporate agriculture into the climate change negotiations and legal frameworks.

Many agricultural practices have small net emission reduction potentials per hectare, but if the 404 million smallholder farms worldwide (Nagayets 2005) across different farming systems adopted these mitigation practices, a substantial impact on global emissions could be made. Given the range of data presented of the different farming practices it is hard to make one recommendation to prioritize efforts towards one of these, as they will differ with respect to the agro-ecological zones, the soils, the climate, but also according to the economic potential and the feasibility of the farmers to adopt new practices. We would stress the importance of finding technical solutions which are practical and less time consuming for the farmers to implement from a technical, an economic, as well as a cultural point of view.

As the overview of projects indicates, there is still a vast potential for agricultural mitigation to be tapped, with an understanding of the challenges and opportunities unique to agriculture in measuring and quantifying mitigation impacts. Mitigation can never be the main goal of agricultural production, but it can be an important co-benefit. Agriculture is always aimed to produce food and other necessities and support livelihoods. However, if we find a way to engage the small farms that occupy about 60% of the arable land worldwide (IAASTD, 2009), a significant contribution to climate change mitigation can be made.

To develop viable agricultural mitigation programs, there is a need for better data on the mitigation impacts of different farming practices. The data in this chapter are based on cases that used different methodologies, making it difficult to compare across studies. For many farming practices in developing countries more data is needed to represent the diversity of agro-ecological zones, as well as provide input for site-specific mitigation strategies. Harmonized data would support the calculation of emission factors for meaningful bundles of farming practices, which would better reflect real conditions.

It is difficult to imagine a financing system that would work for smallholders based on monitoring of exact measurements, as the transaction costs would be impractically high. Therefore, we need to establish practice-based approaches and build in the necessary buffers to guarantee that emissions reductions are real. The emission factors can be used in deriving the net impact of emissions when adopting a defined bundle of farming practices. This would mean monitoring the adoption of practices, rather than the exact emission reduction, to be used as a basis for mitigation payments.

In the AFOLU context, the interactions between agricultural production and forestry need to be considered. Agriculture is a significant driver of deforestation in many parts of the world, and agricultural intensification on existing areas might be required to be part of a mitigation package. Therefore a comprehensive landscape approach is needed to manage REDD and agriculture and to ultimately achieve the goal of reduced emissions.

## Note

1 AFOLU and terrestrial carbon are used interchangeably.

## References

Conant, R.T., Paustian, K. and Elliott, E.T. (2001) Grassland management and conversion into grassland: effects on soil carbon, *Ecol. Appl.*, vol. 11, no. 2, pp. 343–355.

Conant and Paustian (2002) Potential soil carbon sequestration in overgrazed grassland ecosystems, *Global Biogeochemical Cycles*, vol. 16, no. 4, p. 1143.

Eagle, A.L, Henry, L.R., Olander L.P., Haugen-Kozyra, K., Millar, N. and Robertson, P. (2010) *Literature review: greenhouse gas mitigation potential of agricultural land management activities in the U.S.*, Draft, Nicholas Institute for Environmental Policy Solutions, Duke University, Durham, NC.

FAO (2010a) 'Agriculture, Forestry and Other Land Use Mitigation Project Database – An assessment of the current status of land-based sectors in the carbon markets', by Varming, M., Seeberg-Elverfeldt, C. and Tapio-Biström, M. Mitigation of Climate Change in Agriculture Project, FAO, Rome, Italy.

FAO (2010b) 'Review of Evidence on Drylands Pastoral Systems and Climate Change', by Neely, C., Bunning, S. and Wilkes, A., FAO, Rome, Italy.

Forest Trends, The Katoomba Group, EcoAgriculture Partners & Climate Focus (2010) 'An African Agricultural Carbon Facility–Feasibility Assessment and Design Recommendations'.

Gockowski, J. and Sonwa, D. (2011) Cocoa intensification scenarios and their predicted impact on $CO_2$ emissions, biodiversity conservation and rural livelihoods in the Guinea Rainforest of West Africa, *Environmental Management*, vol. 48, no 2. pp 307-321

Govaerts, B., Verhulst, N., Castellanos-Navarrete, A., Sayre, K.D., Dixon, J. and Dendooven, L. (2009) Conservation Agriculture and Soil Carbon Sequestration; Between Myth and Farmer Reality, *Critical Reviews in Plant Science*, vol. 28, pp. 97-12.

Grace, J., San Jose, J., Meir, P., Miranda, H. and Montes, R. (2006) Productivity and carbon fluxes of tropical savannas, *Journal of Biogeography*, vol. 33, pp. 387-400.

IAASTD (2009) *Agriculture at a Crossroads: Global Report*. Island Press, Washington, DC.

IPCC (2007) 'Agriculture' in *Climate Change 2007: Mitigation. Contribution of Working Group III to the Fourth Assessment Report of the Intergovernmental Panel on Climate Change*. Cambridge University Press, Cambridge, United Kingdom and New York, NY.

Nair, P.K.R., Kumar, B.M. and Nair, V.D. (2009) Agroforestry as a strategy for carbon sequestration, *Journal of Plant Nutrition and Soil Science*, vol. 172, pp. 10-23.

Nagayets, O. (2005) *Small farms: current status and key trends,* Information brief, Prepared for the Future of Small Farms Research Workshop, Wye College, 26–29 June 2005.

Niggli, U., Fliessbach, A., Hepperly, P. and Scialabba, N. (2009) 'Low Greenhouse Gas Agriculture: Mitigation and Adaptation Potential of Sustainable Farming Systems'. FAO, Rome, Italy.

Ortiz-Monasterio, I., Wassmann, R., Govaerts, B., Hosen, Y., Katayanagi, N. and Verhulst, N. (2010) 'Greenhouse gas mitigation in the main cereal systems: rice, wheat and maize', in M.P. Reynolds (ed.) *Climate change and crop production*. CAB International, Wallingford.

Tennigkeit, T. and Wilkes, A. (2008) *An assessment of the potential for carbon finance in rangelands*. ICRAF Working Paper No. 68.

UNEP Risoe Centre (2010) *CDM pipeline overview* [Online]. Available at: http://cdmpipeline.org/publications/CDMpipeline.xlsx (Accessed: 15 August 2010).

UNFCCC (2010) *Glossary of climate change acronyms* [Online]. Available at: http://unfccc.int/essential_background/glossary/items/3666.php#C (Accessed: 3 December 2010).

Wassmann, R., Lantin. R.S., Neue, H.U., Buendia, L.V., Corton, T.M. and Lu, Y. (2000) Characterization of methane emissions from rice fields in Asia. III. Mitigation options and future research needs, *Nutrient Cycling in Agroecosystems*, vol. 58, pp. 23–36, 2000.

West, T.O., and Post, W.M. (2002) Soil organic carbon sequestration rates by tillage and crop rotation: a global data analysis, *Soil Science Society of America Journal*, vol. 66, pp. 1930–1946.

Yagi, K., Tsuruta, H. and Minami, K. (1997) Possible mitigation options from rice cultivation, *Nutrient Cycling in Agroecosystems*, vol. 49, pp. 213–220.

Yan, X., Akiyama, H., Yagi, K. and Akimoto, H. (2009) Global estimations of the inventory and mitigation potential of methane emissions from rice cultivation conducted using the 2006 Intergovernmental Panel on Climate Change Guidelines, *Global Biochemical Cycles*, vol. 23, pp. 1–15.

# 3

# EVALUATING SYNERGIES AND TRADE-OFFS AMONG FOOD SECURITY, DEVELOPMENT, AND CLIMATE CHANGE

*Nancy McCarthy, Leslie Lipper, Wendy Mann,*
*Giacomo Branca, and Jeronim Capaldo*

## Introduction

Meeting the food demands of a global population expected to increase to 9 billion by 2050, and improving incomes and livelihoods, will require greater resilience to the impacts of climate change and other major improvements in agricultural production systems. While producing food, fuel and fibre, agriculture can also lower emissions and remove carbon from the atmosphere. The questions addressed in this chapter are (1) to what extent are there potential synergies or trade-offs between the changes in production required by food security objectives and those needed to increase mitigation; and (2) identifying means of enabling actions that contribute to both objectives.

## Global food needs to 2050

Although the growth rate of global population is declining, the United Nations (UN) projects that total population will increase by more than 50% by 2050 (UN 2009), from the current 6 billion to approximately 9 billion. Most of the increase is projected to occur in South Asia and sub-Saharan Africa. Both regions have a large share of the world's poor and food insecure, whose livelihoods depend on agriculture.

FAO projects that by 2050 global agricultural production will need to grow by 70% in order to feed the greater population and reduce rates of undernutrition (Bruinsma 2009). Projections' results are based on several assumptions about population and income growth as well as dietary patterns. Under the baseline scenario,[1] per capita calorie availability is projected to increase 11% by 2050 to an average of 3130 kcal/per capita.

Under this scenario, 4% of the developing world population would still be food insecure.[2]

There are three main ways of increasing agricultural production: (1) bringing new land into agricultural production; (2) increasing the cropping intensity on existing agricultural lands; and (3) increasing yields on existing agricultural lands. Adopting any one of these strategies will depend upon local availability of land and water resources, agro-ecological conditions, technologies used for crop production and the state of infrastructure and institutions.

Bruinsma (2009) investigates present and future land and yield combinations for 34 crops under rainfed and irrigated conditions in 108 countries and country groups. Results show that 75% of projected crop growth in developing countries comes from yield gains and 16% from cropping intensity. Arable land expansion is found to be an important source of growth in sub-Saharan Africa and Latin America. These results suggest that potential tensions may arise between the need to increase food production and the transition towards sustainable, low-emission agriculture strategies, unless viable strategies to meet both goals are developed.

## Agriculture mitigation options

The remainder of this chapter focuses on changes in agricultural practices and their implications for mitigation, food security and adaptation following FAO (2009). We refer to four of the five categories of agricultural practices identified by Smith *et al.* (2007) for terrestrial mitigation options from agriculture.[3] As we show below, it is striking that changes in agricultural practices needed to generate mitigation are often also required to improve food security and adaptation.

The major sources of terrestrial mitigation from agriculture, following Smith *et al.* (2007a) are described below:

### *Cropland management*

* Improved agronomic practices generate higher inputs of carbon (C), leading to potential increased soil C storage (Follett *et al.* 2001). Such practices include using improved crop varieties, extending crop rotations, avoiding use of bare fallow and using cover crops.
* Integrated nutrient management, which can reduce emissions by reducing leaching and volatile losses, improving nitrogen (N) use efficiency through precision farming and improving fertilizer application timing (Lal and Bruce 1999). Off-site emissions may also decrease to the extent that demand for N fertilizer diminishes.
* Increasing available water in the root zone through water management, which can enhance biomass production, increase the amount

of above-ground and the root biomass returned to the soil, and improve soil organic C concentration.

- Tillage that minimally disturbs soil and incorporates crop residue, which decreases soil C losses through enhanced decomposition and erosion.[4] Systems that retain crop residues tend to increase soil C because these residues are the precursors of soil organic matter.
- Agroforestry systems increase above-ground C storage and may also reduce soil C losses stemming from erosion.

### Improved grassland management

- Improved productivity through increasing nutrients for plant uptake and reducing the frequency or extent of fires.
- Improved grazing management by controlling intensity and timing of grazing (e.g., stocking rate management,[5] rotational grazing, and enclosure of grassland from livestock grazing).

### Restoration of degraded lands

- Carbon storage in degraded lands can be partly restored by practices that reclaim soil productivity (Lal 2004) (e.g. revegetation; applying nutrient amendments and organic substrates such as manures, biosolids, and composts; reducing tillage and retaining crop residues; and conserving water).

The technical mitigation potential in agriculture, considering all gases, is estimated to be between 4,500 (Caldeira *et al.* 2004) and 6,000Mt $CO_2e$/year (Smith *et al.* 2008) by 2030. Figure 3.1 provides rough estimates by region.

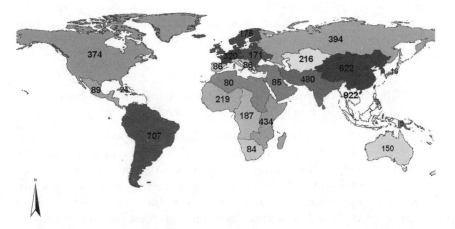

*Figure 3.1* Technical mitigation potentials (all practices, all GHGs: $MtCO_2$-e/yr) by 2030, by region (source: Smith *et al.* 2007).

## Food security and adaptation options

This section assesses the impact of agricultural practices on the two components of food security directly linked to land use and management: [6] the level and stability of food production (FAO 2009). Table 3.1 provides details on examples of activities and investments that have been proposed to increase agricultural production and decrease the variability of output caused by climate variability and extreme climate shocks.

An important point captured in the table is that short-term impacts on food production can differ substantially from long-term impacts. For several options, short-term impacts may be negative depending on underlying agro-ecological conditions, previous land-use patterns, and current land use and management practices. However, long-term impacts on production are expected to be positive both in terms of average levels and of stability. For instance, crop and grassland restoration projects often take land out of production for a significant period of time, reducing cultivated or grazing land available in the short run, but leading to overall increases in productivity and stability in the long run.

A different type of trade-off may occur when incorporating crop residues that are expected to increase soil fertility and water retention capacity, thereby increasing yields at least over the medium–long term. In African countries where livestock are an important component of the food production system, trade-offs between livestock and crop use for residues is often cited as a main constraint to the adoption of conservation tillage by smallholders (Giller *et al.* 2009).

While there can be trade-offs in terms of average yearly food production in the short term, there are fewer identified negative impacts on yield variability. However, yield variability can increase in the short term where changes in activities require new knowledge and experience, and farmers unfamiliar with such systems require some time to successfully adopt the practice (e.g., fertilizer application or the construction of water retention structures where incidence and severity of both droughts and floods are expected to increase in the future).

## Mitigation, adaption and food security: where and how can synergies be realized?

A comparison of the mitigation options discussed above with those for food security and adaptation analyzed in Table 3.1 shows that nearly all terrestrial-based agriculture mitigation options contribute as well to adaptation and food security. The potential for synergies is particularly promising for changing food production practices such as improved crop varieties; avoiding bare fallow and changing crop rotations to incorporate food-producing cover crops and legumes; adopting precision fertilizer management in other regions; seeding fodder and improving forage quality and quantity on pastures areas; expansion of low-energy irrigation; and expansion of agroforestry and soil and water conservation techniques that do not take significant amounts of land out of food production. Trade-offs are more likely when mitigation options take land out of production, either temporarily or permanently. For instance, restoration of degraded lands often requires that land not be used for production at least in the short term, whereas avoiding draining or restoring wetlands would directly take land out of production permanently.

By combining information from Table 3.1 with estimates of mitigation potential by practice, it is possible to derive a chart of the synergies and trade-offs between mitigation and food security for specific practices and regions, as shown in Figure 3.2. Because impacts will vary by agro-ecological conditions, historical land use and current production systems, the chart is only illustrative. Additionally, to simplify the chart, we over-look short-term benefits to food security, although it is important to recognize that trade-offs can occur.

## 2.5 Costs

The analysis above essentially looks at the synergies and trade-offs in terms of benefits for mitigation, adaptation and food security. However, in order to prioritize actions, it is necessary to calculate the full costs of achieving these benefits. There are wide ranges in cost estimates by different sources, reflecting the large differences among regions but also reflecting which costs were considered in the analyses. For instance, McKinsey (2009) pro-vides cost estimates for mitigation from crop and grassland management,

| | Mitigation potential | |
|---|---|---|
| **Food security potential** | Food Security Potential: High<br>C-Sequestration Potential: Low<br><br>• Expand cropping on marginal lands<br>• Expand energy-intensive irrigation | Food Security Potential: High<br>C-Sequestration Potential: High<br><br>• Restore degraded land<br>• Expand low-energy intensive irrigation<br>• Agro-forestry options that increase food/incomes<br>• Zero tillage and residue cover, where limited trade-offs with livestock |
| | Food Security Potential: Low<br>C-Sequestration Potential: High<br><br>• Reforestation/afforestation<br>• Agro-forestry options with limited food/income benefits | Food Security Potential: Low<br>C-Sequestration Potential: Low<br><br>• Bare fallow<br>• Continous cropping without soil amendments<br>• Heavily grazed pastures |

Mitigation potential

*Figure 3.2* Examples of potential synergies and trade-offs

Table 3.1 Food production and resilience impacts of changes in selected agricultural production systems

| Changes in agricultural systems | Impacts on food production | | Impacts on yield variability and exposure to extreme weather events | |
|---|---|---|---|---|
| | Positive | Negative | Positive | Negative |
| **Cropland management** | | | | |
| Improved crop varieties | Increased crop yield | | Reduced variability at plot level; potentially reduced variability at the local/national levels | |
| Use of cover crops | Higher yields due to reduced on-farm erosion and reduced nutrient leaching | May conflict with custom of using cropland for grazing post-harvest | Reduced variability due to increased soil fertility, water-holding capacity | |
| Increased efficiency of fertilizer | Higher yields through more efficient use of N fertilizer and/or manure | | Lower variability more likely where good drainage and drought infrequent; experience can reduce farm-level variability over time | |
| Incorporation of residues | Higher yields through increased soil fertility, increased water-holding capacity | Potential trade-off with use as animal feed | Reduced variability due to increased soil fertility, water-holding capacity | |
| Reduced/zero tillage | Higher yields over long run, particularly where increased soil moisture is valuable | May have limited impacts on yields in short term; weed management becomes very important | Reduced variability due to reduced erosion and improved soil structure, increased soil fertility | Potentially greater variability where drought frequent and inexperienced users |

| | | | | |
|---|---|---|---|---|
| Perennials/agro-forestry | Higher yields on adjacent lands in medium–long term, better rainwater management; higher tree-crop incomes | Potentially less food, at least in short term, if displaces intensive cropping patterns | Reduced variability of agroforestry and adjacent crops | |
| Water management | | | | |
| Irrigation | Higher yields, greater intensity of land use | | Reduced variability in well-functioning systems | |
| Bunds and terraces | Higher yields, particularly where increased soil moisture is key constraint | Potentially lower yields when extremely high rainfall | Reduced variability in dry areas with low likelihood of floods and/or good soil drainage | Some designs may increase crop damage under heavy rains |
| Pasture and grazing management | | | | |
| Improving forage quality and quantity | Higher livestock yields due to more and higher quality forage | | Reduced variability where improved forage is adapted to local conditions | Increased variability where improved forage more sensitive to climate than natural pasture |
| Seeding fodder grasses | Higher livestock yields due to greater forage availability | | Reduced variability where seeded fodder is adapted to local conditions | Increased variability where improved fodder more sensitive to climate than natural pasture |
| Stocking rate management | Potential increased returns per unit of livestock | Returns at the herd level may decline, at least in the short term | Potentially lower variability in long term, where forage availability is key factor in livestock output variability | |

*Sources:* Antle *et al.* (2007); Baudeon *et al.* (2007); Dutilly-Diane *et al.* (2003); FAO (2007); Freibauer *et al.* (2004); Giller *et al.* (2009); Kwesiga *et al.* (1999); Lal (2004); Niles *et al.* (2002); Nyende *et al.* (2007); Pretty and Hine (2001); Pretty *et al.* (2003); Rosenzweig and Tubiello (2006); Verchot *et al.* (2007).

restoration of organic soil, and restoration of degraded land. Average costs per tonne of carbon equivalent abated to the year 2030 are computed to be negative for crop and grassland nutrient management and tillage and residue management, indicating that these activities should generate higher benefits than costs over the time frame considered. While this analysis is useful in indicating which practices will be self-sustaining in the long run, it does not include opportunity costs and, by using a 20-year time horizon, masks the magnitude of the initial investments required, which is a main barrier faced by smallholders.

These costs constitute a potential adoption barrier as smallholders often lack resources to make the investments needed to realize higher yields and positive net benefits in the longer run (FAO 2009). And while the McKinsey figures appear to show great promise for many agriculture-based mitigation options, there is also a very large body of literature documenting the costs and barriers to sustainable land management practices (World Bank 2007; Nkonya et al. 2004; Barrett et al. 2002; Otsuka and Place 2001; McCarthy et al. 2004). These barriers include limited credit to finance upfront costs, tenure security; lack of access to effective research and extension services for capacity building and technical assistance; limited access to insurance; and lack of access to markets for inputs and equipment as well as output markets. More extensive and country-specific analysis of costs will be needed in future.

## 3 Conclusion

The main finding of the chapter is that there are a wide range of practices and technologies that have the potential to increase food production and the adaptive capacity of the food production system, as well as reduce emissions or enhance carbon storage in agricultural soils and biomass. However, even where such synergies exist, capturing them may entail significant investment and opportunity costs, particularly for smallholders in the short term. Other findings include:

- Synergies differ across localities, and thus a necessary first step is to identify where the potential synergies and trade-offs occur in specific circumstances.
- Even where significant trade-offs between food security and mitigation might occur from a proposed land-use change, it is important to determine if there are opportunities to minimize such trade-offs.

These results provide a framework for identifying synergies and trade-offs, which can be used to prioritize actions and identify appropriate funding sources (FAO 2009). This is particularly important in the agricultural sector where, after 25 years of relative neglect, there is renewed interest in investing in agriculture for food security and in ensuring that agricultural systems can adapt to climate change (World Bank 2007; Parry *et al.* 2007). Additional financing from mitigation and adaptation funds can be leveraged to stimulate synergies and minimize trade-offs between food security, adaptation and mitigation (FAO 2009).

## Notes

1 Described in FAO (2006).
2 This result is based upon an assumption of shifts in dietary patterns involving an increase in the share of high value foods and meat consumed as incomes rise. Lower calorie content and inefficiency associated with conversion of feed grains to meat calories translates into reduced increases in caloric availability per increase of agricultural production. Actual experience with income growth and dietary transformation could vary from this projection, which would in turn affect the needed supply response from agriculture—however in this report we will use this as the base case.
3 In this chapter, we will not consider further the scope for emissions reductions from organic soils; these soils comprise a relatively small land area, though degraded organic soils also offer the greatest potential increases in soil organic carbon. Options for changing agricultural practices, e.g., in paddy rice, are generally complex (Seeberg-Elverfeldt and Tapio-Biström, this volume, offer a more detailed discussion of options for changing rice cultivation practices).
4 In general terms, conservation tillage (CT) is defined as any method of seedbed preparation that leaves at least 30% of ground covered by crop residue mulch (Lal 1997).
5 The intensity and timing of grazing can influence the removal, growth, C allocation, and flora of grasslands, thereby affecting the amount of C in the soils (Freibauer *et al.* 2004).
6 Adaptation options, which do not directly affect land use, include improved climate forecasting and dissemination, altering planting/harvesting dates, and developing weather-based insurance schemes (FAO 2008; Howden *et al.* 2007).

## References

Antle, J.M., Stoorvogel, J.J. and Valdivia, R.O. (2007) Assessing the Economic Impacts of Agricultural Carbon Sequestration: Terraces and Agroforestry in the Peruvian Andes, *Agriculture, Ecosystems & Environment*, vol. 122, no. 4, pp. 435–445.
Barrett, C.B., Place, F. and Aboud, A.A. (eds) (2002) *Natural Resources Management in African Agriculture: Understanding and Improving Current Practices*, CAB International.
Baudeon, F., Mwanza, H.M., Triomphe, B. and Bwalya, M. (2007) *Conservation agriculture in Zambia: a case study of Southern Province*, African Conservation Tillage Network, Centre de Coopération Internationale de Recherche Agronomique pour le Développement, FAO, Nairobi.
Bruinsma, J. (2009) *The Resource Outlook to 2050*, in *Expert Meeting on 'How to Feed the World in 2050'*, FAO, Rome.
Caldeira, K., Morgan, G., Baldocchi, D., Brewer, P., Chen, C.T.A., Nabuurs, G.-J. Nakicenovic, N. and Robertson, G.P. (2004) A portfolio of carbon management options, in C.B. Field (ed), *Towards CO$_2$ stabilization*, Island Press, New York.
Dutilly-Diane, C., Sadoulet, E. and de Janvry, A. (2003) How Improved Natural

Resource Management in Agriculture Promotes the Livestock Economy in the Sahel, *Journal of African Economies*, vol. 12, pp. 343–370.

FAO (2006) *World Agriculture: towards 2030/2050. Interim report*, FAO, Rome.

FAO (2007) *The State of Food and Agriculture. Paying Farmers for Environmental Services*. FAO: Rome.

FAO (2008) Climate Change and Food Security: A Framework Document. Rome.

FAO (2009) Food Security and Agricultural Mitigation in Developing Countries: Options for Capturing Synergies. Rome.

Follett, R.F., Kimble, J.M. and Lal, R. (2001) *The potential of U.S. grazing lands to sequester soil carbon*, in Follett, R.F., Kimble, J.M. and Lal R. (ed), *The Potential of U.S. Grazing Lands to Sequester Carbon and Mitigate the Greenhouse Effect*, Lewis Publishers, Boca Raton, Florida, pp. 401–430.

Freibauer, A., Rounsevell, M., Smith, P. and Verhagen, A. (2004) Carbon sequestration in the agricultural soils of Europe, *Geoderma*, vol. 122, pp. 1–23.

Giller, K.E., Witter, E., Corbeels, M. and Tittonell, P. (2009) Conservation agriculture and smallholder farming in Africa: The heretics' view, *Field Crops Research*, vol. 114, pp. 23–34.

Howden, S.M., Soussana, J.F., Tubiello, F.N., Chhctri, N., Dunlop, M. and Meinke, H. (2007) Adapting agriculture to climate change, *PNAS*, vol. 104, pp. 19691–19696.

Kwesiga, F.R., Franzel, S., Place, F., Phiri, D. and Simwanza, C.P. (1999) Sesbania sesban improved fallows in eastern Zambia: Their inception, development and farmer enthusiasm, *Agroforestry Systems*, vol. 47, pp. 49–66.

Lal, R. (1997) Residue management, conservation tillage and soil restoration for mitigating greenhouse effect by $CO_2$-enrichment, *Soil and Tillage Research*, vol. 43, pp. 81–107.

Lal, R. (2004) Carbon emissions from farm operations, *Environment International*, vol. 30, pp. 981–990.

Lal, R. and Bruce, J.P. (1999) The potential of world cropland soils to sequester C and mitigate the greenhouse effect, *Environmental Science and Policy*, vol. 2, pp. 177–185.

McCarthy, N., Dutilly-Diane, C., Drabo, B., Kamara, A. and Vanderlinden, J.P. (2004) *Managing Resources in Erratic Environments: An Analysis of Pastoralist Systems in Ethiopia, Niger, and Burkina Faso*. International Food Policy Research Institute: Washington, DC.

McKinsey & Company (2009) *Pathways to a Low-Carbon Economy: Version 2 of the Global Greenhouse Gas Abatement Cost Curve*, London, McKinsey & Company.

Niles, J.O., Brown S., Pretty J.N., Ball, A.S. and Fay, J. (2002) Potential carbon mitigation and income in developing countries from changes in use and management of agriculture and forest lands, *Philosophical Transactions of the Royal Society of London A*, vol. 360, pp. 1621–1639.

Nkonya, E., Pender, J., Jagger, P., Sserunkuuma, D., Kaizzi, C. and Ssali, H. (2004) *Strategies for sustainable land management and poverty reduction in Uganda.*, in *Research Report n. 133*, International Food Policy Research Institute, Washington, DC.

Nyende, P., Nyakuni, A., Opio, J.P. and Odogola, W. (2007) *Conservation agriculture: a Uganda case study*, African Conservation Tillage Network, Centre de Coopération Internationale de Recherche Agronomique pour le Développement, FAO, Nairobi.

Otsuka, K. and Place, F. (2001) *Land tenure and natural resource management. A comparative study of agrarian communities in Asia and Africa.* International Food Policy Research Institute: Washington, DC.

Parry, M.L., Canziani, O.F., Palutikof, J.P., van der Linden, P.J. and Hanson, C.E. (eds) (2007) *Climate Change 2007: Impacts, Adaptation and Vulnerability. Contribution of Working Group II to the Fourth Assessment Report of the Intergovernmental Panel on Climate Change*, Cambridge University Press, Cambridge, UK. pp. 391–431.

Pretty, J. and Hine, R. (2001) Reducing Food Poverty with Sustainable Agriculture: A Summary of New Evidence, in *Final Report of the 'SAFE-World' (The Potential of Sustainable Agriculture to Feed the World) Research Project*, University of Essex, Colchester, UK.

Pretty, J.N., Morison, J.I.L. and Hine, R.E. (2003) Reducing food poverty by increasing agricultural sustainability in developing countries, *Agriculture, Ecosystems & Environment*, vol. 95, pp. 217–234.

Rosenzweig, C. and Tubiello, F.N. (2006) Adaptation and mitigation strategies in agriculture: an analysis of potential synergies, *Mitigation and Adaptation Strategies for Global Change*, vol. 12, pp. 855–873.

Smith, P., Martino, D., Cai, Z., Gwary, D., Janzen, H.H., Kumar, P., McCarl, B., Ogle, S., O'Mara, F., Rice, C., Scholes, R.J., Sirotenko, O., Howden, M., McAllister, T., Pan, G., Romanenkov, V., Rose, S., Schneider, U. and Towprayoon, S. (2007) 'Agriculture', in Metz, B., Davidson, O.R., Bosch, P.R., Dave, R. and Meyer, L.A. (eds) *Climate Change 2007: Mitigation*, Contribution of Working Group III to the Fourth Assessment Report of the Intergovernmental Panel on Climate Change. Cambridge University Press, Cambridge and New York, NY.

UN (2009) World Population Prospects: The 2008 Revision. New York, United Nations Population Division.

Verchot, L.V., Van Noordwijk, M., Kandji, S., Tomich, T., Ong, C., Albrecht, A., Mackensen, J., Bantilan, C., Anupama, K.V. and Palm, C. (2007) Climate Change: Linking adaptation and mitigation through agroforestry, *Mitigation and Adaptation Strategies for Global Change*, vol. 12, pp. 901–918.

World Bank (2007) *Agriculture for Development: World Development Report 2008.* The World Bank: Washington, DC.

# 4

# ACHIEVING MITIGATION THROUGH SYNERGIES WITH ADAPTATION

*Pete Smith and Eva Wollenberg*

## Introduction

Agriculture has the potential to contribute significantly to cost-effective greenhouse gas (GHG) mitigation potential (Smith *et al.* 2007a; 2008), comparable at a range of future carbon prices to the economic potential in the industry, energy, transport, and forestry sectors (Barker *et al.* 2007). Based on carbon prices of US$20–100/t $CO_2$e, the agricultural mitigation potential from all gases is ~1,500–4,300 megatonnes (Mt) $CO_2$e/yr, or 25–78% of the technical potential of ~5,500–6,000 Mt $CO_2$e/yr (Smith *et al.* 2007a; Smith and Olesen 2010). The annual economic mitigation potential is estimated to be worth between US$32 thousand million and US$420 thousand million. About 70% of the potential arises from developing countries with a further 10% from countries with economies in transition (Trines *et al.* 2006).

Despite mitigation's significant economic potential, however, most farmers in developing countries are smallholders who will have to prioritize farm practices that enhance their food and livelihood security. This implies that to achieve the economic mitigation potential in developing countries, mitigation measures will need to support improved food production, especially that which is adapted to the changing climate, and profitability. Such practices are often referred to as "win–win" options, and strategies to implement such measures can be encouraged on a "no regrets" basis (Smith & Powlson, 2003), i.e., they provide other benefits even if the mitigation potential is not realized.

For many farmers, mitigation payments will only be co-benefits to the main enterprise of food production. The aim of this chapter is therefore to identify the synergies of mitigation with food security and improved livelihoods to support viable mitigation options for farmers in developing countries. We focus on agronomic adaptation of crops and livestock as an indicator of food security. Synergies between mitigation and food security have been examined elsewhere (Mann *et al.* 2009; see also McCarthy *et al.*). Below we discuss the range of mitigation options, the relationship of mitigation to adaptation and livelihoods, and a framework for further testing of synergies.

## Identifying the best options for greenhouse gas mitigation in agriculture in developing countries

The Intergovernmental Panel on Climate Change (IPCC) AR4 considered approximately 60 greenhouse gas mitigation options in agriculture (Smith *et al.* 2008). These can be grouped into several broad categories including cropland management (such as agronomy, nutrient management, tillage and residue management, and better use of manures), grazing land management, agroforestry and set-aside, the restoration of cultivated organic soils (such as peats), the restoration of degraded lands, livestock management, improved manure storage, and rice management.

Some of these options work better than others in different contexts, so a "one-size-fits-all" recommendation cannot be made. For this reason, mitigation strategies need to be determined according to local conditions. Farmers in different contexts will also have different priorities. Given the importance of food security and economic development options for most farmers in developing countries, the most favourable mitigation options also confer increased resilience and enhanced adaptation to future climate change, and support sustainable development. Focusing on options where synergies occur among mitigation, adaptation, and sustainable development should help to incentivize agricultural mitigation in developing countries.

## Relationship between mitigation and adaptation in agriculture

Climate change mitigation, adaptation and impacts will happen simultaneously and interactions will occur. Mitigation-driven actions in agriculture can have either (a) positive adaptation consequences (such as soil carbon sequestration projects that enhance water-holding capacity and increase drought preparedness) or (b) negative adaptation consequences (for example, if heavy dependence on biomass energy increases the sensitivity of energy supply to climatic extremes).

Nearly 90% of the mitigation potential in annual crop-based agriculture lies in reducing net soil carbon dioxide ($CO_2$) emissions (by restoring cultivated organic soils, for example) or in sequestering $CO_2$ in the soil organic matter of mineral soils (Smith *et al.* 2007a; 2008). It has long been known that increasing soil organic matter content improves soil fertility, nutrient supply, soil structure, water-holding capacity, and a host of other vital soil functions (Smith 2004). These functions increase the resilience of the soil under threat from future climate change. Soil carbon sequestration then is one of the clearest examples of a mitigation measure that also protects against changes in climate and enhances adaptation and the level and sustainability of crop production (Pan *et al.* 2009). See also chapter by Verhulst *et al.* in this volume for a discussion. Other examples include the application of animal manure to soils, which reduces fertilizer use and thus emissions, but also improves soil structure and water-holding capacity; the reduction of tillage intensity with improved residue management, which can increase soil carbon while retaining soil moisture and reduce fossil fuel inputs while increasing productivity per hectare; and the restoration of degraded lands, which can sequester carbon and also enhance livelihoods and the resilience of the soils for sustaining agriculture under a changing climate.

Agroforestry and management of trees or forests across landscapes also should constitute a high proportion of mitigation potential, but sectoral analyses have limited our understanding of integrated estimates for their technical or economic potential (see chapters by Semroc *et al.* and Mendez *et al.* for examples of mitigation in agroforestry). The multiple functions of trees such as *Faidherbia albida* could contribute to adaptation through enhanced soil quality and microclimate. Intensification of agriculture across the landscape that also reduces conversion of carbon-rich forests and grasslands should also enhance mitigation (Burney *et al.* 2010).

Adaptation-driven actions also may have both (a) positive mitigation consequences (as when residue returned to fields to improve water-holding capacity also sequesters carbon), or (b) negative mitigation consequences (for example, an increased use of nitrogen fertilizer to overcome falling yield that leads to increased nitrous oxide emissions and fossil fuel use for the production of fertilizers).

Some management changes that are mainly made for the purposes of adaptation also increase carbon sequestration and so enhance mitigation. For example, conservation tillage increases soil water retention in the face of drought while also sequestering carbon below ground. Small-scale irrigation facilities not only conserve water to cope with greater variability, but also to increase crop productivity and soil carbon stocks. Agroforestry systems increase above- and below-ground carbon storage while also increasing water storage below ground, even in the face of extreme climate events. Properly managed rangelands can cope better with drought and sequester significant amounts of carbon. Project- and program-based funding schemes that support adaptation should also be able to draw on mitigation resources.

To understand the synergies between mitigation and adaptation, full life cycle accounting is necessary of mitigation according to GHG emissions, carbon sequestration and changes in land use at the farm and landscape scale over multiple years (Palm *et al.* 2010). These results need to be compared against resulting yields and ecosystem services that support agricultural resilience and adaptation. Mitigation measures can have conflicting impacts on emissions. For example, use of animal manure may decrease needs for fertilizer, but increase nitrous oxide emissions. Residue management may increase sequestration of carbon in soil organic matter, but also result in increased nitrous oxide emissions. Therefore net mitigation impacts across all GHGs need to be examined with the life cycle accounting. Mitigation can also have mixed consequences with adaptation. Precision fertilizer may increase yields, but increase expansion of agricultural land into forest. A variety of livestock may be better adapted to heat stress and therefore requires fewer numbers of animals, but yield more waste. While it will not be possible to determine the net consequences of a multitude of options in all locations, testing a range of likely "win–win" options should begin to indicate patterns.

In many cases, farmers will take actions for reasons that have nothing to do with either mitigation or adaptation—for example, actions taken toward enhancing soil fertility or food security. But these events may still have considerable consequences for mitigation or adaptation as seen when deforestation for agriculture results in the loss of both carbon and ecosystem diversity. Understanding the mitigation and adaptation consequences of predominant trends among farmers will be important to developing policies for improved mitigation.

Climate change impacts also influence mitigation. In terms of mitigation, the accumulation rates for sequestered soil carbon, the growth rates for bioenergy feedstocks, the size of livestock herds, and the rates of sequestration are all variables affected by climate change. Depending upon the climatic impact, there are likely to be shifts in, among other things, plant and tree growth, microbial decomposition of soil carbon, livestock growth, and the composition and distribution of plant and animal species. All these factors will alter mitigation potential; some positively and some negatively. For example, lower livestock growth rates could increase herd size and, consequently, emissions from manure and enteric fermentation, while increased microbial decomposition under higher temperatures will lower soil carbon sequestration potential. Interactions also occur with adaptation. Crop mix and changes in both land usage and irrigation are all potential adaptation strategies to warmer climates. All would alter mitigation potential.

## Mitigation measures and livelihoods

There are various potential impacts of agricultural GHG mitigation on smallholder farmers' livelihoods. Smith *et al.* (2007b) reviewed evidence for the economic impacts of 18 mitigation measures and identified nutrient management, rice management (e.g., mid-season drainage and improved fertilization), water management, grazing land management or pasture improvement, land restoration, enhanced energy efficiency, improved livestock feed practices, and improved manure storage to all have positive or likely positive impacts.

In contrast, increased use of agricultural land under trees or forest is likely to cause farmers to lose income from the converted cropland, but could yield other benefits. Use of biosolid wastes from sewage would likely increase costs. Other practices such as agroforestry, reduced tillage, and management of organic soils had uncertain economic impacts. There was insufficient data to evaluate the impacts of animal manure, breeding of livestock, or use of inoculants or additives in livestock practices. The magnitude of impacts will depend upon the scale and intensity of the mitigation measures and where they are undertaken.

Mitigation measures can indirectly affect livelihoods by improving the quality and quantity of natural capital such as soil or water resources in the long run. Environmental aspects of mitigation need to be better acknowledged, quantified, valued and incorporated in decision-making frameworks to understand their potential value to farmers' livelihoods over time. Payments or other compensation for mitigation practices should acknowledge these benefits.

More broadly, macro-economic policies to alleviate poverty and improve well-being in developing countries, through encouraging sustainable economic growth and sustainable development, should also serve to lower barriers to mitigation by removing the threat of poverty and hunger. Ideally policies associated with fair trade, reduced subsidies for agriculture in the developed world, and less onerous interest rates on loans and foreign debt all need to be considered. This may provide an environment in which climate change mitigation in agriculture in developing countries could flourish. The UK's Stern Review (Stern 2006) warns that unless we take action in the next 10–20 years, the environmental damage caused by climate change later in the century could cost between 5 and 20% of global GDP every year.

## A framework for assessing synergies among mitigation, adaptation and livelihood impacts

To summarize, while the general relationship of some mitigation practices with adaptation and livelihoods is known, there is still much that needs to be learned. Uncertainties and lack of data still constrain our understanding of the impacts of some mitigation practices. A full life cycle and greenhouse gas accounting is often lacking for most practices. Impacts will vary among agro-ecosystems and socio-economic contexts, and the magnitudes of impacts are often unknown for specific sites. Interactions among practices and across the farm and landscape levels still require further study. Understanding whether mitigation practices have clear positive impacts for adaptation and livelihoods will be necessary to design policies and interventions acceptable to smallholder farmers.

We suggest a framework here for analysing the adaptation and livelihood synergies associated with agricultural mitigation.

First, a full life cycle accounting of mitigation practices should be performed to determine net emissions and interactions with other green-house gases during several agricultural cycles and across unit of product (see Baker and Murray, this volume), farm and landscape scales.

Second, net emissions should be compared against different possible positive and negative impacts on agronomic adaptation, including (1) productivity, yields and nutritive content of food produced, both current and expected trends under further climate change; (2) natural capital, especially soil and water quality; (3) other agronomic conditions affecting crop or livestock productivity such as competition, shade, toxicity; and (4) diversity of resulting crop or livestock varieties and species.

Third, net emissions and agronomic adaptation should be compared against their livelihood implications, including (1) increased income or profitability, (2) reduced costs, (3) increased efficiencies (e.g., energy efficiency and recycling of waste), (4) access to land and natural resources, (5) improvements in other economic, social or natural assets, (6) risk management opportunities, such as diversification, insurance schemes or other buffers to perturbations, and (7) the time stream of delivery of these benefits, both across the short- and long term, as well as the stability and frequency of benefit flows. Understanding the different impacts of mitigation practices on benefits for poorer and better-off households, and for men versus women will be important to develop mitigation policies that support social equity.

A comparison of these three elements should help to identify best management practices for mitigation and contribute to sustainable development goals for smallholder farmers in developing countries. Potential trade-offs can be identified to anticipate where compensating actions will be needed.

## Conclusion

Applying best management practices for mitigation would help to reduce greenhouse gas emissions per unit food produced and per unit area (Smith *et al.* 2007b) in developing countries if we are to realize even a proportion of the 70% of global agricultural climate mitigation potential that is available in these countries. Ideally agricultural mitigation measures need to be considered within a broader framework of sustainable development that assesses a comprehensive accounting of net greenhouse gases, and the interactions of practices with adaptation and livelihood outcomes. There is a need to model and empirically test more mitigation practices to understand their relative magnitudes and implications for adaptation and livelihoods. Policies that encourage triple-win arrangements will make agricultural mitigation in developing countries more achievable. In both environmental and economic terms, we must act strongly and quickly.

# References

Barker, T., Bashmakov., I., Bernstein, L., Bogner, J., Bosch, P., Dave, R., Davidson, O., Fisher, B., Grubb, M., Gupta, S., Halsnaes, K., Heij, B., Kahn Ribeiro, S., Kobayashi, S., Levine, M., Martino, D., Masera Cerutti, O., Metz, B., Meyer, L., Nabuurs, G.-J., Najam, A., Nakicenovic, N., Rogner, H.-H., Roy, J., Sathaye, J., Schock, R., Shukla, P., Sims, R., Smith, P., Swart, R., Tirpak, D., Urge-Vorsatz, D., Zhou, D. (2007) 'Summary for Policy Makers', in Metz, B., Davidson, O.R., Bosch, P.R, Dave, R., Meyer, L.A. (eds) *Climate change 2007: Mitigation*, Contribution of Working Group III to the Fourth Assessment Report of the Intergovernmental Panel on Climate Change. Cambridge University Press, Cambridge and New York, NY.

Burney, J.A., Davis, S.J., Lobell, D.B. (2010) Greenhouse gas mitigation by agricultural intensification, *PNAS*, vol. 107, pp. 12052–12057.

Mann, W., Lipper, L., Tennigkeit, T., McCarthy, N. and Branca, G. (2009) 'Food Security and Agricultural Mitigation in Developing Countries: Options for Developing Synergies'. FAO, Rome.

Palm, C.A., Smukler, S.M., Sullivan, C.C., Mutuo, P.K., Nyadzi, G.I. and Walsh, M. (2010) Climate Mitigation and Food Production in Tropical Landscapes Special Feature: Identifying potential synergies and trade-offs for meeting food security and climate change objectives in sub-Saharan Africa', *PNAS*, vol. 107, pp. 19661–19666.

Pan, G., Pan, W. and Smith, P. (2009) The role of soil organic matter in maintaining the productivity and yield stability of cereals in China, *Agriculture, Ecosystems & Environment*, vol. 129, pp. 344–348.

Smith, P. (2004) Soils as carbon sinks – the global context, *Soil Use and Management*, vol. 20, pp. 212–218.

Smith, P. and Olesen, J.E. (2010) Synergies between mitigation of, and adaptation to, climate change in agriculture, *Journal of Agricultural Science*, vol. 148, pp. 543–552.

Smith, P. and Powlson, D.S. (2003) 'Sustainability of soil management practices – a global perspective', in Abbott, L.K. and Murphy, D.V. (eds) *Soil Biological Fertility – A Key To Sustainable Land Use In Agriculture*. Kluwer Academic Publishers, Dordrecht, Netherlands, pp. 241–254.

Smith, P., Martino, D., Cai, Z., Gwary, D., Janzen, H.H., Kumar, P., McCarl, B., Ogle, S., O'Mara, F., Rice, C., Scholes, R.J., Sirotenko, O., Howden, M., McAllister, T., Pan, G., Romanenkov, V., Rose, S., Schneider, U. and Towprayoon, S. (2007a) 'Agriculture', in Metz, B., Davidson, O.R., Bosch, P.R., Dave, R. and Meyer, L.A. (eds) *Climate change 2007: Mitigation*, Contribution of Working Group III to the Fourth Assessment Report of the Intergovernmental Panel on Climate Change. Cambridge University Press, Cambridge and New York, NY.

Smith, P., Martino, D., Cai, Z., Gwary, D., Janzen, H.H., Kumar, P., McCarl, B., Ogle, S., O'Mara, F., Rice, C., Scholes, R.J., Sirotenko, O., Howden, M., McAllister, T., Pan, G., Romanenkov, V., Schneider, U. and Towprayoon, S. (2007b) Policy and technological constraints to implementation of greenhouse gas mitigation options in agriculture, *Agriculture, Ecosystems & Environment*, vol. 118, pp. 6–28.

Smith, P., Martino, D., Cai, Z., Gwary, D., Janzen, H.H., Kumar, P., McCarl, B., Ogle, S., O'Mara, F., Rice, C., Scholes, R.J., Sirotenko, O., Howden, M., McAllister, T.,

Pan, G., Romanenkov, V., Schneider, U., Towprayoon, S., Wattenbach, M. and Smith, J.U. (2008) Greenhouse gas mitigation in agriculture, *Philosophical Transactions of the Royal Society, B,* vol. 363, pp. 789–813.

Stern, N. (2006) *Stern Review: the economics of climate change.* Cambridge University Press, Cambridge, UK.

Trines, E., Höhne, N., Jung, M., Skutsch, M., Petsonk, A., Silva-Chavez, G., Smith, P., Nabuurs, G.J., Verweij, P. and Schlamadinger, B. (2006) *Integrating Agriculture, Forestry, And Other Land Use In Future Climate Regimes: methodological issues and policy options.* A Report for the Netherlands Research Programme on Climate Change (NRP-CC), 188pp.

# Part II

# INSTITUTIONAL ARRANGEMENTS AND INCENTIVES

# 5

# ECONOMIC CHALLENGES FACING AGRICULTURAL ACCESS TO CARBON MARKETS

*Alessandro De Pinto, Claudia Ringler, and*
*Marilia Magalhaes*

## Introduction

While the combined value of markets for greenhouse gas (GHG) emission reductions increased to more than US$100 billion[1] (Capoor and Ambrosi 2009), agriculture has been largely excluded from both formal and informal carbon markets. Key reasons for the marginalization of agriculture include the high transaction costs associated with smallholder agriculture and the high level of uncertainty surrounding carbon sequestration and emission reductions. Transaction costs depend largely on the costs of monitoring, reporting, and verifying changes in the above- or below-ground carbon pool and emissions, and on the cost of aggregating and organizing farmers. Uncertainties in mitigation include the amount of carbon that can be sequestered by agricultural soils, the reduction in net emissions for all GHG obtainable from the agricultural sector, and the length of time that carbon can be stored in the soil. This chapter reviews these challenges and highlights some of the areas that need further research and exploration.

## Economic and behavioral constraints

While agriculture has been widely recognized as a fundamental force in the reduction of poverty, the active role of agriculture in slowing down or even reversing ecosystem degradation is a somewhat new idea. For many years, the problem was framed in terms of a trade-off between agricultural and economic development and environmental degradation. More recently, scientists from different disciplines have posited that the two objectives are not mutually exclusive and that agriculture has the potential to generate both poverty reduction and ecosystem services (Lipper *et al.* 2009). While the economic literature about payment for environmental services was initially mostly focused on forest and water resources, more recently, attention has turned to agricultural landscapes and the rural poor who live in environmentally degraded areas. Climate change mitigation activities are just one of the many environmental services that farmers can provide to the global community and, as such, could be rewarded for this service. Climate change mitigation practices have the potential, if implemented appropriately, to support farmers' resilience to climate shocks. There is growing literature that analyzes the economics of farmers' participation in regulatory and voluntary carbon markets. Most of the empirical literature concentrates on cases in the United States and Europe. However, some of the findings are general and potentially applicable to small farmers in developing countries.

### Conditions for adoption of mitigation practices

The economic literature that looks at the conditions for adoption of mitigation practices is relatively simple. Stavins (1999), Antle (2002), and Gonzáles-Estrada *et al.* (2008), among many, assume that a risk-neutral farmer will try to maximize the present value of net benefits deriving from farming land. Therefore, a farmer will adopt mitigation practices when the net present value of farming with these practices is greater than that of alternatives. Still, farmers might incur additional costs or there might be a temporary decrease in productivity when adopting mitigation practices. In these cases, some form of payment could be made available to farmers to overcome the reduction in benefits. Even though a considerable amount of research has addressed the impact of risk, uncertainty, and risk aversion on farmers' adoption of technology, particularly in developing countries (Sunding and Zilberman 2001), the literature that concentrates on climate change mitigation activities has so far ignored these issues.

### Costs of adoption and barriers

Many studies have written about the substantial barriers that hinder the adoption of climate change mitigation practices and sustainable land management practices in general (e.g., Otsuka and Place 2001, Barrett *et al.* 2002, Nkonya *et al.* 2004; see also McCarthy *et al.*, this volume). These barriers may be due to lack of knowledge, imperfectly functioning markets and consequent lack of credit, or even a drop in yields during the first years of adoption. At the project level, there are important costs that need to be considered (see also Shames *et al.*). Negotiation, organization, management, monitoring, and enforcement act as potential barriers to the implementation of projects. Although rapidly changing, this is an area still characterized by a considerable lack of data; the project-level data available show upfront costs that range from US\$12 to US\$600 per hectare (FAO 2009). In a review of the literature that reports Clean Development Mechanism (CDM) transaction cost estimates, Cacho (2009) finds that ex-ante fixed costs vary from US\$34,000 to US\$280,000 (negotiation and project approval) and that ex-post costs vary from some US\$6,000 to US\$280,000 (project monitoring, verification, and insurance). See Tennigkeit *et al.* for a more detailed discussion of project costs for the case of sub-Saharan Africa.

Economists differentiate between two types of costs associated with the implementation of contracts for the provision of an environmental service: farm opportunity costs and transaction costs. Farm opportunity costs are the costs of resources used on the farm to provide the service. These include the forgone returns from possibly more profitable activities. Transaction costs are costs associated with negotiating and implementing contracts, which also include brokerage fees and monitoring of compliance with the terms of the contract. Contracts are likely to involve a large number of farmers and institutional structures that work as intermediaries between sellers and buyers. These intermediaries will need to group contract agreements from large numbers of farmers to construct a commercially viable contract (for example, the unit of trade for the Chicago Climate Exchange (CCX) is the Carbon Financial Instrument (CFI), representing $100 \, Mt \, CO_2e$). Since these intermediaries act as go-betweens, they reduce the payments that farmers receive per ton of sequestered carbon. Cacho (2009) provides a theoretical demonstration of how increasing the total project area could allow higher payments to farmers. However, this result relies on keeping transaction costs "relatively fixed," while the costs of negotiation could be increasing with the number of farmers participating in the contract. As Antle (2002) points out, negotiation costs will also be affected by the total amount of carbon sequestered and sold on the market, the type of soil where the sequestration activity is undertaken, and the institutional setting in which the transactions take place.

Antle and Stoorvogel (2008) use simulated data to show that the portfolio of profitable mitigation practices is strongly dependent of transaction costs.

### Types of contracts

The recent literature on carbon sequestration in both agriculture and forestry sectors supports the view that it would be more efficient to pay farmers per ton of carbon sequestered rather than for the adoption of specified prescribed management practices (see also chapter by Baker and Murray). Parks and Hardie (1995) use the opportunity costs of forgone agricultural output to show that the least-cost policy for sequestering carbon by converting agricultural lands to forest is associated with bids offered on a per-ton basis rather than on a per-acre basis. Pautsch *et al.*

(2001) analyze the potential for carbon sequestration using different tillage practices and reach a conclusion similar to that of Parks and Hardie. Antle *et al.* (2003) estimate that the costs to implement the per-ton contracts are at least an order of magnitude smaller than the efficiency losses of the per-acre contract. The increased cost of measuring carbon sequestration is counterbalanced by a higher return per dollar spent.

The length of a contract can also play a significant role (Lewandrowski *et al.* 2004). The authors advocate the use of shorter but renewable contracts, in which payments are based on the annualized value of a permanent reduction of a ton of carbon. These contracts would more closely reflect the changing opportunity cost of adoption of mitigating practices and thus encourage farmers' participation and reduce the incentives of cheating. Nelson *et al.* (2009) suggest using the type of contract used in the U.S. Conservation Reserve Program (CRP), in which farmers submit a bid to adopt a set of conservation practices and the bids with the lowest cost per unit of environmental service are accepted. The advantage of this approach is that the program administrator does not have to figure out what the "correct" payment must be and can concentrate on assessing the benefits associated to each suggested practice. Even though the implementation of the CRP has not delivered in full the expected environmental benefits (Claassen *et al.* 2008) this type of contracts has the potential to lower the implementation costs.

### *Role of marginal land*

Marginal lands are not necessarily more economically efficient at sequestering carbon than fertile land. In fact, fertile land may also have the highest potential for carbon sequestration.[2] For example, land that produces greater quantities of crop residues provides farmers with larger amounts of organic material that can be incorporated into the soil.[3] Tschakert (2004) and Tschakert *et al.* (2006), using a cost–benefit analysis approach, find that initial resource endowment has a strong effect on the profitability of recommended carbon sequestering practices. Graff-Zivin and Lipper (2008) reach a similar conclusion using a purely theoretical approach. Their results indicate that farmers who live on land of intermediate quality will generate the highest increment in soil carbon. This has important implications regarding which farmers are good candidates for cost-effective carbon-payment schemes and in particular on the role of poor farmers, who most often work on land of poor quality (Lipper 2001) (see also Lee and Newman in this volume for the observation that the carbon projects they surveyed focused on mid-range farmers rather than the poorest). These findings need further verification through empirical studies, but they suggest that farmers on marginal lands would not be good candidates for cost-effective carbon sequestration programs.

## Monitoring and compliance

An issue that has attracted considerable attention is the design of compliance monitoring strategies when compliance is not self-enforcing. Payments for climate change mitigation services generally have two common features. First, they are voluntary, and second, participation involves a contract between the buyer, an intermediary agent, and the landowner.

Two major issues need to be dealt with when implementing payments for environmental services such as climate change mitigation activities: adverse selection and moral hazard. Adverse selection arises when negotiating the contract. Landowners have better information than the buying agent about the opportunity costs of supplying environmental services. Landowners can thus secure higher payments by claiming their costs are higher than they actually are. These payments will therefore be above the minimum payment necessary to engage the landowner in a program. Adverse selection has been the subject of theoretical analyses in the context of agri-environmental payment schemes, but has not been directly applied to climate change mitigation (Fraser 1995; Moxey et al. 1999; Ozanne et al. 2001; Peterson and Boisvert 2004).

The problem of adverse selection is important because the buying agents might obtain fewer environmental services per dollar spent than they would in a world with perfect information. In contrast, the moral hazard problem arises after a contract has been negotiated and it is due to imperfect information about compliance. The certification process requires that farmers are monitored for contract compliance, but the intermediary may find monitoring costly and thus may be unwilling to verify compliance with certainty. Thus, the farmer has an incentive to avoid fulfilling contractual responsibilities. Cheating behavior in agri-environmental payment schemes has also been the subject of theoretical analyses (Choe and Fraser 1998, 1999; Ozanne et al. 2001; Fraser 2002; Hart and Latacz-Lohmann 2004). Most of the economic literature on the subject concentrates on cases in the United States and Europe. Choe and Fraser (1999) derive optimal monitoring strategies and incentives. Kampas and White (2004) examine the impacts of monitoring costs on the relative efficiency of alternative agri-environmental policy mechanisms and Fraser (2002) investigates the effects of penalties for non-compliance. More recent studies have applied the results of this type of economic analysis to the problems of payments for environmental services, including carbon sequestration (Grieg-Gran et al. 2005). However, several important empirical issues remain unaddressed and should be tested. For example, theoretical analysis by Ozanne et al. (2001) and Fraser (2002) find that risk aversion among farmers can reduce the moral hazard problem in relation to policy compliance. The implications of their findings are still untested in developing countries.

## Considerations and conclusions

The most likely effect of climate change on the rural poor is the increased risk and reduced livelihoods opportunities with consequent stress on existing social institutions. The long-term success of a carbon-payment scheme depends not only on farmers' receiving payments for the environmental services provided (whether it be carbon sequestration or ecosystem conservation), but also on its capacity to improve people's well-being. Given the considerable overlap between adaptation and mitigation practices, carbon payments have the potential to contribute to farmers' food security and resilience to climate change.

Much work is still necessary before farmers can reliably receive payments for climate change mitigation activities. The need for an accurate baseline, demonstrable permanence in emission reductions, and the role of institutions and governance systems are just a few of the areas that require additional investigation. From a purely economic standpoint, work on the role of uncertainty, risk, and risk aversion is necessary particularly if the contribution to mitigation by agriculture in developing countries is to be substantial. Uncertainty still existing in management costs, in the amount of mitigation that each particular soil and agronomic practice can deliver (particularly when a full accounting of GHG is attempted), and uncertainty embedded in basic data on soil, yields, and prices make the design of a cost-effective payment scheme still difficult. Compounded uncertainty strongly reduces the attractiveness of agriculture for investors as a viable option to offset emissions. This uncertainty needs to be resolved by implementing pilot projects and investments in data acquisition.

Risk and farmers' risk attitudes strongly influence the type of financial mechanisms and contracts, and the type and intensity of monitoring needed. Financial incentives might have to compensate for the risk associated with adoption of mitigation practices and contracts must reflect the role of risk on the possibility of defaulting on contractual obligations. Measuring risk will also inform on the need and type of insurance products that might be needed.

## Notes

1 All dollar amounts in this article are expressed in U.S. dollars.
2 Antle and Diagana (2003) show how the opportunity cost of alternative land uses is inversely proportional to the soil's potential for carbon sequestration.
3 Studies on the adoption of soil fertility management practices have returned similar results (Nkonya *et al.* 2004).

## References

Antle, J.M. (2002) 'Economic analysis of carbon sequestration in agricultural soils: An integrated assessment approach', *A Soil Carbon Accounting and Management*

*System for Emissions Trading,* Soil Management Collaborative Support Research Program Special Publication SM CRSP 2002–2004. University of Hawaii, Honolulu, HI.

Antle, J.M. and Diagana, B. (2003) Creating incentives for the adoption of sustainable agricultural practices in developing countries: The role of soil carbon sequestration, *American Journal of Agricultural Economics,* vol. 85, no. 5, pp. 1178–1184.

Antle, J.M. and Stoorvogel, J.J. (2008) Agricultural carbon sequestration, poverty, and sustainability, *Environment and Development Economics,* vol. 13, pp. 327–352.

Antle, J.M., Capalbo, S., Mooney, S., Elliott, E., and Paustian, K. (2003) Spatial heterogeneity, contract design, and the efficiency of carbon sequestration policies for agriculture, *Journal of Environmental Economics and Management,* vol. 46, pp. 231–250.

Barrett, J., Vallack, J., Jones, A., and Haq, G. (2002) *A Material Flow Analysis and Eco-footprint of York,* Technical Report. Stockholm Environment Institute, Stockholm, Sweden.

Cacho, O.J. (2009) 'Economics of carbon sequestration projects involving smallholders', in Lipper, L., Sakuyama, T., Stringer, R., and Zilberman, D. Springer and FAO (eds.) *Payment for Environmental Services in Agricultural Landscapes,* Rome, Italy.

Capoor, K. and Ambrosi, P. (2009) *State and Trends of the Carbon Market 2009.* World Bank, Washington, DC.

Choe, C. and Fraser, I. (1998) A note on imperfect monitoring of agri-environmental policy, *Journal of Agricultural Economics,* vol. 49, no. 2, pp. 250–258.

Choe, C. and Fraser I. (1999) Compliance monitoring and agri-environmental policy, *Journal of Agricultural Economics,* vol. 50, no. 3, pp. 468–487.

Claassen, R., Cattaneo, A., Johansson, R. (2008) Cost-effective design of agri-environmental payment programs: U.S. experience in theory and practice, *Ecological Economics,* vol. 65, no. 4, pp. 737–752.

FAO (Food and Agriculture Organization of the United Nations) (2009) *Food security and agricultural mitigation in developing countries: Options for capturing synergies* [Online]. Available at: www.fao.org/docrep/012/i1318e/i1318e00.pdf (Accessed: 20 January 2010). Rome, Italy.

Fraser, I.M. (1995) An analysis of management agreement bargaining under asymmetric information, *Journal of Agricultural Economics,* vol. 46, no. 1, pp. 20–32.

Fraser, R. (2002) Moral hazard and risk management in agri-environmental policy, *Journal of Agricultural Economics,* vol. 53, no. 3, pp. 475–487.

González-Estrada, E., Rodriguez, L.C., Walen, V.K., Naab, J.B., Koo, J., Jones, J.W., Herrero, M. and Thornton, P.K. (2008) Carbon sequestration and farm income in West Africa: Identifying best management practices for smallholder agricultural systems in northern Ghana, *Ecological Economics,* vol. 67, pp. 492–502.

Graff-Zivin, J. and Lipper, L. (2008) Poverty, risk, and the supply of soil carbon sequestration, *Environment and Development Economics,* vol. 13, pp. 353–373.

Grieg-Gran, M., Porras, I., and Wunder, S. (2005) How can market mechanisms for forest environmental services help the poor? Preliminary lessons from Latin America, *World Development,* vol. 33, no. 9, pp. 1511–1527.

Hart, R. and Latacz-Lohmann, U. (2004) Combating moral hazard in agrienvironmental schemes: A multiple-agent approach, *European Review of Agricultural Economics,* vol. 32, no. 1, pp. 75–91.

Kampas, A. and White, B. (2004) Administrative costs and instrument choice for stochastic non-point source pollutants, *Environmental & Resource Economics*, vol. 27, no. 2, pp. 109–133.

Lewandrowski, J., Peters, M., Jones, C., House, R., Sperow, M., Eve, M., and Paustian, K. (2004) 'Economics of sequestering carbon in the U.S. agricultural sector', *Technical Bulletin TB1909*. USDA, Economic Research Service, Washington, DC.

Lipper, L. (2001) 'Dirt poor: Poverty, farmers and soil resource investment', in *Two Essays on Socio-Economic Aspects of Soil Degradation*, Economic and Social Development Paper Series 149. FAO, Rome, Italy.

Lipper, L., Sakuyama, T., Stringer, R., and Zilberman, D. (eds.) (2009) *Payment for Environmental Services in Agricultural Landscapes*. FAO and Springer Science, Rome, Italy.

Moxey, A., White, B., and Ozanne, A. (1999) Efficient contract design for agri-environmental policy, *Journal of Agricultural Economics*, vol. 50, no. 2, pp. 187–202.

Nelson, G.C., Robertson, R., Msangi, S., and Zhu, T. (2009) 'Greenhouse gas mitigation: Issues for Indian agriculture', Discussion Paper 177. International Food Policy Research Institute, Washington, DC.

Nkonya, E., Pender, J., Jagger, P., Sserunkuuma, D., Kaizzi, C.K., and Ssali, H. (2004) 'Strategies for sustainable land management and poverty reduction in Uganda', Research Report 133. International Food Policy Research Institute, Washington, DC.

Otsuka, K. and Place, F. (2001) 'Land tenure and natural resource management', *Food Policy Statements 34*. International Food Policy Research Institute, Washington, DC.

Ozanne, A., Hogan, T., and Colman, D. (2001) Moral hazard, risk aversion and compliance monitoring in agri-environmental policy, *European Review of Agricultural Economics*, vol. 28, pp. 329–347.

Parks, P.J. and Hardie, I.W. (1995) Least cost forest carbon reserves: Cost-effective subsidies to convert marginal agricultural land to forest, *Land Economics*, vol. 71, no. 1, pp. 122–136.

Pautsch, G.R., Kurkalova, L.A., Babcock, B.A., and Kling, C.L. (2001) The efficiency of sequestering carbon in agricultural soils, *Contemporary Economic Policy*, vol. 19, no. 2, pp. 123–134.

Peterson, J.M. and Boisvert, R.N. (2004) Incentive-compatible pollution control policies under asymmetric information on both risk preferences and technology, *American Journal of Agricultural Economics*, vol. 86, no. 2, pp. 291–306.

Stavins, R.N. (1999) The costs of carbon sequestration: A revealed-preference approach, *American Economic Review*, vol. 89, no. 4, pp. 994–1009.

Sunding, D. and Zilberman, D. (2001) 'The agricultural innovation process: Research and technology adoption in a changing agricultural sector', in *Handbook of agricultural economics*, ed. Gardner, B.L., and Rausser, G.L., vol. 1A. Elsevier Science B.V., Amsterdam, North-Holland.

Tschakert, P. (2004) The costs of soil carbon sequestration: An economic analysis for small-scale farming systems in Senegal, *Agricultural Systems*, vol. 81, no. 3, pp. 227–253.

Tschakert, P., Coomes, O.T., and Potvin, C. (2006) Indigenous livelihoods, slash-and-burn agriculture, and carbon stocks in Eastern Panama, *Ecological Economics*, vol. 60, no. 4, pp. 807–820.

# 6

# REDUCING COSTS AND IMPROVING BENEFITS IN SMALLHOLDER AGRICULTURE CARBON PROJECTS

## Implications for going to scale

*Seth Shames, Louise E. Buck, and Sara J. Scherr*

## Introduction

The mitigation of climate change through project-based smallholder agriculture faces inherent complexity, high costs of project development and challenges of risk management and securing benefits for smallholder farmers (Bracer *et al.* 2007). Large groups of farmers can be difficult to organize and rural institutions supporting these projects often are weak. Without adequate social, financial and property protections, farmers will be at risk of exploitation in the rush by carbon developers to capitalize on international carbon market opportunities.

So far, the number and scale of these projects has been relatively small in both regulated and voluntary carbon markets. Even in cases where projects are established, their size is minuscule compared to the potential for climate change mitigation and farmer participation. For these projects to achieve more significant strides towards climate and rural livelihood objectives, and to play a transformative role in rural landscapes, they must be organized at scales that are orders of magnitude greater. To achieve this transition to a new generation of larger and pro-poor agriculture carbon projects, the constraints of high cost and risk management to assure farmer benefits must be addressed.

This chapter proposes design principles that can be applied to overcome these barriers. The principles are elaborated through discussions of various elements of project design that can be applied by project developers and farmers in agriculture carbon projects. We draw on insights from EcoAgriculture Partners' work as a non-governmental organization (NGO)

69

focused on policy and advocacy related to agricultural carbon projects, mostly in Africa (Bracer *et al.* 2007; Scherr *et al.* 2007; Forest Trends *et al.*, Climate Focus and EcoAgriculture Partners, 2010; Shames and Scherr, 2010). To reduce costs, we consider the principles of leveraging pre-existing institutional capacity and working at scale. These principles are applied to farmer aggregation, land management extension services, carbon measurement, and farmer compensation. To reduce risk and improve benefits for farmers, we propose the principles that farmers benefit principally from yield improvements resulting from climate-friendly interventions and that they partner with groups who are sensitive to their needs and play active roles in project development and implementation. These principles are applied to project design and monitoring, negotiations and contracting, land tenure and carbon rights, and project financing for farmers.

Our analysis suggests that there are signposts for projects to move in larger, farmer-friendly directions. However, project designs and rules for project development will need to embrace approaches that can integrate agro-ecological and social heterogeneity at large scales and provide space for significant farmer participation.

## Principles for cost reduction

Relative to other carbon mitigation opportunities, smallholder agriculture projects face greater institutional complexity, measurement and monitoring challenges and therefore, higher costs (see also chapters by de Pinto *et al.* for a discussion of costs, Lager and Nyberg for an example of project-level lessons). It is difficult to estimate the specific costs of developing an agricultural carbon project and few empirical data exist. Even where data exist, many projects have received co-financing from other sources for other purposes—often as conventional sustainable land management projects—so that these costs are not necessarily passed on to carbon offset buyers. We identify two general principles for reducing costs in these projects: 1) leveraging pre-existing institutional capacity and 2) working at scale.

Given the relatively low price of carbon, the principle benefits of carbon projects to farmers will typically be in long-term yield increases, cost reductions or new income sources, rather than the cash payments from carbon offsets (Smith and Scherr, 2003; Tennigkeit *et al.* 2010). Thus, project developers and managers who are best placed to work with farmers on these projects are those who have pre-existing relationships with large groups of farmers and already possess the institutional capacity and infrastructure required to implement what, in many ways, will be a typical agricultural development intervention. These established institutions are the ones which are stepping in to play critical project roles in institution-poor environments where many smallholder agriculture projects are based (Shames and Scherr, 2010). Generally speaking, the stronger these links are before the carbon project is initiated, the less effort will be required to establish these relationships for the purpose of the carbon projects, and the lower the project's costs will be.

The principle of scale reflects economies of scale in project design, extension support, and measurement, reporting and verification (MRV). (See also Havemann for a discussion of the importance of scale in finance). For all of these, project costs per carbon credit tend to go down as the number of aggregated farmers in a project increases. Working at scale also provides opportunities for projects to link with landscape-scale planning processes and generate more significant livelihood and ecosystem co-benefits. In certain cases, the benefits of scale can be realized through institutional mechanisms such as contract farming or crop-specific extension programming that delivers relatively uniform financial and technical inputs that are required to realize predictable outputs and capacities within a tightly controlled management structure. More commonly, however, where smallholder farmers predominate, agro-ecological, socio-economic and institutional heterogeneity preclude this approach. Under these conditions a landscape approach is likely to be a more suitable strategy for realizing the benefits of scale (Milder *et al.*, in press; Scherr and Sthapit, 2009). Key characteristics of a landscape approach to bringing agriculture carbon projects to scale include: 1) Managing land-use diversity through landscape design to optimize the location of mitigation enhancement activities in ways that capture ecosystem service co-benefits, advance the potential for linking agriculture carbon projects with Reducing Emissions from Deforestation and forest Degradation (REDD) initiatives, and reduce the need for rigid permanence rules; 2) Linking projects in ways that lower the cost per unit MRV and stimulate investment in landscape MRV; and 3) Engaging in transparent multi-stakeholder forums for decision-making and agreement-setting to coordinate land-use management and institutional support.

The following sub-sections illustrate mechanisms based on these principles to reduce costs: 1) Work with existing aggregators of farmers; (2) Build carbon projects on existing sustainable land management (SLM) extension systems; (3) Streamline carbon measurement techniques; and (4) Transfer cash efficiently to farmers.

### *Work with existing aggregators of farmers*

When pre-existing networks of farmers are harnessed for carbon projects, developers can save resources on identifying farmers to work with and develop the communications infrastructure and trust among project actors that will be critical to efficiency and success. Farmers can be reached through a variety of actors. For example, in the Community Markets for Conservation (COMACO) case in the Luangwa Valley of Zambia, the project is managed by COMACO, a community-owned and run organization with a market-based approach to rural livelihoods, food security, and biodiversity conservation. COMACO works with 50,000 farmers and has established the infrastructure and network of extension services along with payment mechanisms necessary to bring markets to remote rural communities, and value-added agricultural commodities into regional centers. Farmers can also be reached through international NGOs working on conservation and/or development issues with a long-term presence in a given place, as is the case with CARE in western Kenya. Agriculture supply chain infrastructure may also be the organizer as is being proposed in the Cocoa Carbon Initiative in Ghana. In Uganda, the national government via the National Forestry Authority has served as aggregator organized the first African Clean Development Mechanism (CDM) project that works with farmers.

### Build carbon projects on existing SLM extension systems

SLM extension activities are the core of an agriculture carbon project. The elegance of these projects is that in most cases, the climate-friendly interventions are identical to those that would be introduced within a conventional SLM program. The challenge for cost reduction in these activities will be to exploit, to the greatest extent possible, systems already in place that can deliver extension services, or to work within established institutional structures to deliver this training. In areas with strong public agricultural extension services, government could play a key role in these projects. Where government presence is limited, development and conservation NGOs may be appropriate. Contract farming and certification systems also serve as platforms for carbon projects.

### Streamline carbon measurement techniques

Carbon project development costs include the establishment of the carbon baseline and long-term MRV required to certify for buyers that emissions are reduced or carbon sequestered. A full technical discussion of ways that MRV costs can be reduced is beyond the scope of this chapter, but there are strategies that developers can begin to use now that will help reduce these costs. First, they can explore opportunities to either use a pre-existing carbon measurement methodology, or find partners to finance one that they can use. Land-based carbon project methodologies within the CDM are rare and complicated to implement. While agricultural methodologies are more readily available within the voluntary markets, primarily under the Voluntary Carbon Standard (VCS) and Plan Vivo, more are needed. The cost of methodology development, however, is sometimes taken on by outside actors with an interest in the development of the sector. For example, the World Bank BioCarbon Fund financed the development of the Sustainable Agricultural Land Management (SALM) Methodology, soon to be certified under the VCS.

Regardless of the MRV system, if direct measurements need to be taken, efficiencies can be found by working with farmers to collect data. For instance, the MRV system for the International Small Group and Treeplanting Program (TIST) operating in India, Kenya, Uganda, and Tanzania is based on representatives from local communities. These community quantifiers register information about every tree planted with handheld GPS devices.

### Transfer cash efficiently to farmers

While the majority of the benefits for farmers will come in the form of yield increases, cash can be a significant incentive for farmers. Projects, therefore, must consider how to set up efficient systems that transfer the cash benefits of offset sales to farmers. Creative methods are emerging which can reduce these costs. For example, TIST pays farmers in Kenya via M-PESA, the Kenyan mobile-phone based money transfer system whereby payment is sent in voucher form to a mobile phone which can be reimbursed at a bank. In cases where direct farmer payment is simply too expensive for the project, community funds can be set up, and controlled by local groups which invest in public community projects. The payment schemes may also mix the direct payment and community fund models. (See also chapter by Wernick on use of microfinance channels to reduce transaction costs and Havemann on finance mechanisms available to smallholders).

## Principles to reduce risk and improve benefits for farmers

Small farmers have a limited asset base to absorb carbon project risks such as periods of lower returns or higher labor requirements in exchange for a promise of future carbon payments. Also, they generally do not have long-term capital to invest. Therefore, smallholders will only have incentives to participate in carbon projects if project interventions minimize risks and improve their livelihoods. These conditions can be met when project design and implementation is based on the following principles: 1) Farmers receive both short-term and long-term benefits from yield improvements resulting from climate-friendly interventions; 2) They are protected against risks by playing active roles within the project and partnering with groups that are sensitive to their needs.

The key to assuring farmer benefits from carbon projects lies in the fact that many climate-friendly land management interventions create production co-benefits. An analysis of 41 sustainable land management interventions globally showed significant yield increases with 24 interventions showing a yield increase greater than 100% (Pender 2008). Another 45 sustainable land management interventions examined in sub-Saharan Africa found that cereal yields increased between 50% and 100% in almost all of the cases. Most of these land-use practices also showed significant profitability for farmers (Pretty *et al.* 2006).

Farmers can also improve their position in all areas of project development if they are seen as true partners by project developers and managers. Farmers' insights into how best to organize projects and to implement climate-friendly practices will be critical to project success as well as their own protection. Other project actors should be sensitive to farmer needs. (See also the chapter by Lee and Newman for evidence of how institutional structures of projects have been used to engage farmers).

Below we suggest applications of the principles for improving farmer benefits and reducing risk. These suggestions include: (1) Design flexibility for farmers into projects; (2) Empower farmers in negotiations and contracting; (3) Strengthen land tenure and carbon rights; and (4) Support upfront project financing opportunities for farmers.

### Design flexibility for farmers into projects

Despite the opportunities for livelihood improvements from climate-friendly agricultural interventions, carbon projects create risks for farmers if not carefully designed. Projects that require farmers to avoid clearing, or using large pieces of land or forest for long periods of time, present dangers. In these cases, farmers' opportunities for new economic activities will likely be reduced, and new opportunity costs may grow over time as productive land becomes more scarce with population growth or development. Furthermore, poor selection, spacing or management of tree species in agroforestry systems may suppress, rather than increase the productivity of the system.

Risks can be reduced when farmers play key roles in project design so that they can devise their own least-cost solutions to deliver climate change mitigation, and also maximize livelihood and ecosystem service co-benefits. Farmers will have critical knowledge of which places within an agro-ecosystem may have the greatest benefits for land-based livelihoods.

### Empower farmers in negotiations and contracting

The carbon supply chain in a smallholder agricultural project can be long, and there may be pressure on developers to improve their narrow profit margins by reducing payments to farmers. Community or farmer groups engaging in carbon contracts should have a pre-existing and robust organization that supports social cohesion, has broad member input, and is recognized within the community as a forum for legitimate decision-making to ensure a strong negotiating base. Before communities enter into contracts, these organizations should evaluate their objectives for their local economy, society, farming systems, ecosystems, and consider how a carbon project can help achieve these.

Symmetry in negotiations is a key challenge as carbon deals are often driven by experienced commercial project developers with the support of highly specialized legal and technical advisors driving hard bargains. By contrast, most farmer groups are participating in carbon markets for the first time with more limited information and resources, resulting in contractual conditions that are disadvantageous to the farmers. Without clear explanations of the contracts or negotiation process, farmers may not fully understand contractual commitments, or may be pressured to undertake activities on the farm that make little economic sense for them. Greater symmetry in information and access to expert advice and support is essential for negotiations to be fair for farmers, and they may need their own independent legal advisors. This is important while the contract is negotiated, and as it is enforced. (See also discussion of contracts in De Pinto *et al.*)

Contracts should include an element of flexibility. Provisions are needed that allow for changing circumstances in local economics and climate. They should enable projects to incorporate lessons from their own experiences as well as those from elsewhere. These agreements should be re-examined periodically for opportunities to improve them.

### *Strengthen land tenure and carbon rights*

In areas where smallholder land tenure is not secure, there is a risk that powerful interests will claim land that becomes more valuable as a result of the financial potential generated by carbon markets. Once communities restore degraded lands, large landowners or government agencies may come to claim them. The risk is compounded by the fundamental question of who owns rights to the carbon, and whether buyers for it will be found where tenure rights to the land it is on are not firmly established. In many regions vulnerable or low-status populations within communities or landscapes, including women, are at even further risk in the partitioning of valuable carbon benefits. To address questions of equity and avoid tenure conflicts, carbon projects must engage members of local communities in the design and negotiation of schemes to secure and distribute tenure benefits (Freudenberger and Miller 2010).

### *Support upfront project financing opportunities for farmers*

While farmers should be better off as a result of climate-friendly interventions, even without a cash payment, they may still incur significant costs in the transition to the new system. These may include input costs (e.g., tree seedling, herbicides, fertilizer) as well as labor. Some agricultural carbon standards, such as Plan Vivo and Carbon Fix, offer ex-ante credits which allow carbon offset cash to be collected after practices have been implemented, but before the sequestration or emission reduction has actually occurred. This system creates a source of financing to cover these transition and initial investment costs. However, most programs, including VCS, issue carbon credits only after the carbon benefit has been verified. When ex-ante credits are not available, projects need to explore microfinance or revolving fund options to cover these upfront costs. (See also chapters by Baroudy and Hooda, and Wernick.)

## Conclusion

The primary challenges for smallholder agricultural carbon projects are to control project costs and to ensure that participating farmers benefit and that they are protected. The development of models for these projects are still in the early stages, and initiatives have been of highly variable quality, but design and management principles can be drawn from these early experiences. To manage costs, projects should work to leverage preexisting institutional capacity and to work at scale, particularly by employing landscape approaches. To improve the position of farmers in these projects, they should receive both short-term and long-term benefits from yield improvements resulting from climate-friendly interventions. Farmers also should be enabled to play active roles throughout the project and to partner with groups that are sensitive to their needs.

There has been limited capacity in rural institutions in developing countries to support managerial and technical elements of agricultural projects. External expertise is expensive, and unless local capacity improves, these projects will continue to be financially challenged. To build this critical capacity for community groups, and for project developers and managers, lessons from early project experiences must continue to be synthesized. The most effective method for research in this space has been *learning by doing*, and this approach should continue. A variety of groups is pioneering this approach including Climate Change, Agriculture and Food Security (CCAFS) and the Food and Agriculture Organization (FAO)'s Mitigating Climate Change through Agriculture (MICCA) program (Seeberg-Elverfeldt and Tapio-Biström, 2010), and these should be supported.

Even as we review the principles and strategies that can produce cost-effective projects that benefit farmers, and as initiatives emerge to build capacity to effectively and equitably implement these projects, certain limitations remain in the underlying model of agricultural carbon projects. As models for climate-friendly smallholder agriculture move forward, opportunities should be considered which allow initiatives to move beyond the farm-based projects towards landscape-scale initiatives, and to integrate into Nationally Appropriate Mitigation Actions (NAMAs). Ultimately, lessons from the first generation of agriculture carbon projects should be harvested to inform more sophisticated and comprehensive models that will have more significant impacts on climate and livelihoods than even the largest projects currently under way.

# References

Bracer, C., Scherr, S., Molnar, A., Sekher, M., Ochieng, B.O. and Sriskanthan, G. (2007) 'Organization and Governance for Fostering Pro-Poor Compensation for Environmental Services', World Agroforestry Centre CES Scoping Study Issues Paper no. 4, ICRAF Working Paper no. 39. Nairobi, Kenya.

Forest Trends, Climate Focus, EcoAgriculture Partners, Katoomba Group. (2010) 'An African Agricultural Carbon Facility: Feasibility Assessment and Design Recommendations'. Washington, DC.

Freudenberger, M. and Miller, D. (2010) 'Climate change, property rights, & resource governance: Emerging implications for USG policies and programming', Property Rights and Resource Governance Briefing Paper #2. USAID, Washington, DC.

Milder, J.C., Buck, L.E., DeClerck, F. and Scherr, S. (in press) 'Landscape approaches to achieving food production, natural resource conservation, and the Millennium Development Goals', in de Clerck, F.A.J., Ingram, J.C., and Rio, C.R. (eds) *Integrating Ecology into Poverty Alleviation and International Development Efforts: A Practical Guide.* Springer, New York, NY.

Pender, J. (2008) 'The world food crisis, land degradation and sustainable land management: linkages, opportunities and constraints'. International Food Policy Research Institute, Washington, DC.

Pretty, J.N., Noble, A.D., Bassio, D., Nixon, J., Hine, R.E., Penning de Vries, F.W.T., and Morison, J.I.L. (2006) Resource conserving agriculture increases yields in developing countries, *Environmental Science and Technology*, vol. 40, no. 4, pp. 1114–1119.

Seeberg-Elverfeldt, C. and Tapio-Biström, M.L. (2010) *Global survey of agricultural mitigation projects* [Online]. Available at: www.fao.org/docrep/012/al388e/al388e00.pdf. Food and Agriculture Organization of the United Nations (FAO) Mitigation of Climate Change in Agriculture (MICCA) Project, Rome, Italy.

Scherr, S., Milder, J., and Bracer, C. (2007) 'How important will different types of compensation and reward mechanisms be in shaping poverty and ecosystem services across Africa, Asia and Latin America over the next two decades?'. World Agroforestry Center, Nairobi, Kenya.

Scherr, S. and Sthapit, S. (2009) 'Mitigating climate change through food and land use', Worldwatch Report 179. Ecoagriculture Partners and Worldwatch Institute, Washington, DC.

Shames, S. and Scherr, S. (2010) *Institutional models for carbon finance to mobilize sustainable agricultural development in Africa* [Online]. Available at: www.ecoagriculture.org/documents/files/doc_335.pdf. EcoAgriculture Partners, Washington, DC.

Smith, J. and Scherr, S. (2003) Capturing the value of forest carbon for local livelihoods, *World Development*, vol. 31, no. 12, pp. 2143–2160.

Tennigkeit, T., Kahrl, F., Wölcke, J., and Newcombe, K. (2010) 'Agricultural carbon sequestration in Sub-Saharan Africa: economics and institutions'. World Bank, Washington, DC

# 7

# HOW TO MAKE CARBON FINANCE WORK FOR SMALLHOLDERS IN AFRICA

## Experience from the Kenya Agricultural Carbon Project

*Bo Lager and Ylva Nyberg*

## Introduction

The World Bank BioCarbon Fund and Vi Agroforestry signed the first Emission Reduction Purchase Agreement (ERPA) in 2010 for an African agricultural and soil carbon finance project. This event was a benchmark in the development of terrestrial carbon finance in the world. The project, Kenya Agricultural Carbon Project (KACP), will be an important example of a practice-based agricultural carbon credit project. It will offer a dual opportunity for climate change mitigation and adaptation, making it possible for smallholder farmers in Kenya to benefit from carbon finance. The project is groundbreaking both in its methodology and mode of implementation.

The development of this project started in 2007. Because of the complexity of the project, both in terms of implementation and new methodology, several institutions have been involved. While Vi Agroforestry has been the project developer, Joanneum Research, Austria, and Unique Forestry consultants, Germany, were instrumental in the development of the methodology. The World Bank Carbon Finance Unit and the Swedish International Development Cooperation Agency (Sida) have provided crucial financial support (see also chapter by Baroudy and Hooda on the experience of the BioCarbon Fund).

## Kenya agricultural carbon project

The project area, consisting of both cropland and grassland, is characterized by constant or increasing agricultural pressure on lands, decreasing use of agricultural inputs such as fertilizers and decreasing forest land. Smallholder farmers participating in the project practice mixed agriculture dominated by maize, beans and livestock. The majority of farmers live in poverty and suffer from food insecurity. In addition to the advisory services provided by the project, agricultural productivity in the area is promoted through extension services provided by government and other civil society organizations.

The purpose of KACP is to promote Sustainable Agricultural Land Management (SALM) practices for regeneration of degraded lands, mitigation of greenhouse gas (GHG) emission and building of adaptive capacity to enable farmers to cope with the impacts of climate change. Vi Agroforestry's field officers are sensitizing, mobilizing and training farmers on SALM practices through participatory group and organization development approaches. Farmers who adopt SALM practices will generate carbon stocks in agricultural systems, increase staple food production and access carbon market annual revenues until 2029.

## Project roll-out

The KACP is designed to implement community activities that seek to enhance the resilience of communities, or the ecosystems on which they rely. The project emphasizes skills to address community adaptation, instead of only focusing on household adaptive capacity. During the project's intensive phase (year 1 to year 4), 27 field advisers will each work with and recruit 600 farmers per year (approximately 10,800 farmers total annually). The average land area that potentially can be under SALM per household is assumed to be approximately 0.5 ha. During the extensive phase, after year 4, recruitment will be lower or even zero and fewer field advisers will be needed within the area. The main intention during the extensive phase is to create an enabling environment which fosters a farmer-led sustainable implementation system and leadership under advisory services from the project. Local community-based organisations, service providers and stakeholders with strong leadership facilitation from Vi Agroforestry will sustain the project in a cost-effective way.

The project uses participatory planning, monitoring and evaluation of the farmer-led implementation system. The field advisers sensitize as many farmers as possible through existing traditional institutional structures such as *barazas* (community information meetings) and other organized meetings or groups (e.g., schools and local non-governmental organizations (NGOs)). The field adviser will contract farmer groups on adoption of defined farming practices and the contract is signed between the farmer groups and Vi Agroforestry. The contract stipulates the parties' obligations and rights in relation to the implementation of the carbon project and in particular assigns the rights of the Verified Emission Reductions from the farmer to Vi Agroforestry.

The project considers four integrated strategies within an enabling environment: building resilient livelihoods, developing local capacity, disaster risk reduction and tackling drivers of vulnerability. Advisory services promote specific practical aspects such as on-farm diversification; capacity building on appropriate SALM practices like contour planting, composting, terracing and residue management; tree planting; and on- and off-farm business engagement.

## Participatory approach

Since the aim of KACP is threefold (improved living conditions, adaptation and mitigation), the role of the farmers and their groups is crucial for success. Participation is needed at the planning, implementation, monitoring and payment stages. If farmers do not understand what to do, the nature of the contract or why SALM practices should be adopted, the project will fail.

## The methodology

KACP has been developing a new methodology for soil and biomass carbon accounting using carbon stock modeling (RothC model), which monitors changes in land-use management practices that have an influence on carbon stocks. The methodology is in the approval process by the Verified Carbon Standard (VCS).

Within smallholder farmer groups the project is promoting a package of climate-smart agricultural techniques consisting of SALM practices which reduce emissions of GHGs through carbon sequestration by trees and soil. The emission reduction potential in KACP is calculated by comparing the current or projected net emissions under a given baseline (e.g., business as usual with no mitigation put in place) with the net emissions that would occur if management practices are changed. The area under project intervention must follow special applicability criteria concerning the land use at the start of the project, the present and expected agricultural pressure on the land in absence of the project, use of agricultural inputs, livestock numbers, existence of forest land and woody perennials (trees), agricultural residues, manure and the use of fossil fuels. All applicability conditions are related to a simplification of the baseline, the assumption that the area of a given land-use type is degrading in absence of the project, as well as a security for additionality and permanence of the project results (SALM methodology).

Increased yields, as a result from the SALM practices, will also increase the amount of carbon sequestered by plants and accumulated in the soil during growth, or when incorporating plant residues into the soil. Other practices, including water management and improved locally adapted crop varieties increase resilience, while introduction of legumes into grassland improve the nutrient balance. Reduced or non-tillage practices further maintain or increase the soil carbon. Systems that retain crop residues also tend to increase soil carbon because these residues are the precursors of soil organic matter, the main carbon store in soil. Activities such as agroforestry may be useful to increase soil fertility as well as carbon stocks. The highest benefits can be obtained if degraded soils can be reclaimed (Smith *et al.* 2007).

The growth of woody perennials, amount of agricultural production, fertilizer use and the adoption of the targeted activities will be monitored. The amount of manure production will not be physically measured but estimated using an IPCC Tier 1 approach based on the number of animals, average annual temperature and manure system. Also the increase in soil organic carbon will not be physically measured. This will be modeled based on the documented activity using the available software program, RothC. The model input data consists of soil clay content; monthly mean, minimum and maximum temperature; monthly precipitation and radiation. Data on residue and manure inputs from the crops and animals is based on crop yield data and number of animals collected from a sample of farms in the project area (SALM methodology).

Agricultural practices change slowly with time and therefore the farmers should undertake a Farmer Self Assessment (FSA) on a regular basis (every five years) to identify the actual agricultural management practices adopted on croplands and grasslands (see also chapter by Seebauer *et al.*). The FSA should estimate or record details of each management practice including: area; annual biomass production and percentage left in the fields; number and type of grazing animals and amount of manure and nitrogen fertilizers input; use of N-fixing species; amount of biomass burnt and the biomass of woody perennials (SALM methodology).

Based on the FSA, a representative sample of farm households will be interviewed periodically throughout the project's lifetime about their current as well as future management practices to anticipate (ex-ante) the adoption of improved SALM practices and to monitor the adoption (ex-post). The aim of the questionnaire is to estimate the number of farmers and the area where SALM activities are adopted, establish a transparent baseline and a monitoring system to reward farmer groups for generating carbon assets, and to establish a baseline and monitor the impact of the interventions on farmer livelihoods.

This Activity Baseline Monitoring Survey (ABMS) cannot be used to distribute carbon revenues at farm or group level because only a representative sample is surveyed. Therefore, in addition to the ABMS, a monitoring system at farmer group level is employed to distribute carbon revenues according to a set of criteria. Criteria should be performance based in that they consider the area where SALM practices have been adopted and the yield response as a result of the adoption of SALM practices. In order to mitigate any internal and external disagreements, KACP has developed grievances procedures.

## The costs of implementation and sharing of benefits

The project targets 54,000 households and 45,000 hectares of agricultural land, sequestering 1.2 million tonnes $CO_2e$ which will be translated to carbon credits and sold to the World Bank BioCarbon Fund and other potential investors. Out of total carbon revenues it is expected that 60% will be distributed to farmer groups involved in the project; 35% will be used to provide advisory services and 5% will be used for marketing carbon certificates as well as financial management in Sweden, while the transaction costs are covered by outside funding from Sida and the World Bank. Hence, the project will benefit the poorest communities in Africa by utilizing the international carbon market (see also chapter by Lee and Newman).

It is important to underline that carbon revenues provide the "icing on the cake" and the real benefits lie within improved yield, increased food security, market access for agricultural produce and other co-benefits that accrue to farmers and communities through the advisory services and the SALM practices.

Because of the high transaction cost, public sector or philanthropic funds will be essential for launching terrestrial carbon projects like KACP. The mode of carbon payment of the project follows the principle of ex-post payment, which means the payment of carbon credits will be done after verification of the credits by an independent verifier. The high transaction cost in terms of project implementation, validation and verification is a clear and prominent barrier for the implementation of this kind of carbon project. In addition, the soil carbon pool build-up follows the rate of a farmer's adoption of SALM practices. This means that during the first 3–4 years of the project, when the number of field officers is high and the carbon sequestration level is low because few farmers have adopted SALM, the project cost is highest.

Schematic breakdown of implementation costs for KACP:

- salaries for staff: 50%
- logistics/transport: 15%
- training/capacity building of staff: 15%
- seeds and seedlings: 5%
- other (insurance, office rent, telephone, electricity): 15%

Carbon revenues:

- volume: $1.2\,MtCO_2$
- crediting period: 20 years (2009–2029)
- number of farmers: 54,000
- area: 45,000 hectares
  60% to farmers
  35% to project developer
  5% to administrator

The implementation cost over the whole project cycle (20 years) is estimated to US\$1.5 million which is equal to US\$1.25/tonne $CO_2$. This figure includes cost for monitoring, validation and verification. The cost for marketing of the credits is covered by the US\$0.2 to the administrator which is the Vi Agroforestry head office in Stockholm.

Some estimates of carbon mitigation potential are given by Bloomfield and Pearson (2000). Targeting smallholders for carbon sequestration programs can be a good strategy, both in terms of carbon sequestration and poverty alleviation, but the transaction costs (administration, verification) may be higher for smallholdings (Cacho *et al.* 2003). It is clear that a larger scale makes the costs more economic in a bigger project (target area 750 km$^2$) costing much less (US\$0.25/t $CO_2$) compared to a smaller project (10–60 km$^2$) at a price of around US\$3.5/t $CO_2$.

Mitigation costs are highly context-dependent. An agricultural carbon project that funds extension workers, management staff and MRV is likely to spend approximately US\$2.39 per hectare annually, for a 200,000 hectare commodity chain project. The figure does not include establishment costs, marginal operational costs for improved agricultural practices or the operational costs of 'business as usual' planting and harvesting (Streck *et al.* 2010).

## Opportunities, synergies and win–win–win scenarios

Research has shown that African farmers can reduce GHG emissions, increase carbon sequestration and maintain above- and below-ground carbon stocks at relatively low costs, while improving food production and livelihoods. KACP has focused on these objectives and provides a "triple win" of environmental and social services for the small-scale farmer through reduction of GHG emissions, increase in agricultural productivity, and climate change adaptation for farmers (Figure 7.1).

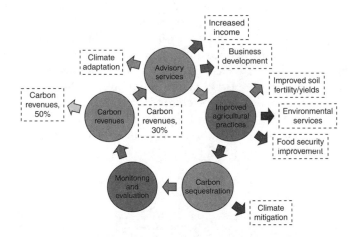

*Figure 7.1* Activity flow and benefits of soil carbon

An additional benefit is environmental, as the biodiversity of both higher (birds, plants and insects) and lower flora and fauna (micro-biological) is increasing in the agro-ecological environment with the implementation of SALM practices.

## Challenges

Despite the potential of the carbon market, traditional carbon funds have largely excluded both terrestrial projects and Africa from carbon finance opportunities. The obstacles are many and have included a lack of credible methodologies, permanence concerns, scarce regional technical expertise, uncertainty of measuring and monitoring of emission reductions, and difficulties in coordinating large numbers of smallholder farmers. These issues are presented in more detail in Table 7.1, along with potential solutions.

## Conclusion

There is a huge potential for smallholders in Africa to get involved in and benefit directly from the global carbon finance system. However, the precautions necessary for a successful project is to use an all-through participatory approach with a holistic livelihood focus including proper understanding of the whole process from both the target group and the donor side. Initial funding will be pivotal and therefore it is important to involve financially solid investors and implementing organizations that have a well-developed network at local and regional levels as well as adequate experience from working directly with farmers. Unfortunately, the high permanence buffer is a hindrance to cover the initial investment needed in the starting up of a carbon project. Fast adoption requires intensive provision of advisory services (high cost) in the initial phase of the project and is directly related to the experience farmers have from the implementing organization at the outset of the project. Women and men in the target group need to be actively encouraged to be part of the same groups, meetings and training for sustainable results and a long-term commitment from the implementing organization in an area is the key to successful carbon sequestration. Today's price for terrestrial carbon is too low and is not adequately taking into consideration the true value of adaptation to climate change and therefore there is a need in the future to tie carbon finance together with adaptation.

*Table 7.1* Challenges and potential solutions for smallholder carbon credit projects

| Challenge | Solution |
|---|---|
| Lack of credible methodologies slows the development of terrestrial carbon projects. | • Invest in methodology development favoring terrestrial carbon. |
| High permanence buffer delays payment to farmers in the early stages of project. | • Advocate for soil and agricultural carbon to be eligible in the post-Kyoto Protocol commitment period so as to tap the regulatory carbon market. |
| Knowledge barrier among small-scale farmers and scarce regional technical expertise. | • Develop insurance system/facility that can cover the project proponent against permanence risks, increasing the confidence of private markets for making investments in emission reductions from the voluntary carbon market. |
| | • Build farmers' knowledge and awareness about climate-smart agricultural practices, climate variability, carbon sequestration, mitigation and carbon finance. |
| | • Build sustainable, technical field advisory systems at local, regional and national level to achieve emission reductions while increasing productivity, boosting adaptive capacity and increasing the number and range of Africans engaged with project design. |
| | • Develop guidelines and handbooks on SALM, based on experience, knowledge-sharing, and capacity building supported by field testing. |
| Difficulties coordinating large numbers of smallholder farmers. | • Careful planning and use of participatory approaches in interaction with communities allowing a fair amount of time for sensitization. |
| | • Develop strong partnerships with capable organizations that have existing dialogue within and among communities in the area. |
| | • Initially favor farmers that are members in groups and organizations and those with on-farm resources that can be beneficial in the process. |
| Carbon market has been biased toward emissions in industrial and energy sectors and buyers' short-term compliance needs rather than long-term mitigation potential. | • Advocate for EU-ETS to include land-use carbon and increase understanding and interests in temporary carbon credits. |
| | • Continue to lobby the Conference of the Parties to the UNFCCC to support terrestrial carbon finance. |
| Modest sequestration rates per farmer, and cost of measuring and monitoring of emission reductions makes the financial model weak with low incentive for farmers to adopt it. | • Develop systems that provide cost-effective carbon monitoring and measurement techniques as well as proven farmer outreach approaches across large landscapes in order to aggregate large numbers of smallholders. |
| | • Increase awareness of the co-benefits of terrestrial carbon. Carbon payment should be seen as an additional benefit, but not the main reason for carrying out the SALM practices. |

*continued*

*Table 7.1* continued

| Challenge | Solution |
|---|---|
| Lack of secure upfront finance for initial cost is a hurdle for project developers. | • Provide payments to groups instead of individuals to manage transaction costs in the distribution of money. Otherwise, at today's low prices of carbon, each farmer will get a very small amount of money, because farms are small.<br>• Donors/Funders need to offer upfront payments against credits delivered during the first 2–5 years, the most critical period to overcome.<br>• Combine public and private finance to bridge the interim financing gap for sustainable business models around carbon finance. |
| Lack of holistic livelihood approach in carbon finance. | • Improved yield and other co-benefits that accrue to farmers and communities could serve as important incentives for carbon projects.<br>• A project of this kind needs a financially stable and reliable organization with a large on-ground network of farmers and established trust with both buyers and sellers of the carbon.<br>• Relatively high investment costs will be a fact during the first years of planning and establishment of the project, as the funds from carbon emission reduction agreements will not be available until upon delivery of the $CO_2$.<br>• Innovative set of transaction models can be used by governments and non-state actors to access carbon and other climate finance sources for climate mitigation and agricultural adaptation that helps African smallholder farmers.<br>• Commodity supply chain actors could provide finance and technical assistance to their producers. |
| Discrimination against women in carbon finance payment to landowners only. | • Target both men (decision-makers and landowners) and women (practitioners) in the advisory services in order to come to a common understanding and decide on the best options for their farm and the payments.<br>• Advocate for government policies related to land titles that promote women as landowners.<br>• Design contracts where more than one person per household can sign.<br>• Promote women and gender equality in important positions within farmer organizations and groups. |
| There are risks in carbon finance of attracting insincere actors as project developers. | • The terrestrial carbon market must have a filtering role to identify well managed field projects that are in a position to absorb the technical capacity support provided by the facility and its partners, and that has credibility as representing the interests of African smallholder carbon offset sellers, rather than international buyers or the international carbon finance expert community. |
| Lifetime of land-based funding programmes are generally short. | • Long-term investments by funders are essential to identify and establish partnerships to ensure that full carbon offset benefits are produced, monitored and delivered to buyers. |

# References

Bloomfield, J. and Pearson, H.L. (2000) Land use, land-use change, forestry, and agricultural activities in the clean development mechanism: estimates of greenhouse gas offset potential, *Mitigation and Adaptation Strategies for Global Change*, vol. 5, pp. 9–24.

Cacho, O.J., Hean, R.L. and Wise, R.M. (2003) Carbon-accounting methods and reforestation incentives, *The Australian Journal of Agricultural and Resource Economics*, vol. 47, no. 2, pp. 153–179.

Smith, P., Martino, D., Cai, Z., Gwary, D., Janzen, H., Kumar, P., McCarl, B., Ogle, S., O'Mara, F., Rice, C., Scholes., B. and Sirotenko, O. (2007) 'Agriculture', in Metz, B., Davidson, O.R., Bosch, P.R., Dave, R. and Meyer, R.L. Climate Change 2007: Mitigation, Contribution of Working Group III to the Fourth Assessment Report of the Intergovernmental Panel on Climate Change, Cambridge University Press, Cambridge, UK.

Streck, C., Coren, M., Scherr, S.J, Shames, S., Jenkins, M. and Waage, S. (2010) 'An African Agricultural Carbon Facility-Feasibility Assessment and Design Recommendations'. Forest Trends, The Katoomba Group, Ecoagriculture Partners, and Climate Focus with support from The Rockefeller Foundation.

# 8

# DESIGNING AGRICULTURAL MITIGATION FOR SMALLHOLDERS IN DEVELOPING COUNTRIES

## A comparative analysis of projects in East Africa

*Jean Lee and Jess Newman*

### Introduction

In the past decade, the global community has witnessed a rise in interest and funding for "win–win," pro-poor carbon sequestration projects as part of a wider sustainable development paradigm (Engel *et al.* 2008). Many are hopeful that agricultural carbon mitigation holds promise for meeting dual goals of alleviating poverty and improving environments in rural areas. Programs such as the Clean Development Mechanism (CDM), Verified Carbon Standard (VCS), and the Reducing Emissions from Deforestation and forest Degradation (REDD) in Developing Countries Programme seek to reduce carbon emissions while ensuring sustainable development and livelihoods for local people. In spite of the potential for "win–win" situations, smallholder farmers do not always benefit from carbon mitigation programs due to various barriers to participation and benefits, such as secure land tenure and certification costs (Boyd *et al.* 2007; Corbera *et al.* 2007).

An understanding of institutional structures and arrangements is key to enabling smallholder participation and delivering meaningful livelihood benefits (Brown and Corbera, 2003; Bracer *et al.* 2007). Institutions play a critical role in the implementation of these programs, and the rules and regulations they establish govern a wide range of issues, from contracts regarding carbon prices to the type of participants targeted and the actual frequency and method of payment distribution. Institutional arrangements that affect participation include land tenure rights, law enforcement, and rule-making authorities (Hayes and Persha, 2010). Strong institutional arrangements provide the necessary structures and incentives for smallholder participation by reducing transaction costs and minimizing risks (see also the chapter by Shames *et al.*). In addition, achieving the pro-poor goals of carbon mitigation programs requires institutions to pay particular attention to how payments are distributed and whether they positively affect marginalized groups (Brown *et al.* 2003).

While several studies have documented the institutional capacity and equity of carbon mitigation projects in Central America (Boyd *et al.* 2007; Brown and Corbera, 2003; Corbera *et al.* 2007; Siegel, 2005), fewer studies have addressed concerns in sub-Saharan Africa, where agricultural carbon mitigation potential is lower due to soil and climate conditions. Literature on institutional arrangements, project design, and incentives for participation remains somewhat theoretical and lacks rigorous comparative analyses focused on the poverty alleviation side of the "win–win" equation. This chapter presents the results of exploratory data collection about project design and institutional arrangements during visits to seven self-professed "pro-poor" carbon mitigation projects in Kenya, Uganda and Tanzania. It serves as a jumping-off point for further research on designing poor-targeted carbon projects that secure the participation of their target audiences and deliver meaningful benefits to them.

## Methods

Our work draws on more than 100 interviews across seven pro-poor, agricultural carbon sequestration projects in East Africa to foster a comparative discussion of project design elements relevant to smallholder participation and benefits. Each project provides either employment on forest plantations or incentives for biomass/soil carbon sequestration on smallholder farms. We spoke with project designers, managers, and participants, and averaged responses across interviewees when necessary. Our research took place during the summer of 2010 and does not contain all project developments since that time.

Our projects are fairly representative of the agricultural carbon project population in East Africa (Ethiopia, Kenya, Uganda and Tanzania). Our sample included 35% of existing projects. Table 8.1 showcases the diversity across key project variables such as size, carbon sequestration practice, benefit type, standard, agro-ecological system, and developer.

During each site visit, we collected qualitative and quantitative data about project design, operations, and participant benefits through 30–90-minute individual and group interviews with project staff, local stakeholders, and participants. Our data collection focused on:

- farmer engagement strategies and organization (structures and leadership through which farmers engage with projects)
- aggregation strategies to reduce transaction costs

Table 8.1 Project characteristics

| Name | Location | Developer | Validation standards | Third-party verification standards | Practices to seq. carbon | Size (ha) | Participants | Benefits |
| --- | --- | --- | --- | --- | --- | --- | --- | --- |
| Forest Again: Compassionate Carbon Offsets | KE | Eco2librium (for-profit) | | CCBA Gold | Permanent reforestation of Kenyan Forest Service reserve | 500 | 60 rotating employees, 80 small group members | Wages, eventual carbon payments to small group |
| TIST Kenya | KE | Clean Air Action Corporation (for-profit) with the Institute for Environmental Innovation (non-profit) | TIST | VCS CCBA | On-farm afforestation | 33,500 | 6,700 small groups | Quarterly cash payments per tree, training, co-benefits; 70% share of carbon profits when trees mature |
| Treeflights Kenya | KE | Treeflights (for-profit) | | | On-farm afforestation | 650 | 46 households | Free seedlings, one-time cash payment, cash crop (cashews) |
| The Western Kenya Smallholder Agricultural Carbon Project (WKSACP)[1] | KE | SCC Vi-Agroforestry-Kitale (NGO) | Developing soil carbon standards with the World Bank | | On-farm improved land management practices (soil carbon) | 45,000 | 7,400 households | Annual cash payments, agricultural extension services, co-benefits |

| | | | | | | | | |
|---|---|---|---|---|---|---|---|---|
| Emiti Nibwo Bulora | TZ | SCC Vi-Agroforestry-Kagera (NGO) | Plan Vivo | | On-farm afforestation | 130 | 24 people | 5 cash payments per tree over the course of 10 years |
| Kikonda Forestation | UG | Global-Woods AG (for-profit) | | CarbonFix CCBA | Rotational harvest plantation reforestation | 1,000 | 500–1,000 employees, 260 small group members | Wages, free pine seedlings |
| Trees for Global Benefits | UG | ECOTRUST (NGO) | Plan Vivo | | On-farm afforestation | 692 | 514 people | 5 cash payments per tree over the course of 10 years |

- incentives for on-farm mitigation practices
- inclusiveness of poor or marginalized farmers and women
- project attention to community/household well-being indicators.

## Results

This section presents results pertaining to our five research themes. A summary of project activities by theme allows for comparisons of design elements and policies across projects (see Tables 2 to 7).

Farmer engagement strategies and organization refer to the people, institutions, structures, and strategies that enable farmers to participate, communicate, and make decisions in projects. The majority of the projects had extension agents or representatives who served as the main point of contact, connecting participants to project staff and transmitting information. Many projects channeled this communication through farmer group leaders. The International Small Group and Treeplanting Program (TIST) and the Western Kenyan Smallholder Agricultural Carbon Project (WKSACP) had more complex hierarchies of participant leadership that involved them in operations management, decision-making, and project governance.

The majority of projects worked with groups of farmers, either pre-existing or specially formed. The two largest projects organized their groups into umbrella groups, which were groups of groups. Employees in the plantation projects were not organized.

*Table 8.2* Structures that enable farmer engagement

| Project | Project extension agents/ representatives | Participant group leaders | More complex self-governance structures or hierarchies | Participant-run project |
|---|---|---|---|---|
| The WKSACP | X | X | X | |
| Emiti Nibwo Bulora | X | | | |
| TIST Kenya | X | X | X | |
| Treeflights | | | | |
| Trees for Global Benefits | X | | | |
| Forest Again: Compassionate Carbon Offsets | | | | |
| Kikonda Reforestation Project | X | X | | |

*Table 8.3* Project aggregation strategies

| Project | Work with individuals | Work with groups formed for participation | Recruit existing groups | Work with umbrella groups |
|---|---|---|---|---|
| The WKSACP | | X | X | X |
| Emiti Nibwo Bulora | | X | | |
| TIST Kenya | | X | X | X |
| Treeflights | X | | | |
| Trees for Global Benefits | X | | | |
| Forest Again: Compassionate Carbon Offsets | X | | | |
| Kikonda Reforestation Project | X | X | | |

*Table 8.4* Incentives for encouraging practices that enhance mitigation

| Project | Monetary incentive (payment or wage) | Monetary equivalent (vouchers, credit) | Free materials or inputs | Subsidized materials or inputs | Knowledge/ education about future benefits |
|---|---|---|---|---|---|
| The WKSACP | (in progress) | | X | | X |
| Emiti Nibwo Bulora | X | | | | X |
| TIST Kenya | X | | | | X |
| Treeflights | X | | X | | |
| Trees for Global Benefits | X | | | | X |
| Forest Again: Compassionate Carbon Offsets | X | | | | X |
| Kikonda Reforestation Project | X | | X | | X |

All projects but Treeflights offered training or seminars about mitigation activities and their benefits. Three projects offered free materials or inputs, such as seedlings or improved maize seeds. All projects offered a cash incentive for participation. The size, regularity, and delivery mechanism vary by site. Treeflights delivered a one-time payment, but all others delivered regular payments on a contract-bounded schedule. Distribution processes included cash handouts, bank transfers, and transferring money via phone.

All projects but one included female participants. Tanzanian law prohibits women from owning land, meaning women in the Emiti Nibwo Bulora project are unable to sign contracts. While participation is relatively gender equitable across projects, women were often restricted in the level or meaningfulness of their participation by gender norms surrounding tasks, especially in the employment projects (e.g., women can't slash and cut). No projects specifically targeted women for general participation, but many included gender sensitivity in their training and encouraged families to make decisions as a whole. The TIST project also targeted women when hiring staff and mandated a baseline level of female leadership in their participant self-governance system.

While all projects had explicit poverty alleviation goals, only Forest Again specifically targeted low-income participants (however, their process of rotating through needy employees was controversial due to claims of corruption). The other projects sought to reduce barriers to participation. Some circumvented strict proof of land title requirements by allowing local government officials to endorse land ownership in contracts. Others tried to ameliorate high capital requirements, including group membership fees, the cost of seedlings, and maintenance/labor costs. Using land holdings as a direct index for income, we found the majority of projects are securing the participation of relatively middle income community members.

Only the WKSACP collected regular data about socio-economic indicators, and thus attempted to measure its livelihood impact. They used an annual Activity Baseline and Monitoring Survey (ABMS) filled out by each farmer to collect information on household well-being indicators, including, but not limited to land size, crop type, livestock holdings, schooling level of parents and children, diet, roof type, and income.

Table 8.5 Inclusiveness of women

| Project | Percentage of female participants (approx.) | Explicit female empower-ment goals | Targeted recruitment of women or women's groups | Targeted participation of spousal pairs and families | Gender sensitizing | Female project leadership requirements |
|---|---|---|---|---|---|---|
| The WKSACP | 50% | X | | | X | |
| Emiti Nibwo Bulora | 0% | | | | X | |
| TIST Kenya | 50% | X | | X | X | X |
| Treeflights | 20% | | | | | |
| Trees for Global Benefits | 30% | | | X | | |
| Forest Again: Compassionate Carbon Offsets | 60% | X | | | | |
| Kikonda Reforestation Project | 40% | | | | | |

Table 8.6 Inclusiveness of the poor or marginalized

| Project | Median income percentile of participants[2] | Explicit pro-poor goals | Affirmative action in hiring or recruitment | Ways for the landless to participate | Relaxed land tenure requirements | Low costs and barriers to project entry | Sensitizing about social issues |
|---|---|---|---|---|---|---|---|
| The WKSACP | 45 | X | | | X | X | |
| Emiti Nibwo Bulora | 50 | X | | | X | | |
| TIST Kenya | 55 | X | | in progress | X | X | |
| Treeflights | N/A | | | | X | | |
| Trees for Global Benefits | 75 | X | | | X | | |
| Forest Again: Compassionate Carbon Offsets | 45 | | X | X | | X | |
| Kikonda Reforestation Project | 55 | X | | X | | X | |

*Table 8.7* Monitoring of community or household well-being indicators

| Project | None | Informal observations | Baseline surveys only | Irregular, variably structured survey sampling | Regular survey sampling of some kind |
|---|---|---|---|---|---|
| The WKSACP | | | | | X |
| Emiti Nibwo Bulora | X | | | | |
| TIST Kenya | X | | | | |
| Treeflights | X | | | | |
| Trees for Global Benefits | X | | | | |
| Forest Again: Compassionate Carbon Offsets | | | | X | |
| Kikonda Reforestation Project | | X | | | |

## Discussion of research results

### *Farmer engagement strategies and organization*

Few projects involved participants in meaningful decision-making or project governance. At TIST and the WKSACP, self-governing hierarchies of participant groups handled budgets, set goals, demanded accountability, and interacted with project staff at the highest level. Leaders and lower-level staff were selected from among project participants. These projects were the largest and most smoothly run. They also generated the most apparent excitement and commitment among participants. However, they also had trained staff on the ground assisting and overseeing activities. We hypothesize that the level and structure of project self-governance, and thus participants' ability to engage with the project and feel ownership, has a significant effect on the level of overall benefits participants experience.

### *Project aggregation strategies*

We found that the majority of on-farm afforestation projects worked with existing or newly formed groups. Existing groups seem to have established trust that eased financial issues, strengthened accountability and led to more successful implementation. The plantation workers were not aggregated. Kikonda left hiring decisions entirely up to local contractors, and Forest Again rotated through a list of needy community members. There was a lower level of interest in long-term relationships or capacity building with employees, which probably led to a lack of investment in facilitating and streamlining communication with them.

### *Incentives for encouraging on-farm practices that enhance mitigation*

While all projects offered some sort of monetary stipend or payment, payments were not usually the main motivation for farmers' participation. Most farmers in on-farm afforestation projects cited future timber profits as their most important incentive given the small size and occasional irregularity of the carbon payments. Some said the carbon payments successfully offset short-term concerns about tree maintenance costs and reduced crop land, but others felt they were too paltry to have any impact on income levels or even offset costs. In many projects, other co-benefits such as firewood and increased soil fertility were more heavily emphasized by the project than the payments, which were framed as "bonuses."

### *Inclusiveness of women and the poor/marginalized*

Despite explicit poverty alleviation goals, only Forest Again targeted "most need" community members (as defined by the Climate, Community and

Biodiversity Alliance (CCBA)). No projects practiced affirmative action in favor of females or members of specific ethnic groups when recruiting participants, and only TIST targeted women in staff hiring and project leadership. Project participants across the board fell in the middle of their communities' income distributions. This is positive in the sense that benefits were not being disproportionally captured by rural elites, but highlights how the poorest 30–40% of community members were not benefiting from pro-poor projects (project land ownership and material cost barriers consistently exclude the landless and those with very small (>1 ha) farms, who are often the poorest). TIST is seeking to address this by pursuing a relationship with the Kenya Forest Service in which participants can plant indigenous trees in state-owned riparian areas and share in the carbon revenues. It still remains unclear which income groups projects can successfully target and how benefits to specific income groups trickle through communities.

## Synthesizing results and gaps in knowledge

There are still significant gaps in our understanding of the role that institutions (ranging from local community groups to government agencies) play in the pro-poor effectiveness of carbon mitigation projects. After our field visits, the strengths and weaknesses of different institutional arrangements remained fuzzy, but some conclusions and critical knowledge gaps emerged.

### *Institutional arrangements and livelihood linkages*

We need more robust and comprehensive documentation of the full range of project livelihood impacts (monetary and non-monetary benefits) and their relationship to overall income. In projects with equally important social and environmental goals, it is critical that the same rigor is applied to documenting welfare benefits as environmental carbon sequestration. Most projects expressed a desire to track their welfare impacts, but cited resource constraints as their largest barrier. Socio-economic information is critical from the perspective of demonstrating results, attracting donors and moving understanding beyond theory. Once welfare benefits are documented, they can be connected, perhaps even causally, back to specific aspects of project design. This could potentially lead to cycles of adaptive project design and conceptualizing the design process as something more experimental that can be tweaked to maximize benefits.

### *Ecosystem benefits and carbon sequestration*

Many projects are unable to detail and document environmental benefits beyond carbon sequestration estimates. In order to fully capture the overall ecosystem impacts of these carbon mitigation programs, we need rigorous, long-term scientific data collection documenting, for example, diversity of flora and fauna, soil erosion levels, and water quality. These unmeasured environmental impacts are not only important from an ecosystem perspective, but also represent important co-benefits driving farmer participation.

### Carbon standards

All but one of the projects used standing biomass to calculate sequestered carbon; other forms of carbon emission reductions are not considered. The different standards employed (see Table 8.1) emphasized different implementation strategies and benefits and required unique sets of monitoring, reporting, and verification tools. In order for smallholder agricultural carbon mitigation projects to be successful, we need a better understanding of which standards are the most effective at securing both environmental and welfare benefits at different scales and in different agro-ecological regions. There is very little consensus here, as evidenced by the low level of overlap in standards among projects.

### Project scaling

As many projects continue to expand, we need more research on successful strategies for scaling up while maintaining a consistent flow of both environmental and livelihood benefits. Our research indicates that transitioning to project self-governance and capitalizing on and developing local agencies and institutions, such as rural banks and local carbon buyers, is both critical and difficult. Which institutions are necessary to financially and organizationally sustain a large project long term?

## Next steps

This exploratory analysis should provide a jumping-off point for more targeted design of institutional arrangements and monitoring or research on outcomes. Longer-term studies of projects with successful self-governance structures to inform project scaling are needed. Institutional designs should be explored that can better reach the poorest 40% of community members. Regular and rigorous collection of information on benefits to farmers will be necessary to assess projects' progress toward meeting important poverty alleviation goals. Only then can we begin to see and capitalize on the connections between institutional arrangements and delivering welfare benefits.

## Notes

1 See chapters by Lager and Nyberg and Seebauer *et al.* for more detail about this project.
2 These estimates were developed using averaged participant land holdings collected during interviews as a direct index for income. They are merely an approximate measure of how well off the average participant of a certain project is relative to the population in that specific area.

## References

Boyd, E., Gutierrez, M. and Chang, M. (2007) Small-scale forest carbon projects: Adapting CDM to low-income communities, *Global Environmental Change*, vol. 17, pp. 250–259.

Brown, K. and Corbera, E. (2003) Exploring equity and sustainable development in the new carbon economy, *Climate Policy*, vol. 3S1, pp. S41–56.

Bracer, C., Scherr, S., Molnar, A., Sekher, M., Owuor Ochieng, B. and Sriskanthan, G. (2007) 'Organization and Governance for Fostering Pro-Poor Compensation for Environmental Services: CES Scoping Study Issue Paper no. 4,' ICRAF Working Paper no. 39. World Agroforestry Center, Nairobi, Kenya.

Corbera, E., Brown, K. and Adger, W.N. (2007) The equity and legitimacy of markets for ecosystem services, *Development and Change*, vol. 38, no. 4, pp. 587–613.

Engel, S., Pagiola, S. and Wunder, S. (2008) Designing payments for environmental services in theory and practice: An overview of the issues, *Ecological Economics*, vol. 65, no. 4, pp. 663–674.

Hayes, T. and Persha, L. (2010) Nesting local forestry initiatives: Revisiting community forest management in a REDD+ world, *Forest Policy and Economics*, vol. 12, no. 8, pp. 545–553.

Siegel, P. (2005) 'Using an Asset-Based Approach to Identify Drivers of Sustainable Rural Growth and Poverty Reduction in Central America: A Conceptual Framework', World Bank Policy Research Working Paper 3475.

# 9

# USING CERTIFICATION TO SUPPORT CLIMATE CHANGE MITIGATION AND ADAPTATION IN AGRICULTURE

*Jeff Hayward, Gianluca Gondolini, Oliver Bach, and Mark Moroge*

## Introduction

New approaches are needed to enable farmers to adapt to climate change, reduce net greenhouse gas (GHG) emissions and adopt practices that will fortify the resilience of their soils, crops and agro-ecosystems. These approaches must build on existing knowledge of the carbon cycle, climate science and sustainable agriculture practices. To be effective, government and non-governmental organization (NGO) efforts, market-based initiatives, and research and extension programs must incentivize farmers' adoption of these approaches by improving their standard of living through increased revenue, improved productivity or improved resilience.

One means to stimulate preparedness for adaptation strategies and reductions in net GHG emissions on farms is to build upon the success of existing certification of sustainable agriculture. Towards this end, the Rainforest Alliance and the Sustainable Agriculture Network (SAN) have developed criteria for best management practices to foster agricultural mitigation and adaptation. These criteria are part of a voluntary, add-on "climate module" intended to be accessible, practical and understandable (with guidance) to farmers. The purpose of this chapter is to review the role of certification in supporting mitigation and adaptation, using the Rainforest Alliance-SAN initiative as an example.

## Setting standards for climate mitigation and adaptation: building on a solid foundation

### *Overview of agricultural certification systems and climate change*

Voluntary agricultural certification has grown considerably, since the advent of organic certification in the 1960s, and Fair Trade in the 1970s, to Rainforest Alliance Certified™ in the 1990s, and other programs in the market today. Objectives vary depending upon the environmental and/or socio-economic dimensions of the certification program. Several prominent certification systems active in the developing world are now exploring and piloting inclusion of adaptation and/or mitigation elements to their programs. Through partner Café Direct, the Fairtrade Labelling Organisation has piloted the "AdapCC" initiative to help coffee and tea farmers cope with the risks and impacts of climate change (Café Direct and GTZ 2010).[1] The 4C Association is developing a voluntary climate component for verification to the 4C's Code of Conduct,[2] and UTZ CERTIFIED is exploring how to strengthen its system to address climate impacts.[3,4] The International Federation of Organic Agriculture Movements (IFOAM) has written several position papers and case studies on the adaptation and mitigation benefits of organic agriculture (IFOAM 2009a, b). Lastly, the SAN released in February 2011 adaptation and mitigation criteria to complement the SAN Sustainable Agriculture Standard (Sustainable Agriculture Network 2011).

These efforts reflect the trend towards adding guidance for producers on climate change adaptation and mitigation into certification standards, which should confer international recognition, potentially add value and create producer incentives, and facilitate adoption of such climate practices at scale.

### *A climate module to complement a sustainable agriculture standard*

The SAN is a coalition of tropical conservation organizations, including the Rainforest Alliance,[5] that strives to improve commodity production in the tropics through advancing sustainable agriculture certification standards. While the SAN develops the criteria for sustainable farm management that make up the SAN Sustainable Agriculture Standard (SAN Standard), the certified farm products earn the seal, with its emblematic frog, of Rainforest Alliance Certified™. The SAN Standard is built on principles of environmental protection, social responsibility and economic viability. Its 90+ criteria make up a comprehensive, holistic approach to farm management, which has been adopted by more than 200,000 farms in 31 tropical and subtropical countries, on more than 1,000,000 ha of farmland. However, the existing SAN Standard did not clearly set criteria for climate change adaptation and mitigation. In recognition of this gap, since July 2009 the Rainforest Alliance, on behalf of the SAN and in partnership with others, has led the process to research, test and develop a suite of adaptation and mitigation criteria—and corresponding best management practices—that can be used in conjunction with the SAN Standard.

Publically released in February 2011, these adaptation and mitigation criteria make up the "SAN Climate Module" (Climate Module) and are a *voluntary, add-on* to the SAN Standard. By explicitly addressing on-farm adaptation and mitigation issues. The SAN Climate Module strengthens the SAN Standard. As a result of its design as an add-on to the SAN Standard, the Climate Module will be used in conjunction with the SAN Standard, but not independent of it.

To understand how the SAN Standard provided a framework for the additional standards of the Climate Module, the following are illustrative examples:

*   The SAN Standard requires farmers to develop a social and environmental management system, while the Climate Module asks for the management system to include an assessment of climate risks and plans to adapt to these (Climate Module criterion 1.12—see section "Criteria to guide adaptation and mitigation" later in this chapter for a full list of criteria).
*   The SAN Standard requires that all natural ecosystems be protected and restored through a conservation program, while the Climate Module asks that restoration is conducted in a way that reduces climate vulnerabilities (Climate Module criterion 2.10).
*   The SAN Standard sets requirements for production-area shade plantings or establishment of conservation areas, while the Climate Module reinforces this by asking farms to conserve and increase carbon stocks and that they have tree inventories to monitor this (Climate Module criterion 2.11).
*   The SAN Standard requires farms to maintain detailed records of agrochemical use and strive to reduce inputs, while the Climate Module specifically asks that farms achieve nitrogen fertilizer efficiencies (Climate Module criterion 8.10).

## What can be achieved through the SAN Climate Module?

The SAN Climate Module expands the definition of sustainable agriculture by defining the standards for what should constitute climate-friendly or climate-smart agriculture. It sets forth well-defined criteria that can be consistently applied and verified within a proven certification system. It will allow farmers to make credible statements about their efforts to reduce net GHG emissions, increase carbon storage, and build adaptive capacity on their farms.

The module's primary aim is to raise awareness about climate change amongst tropical farmers and foster best management practices that can help farmers and communities to better adapt. Additionally, it is hoped that the module and its implementation will spur commitments within food and beverage company supply chains, including via broad industry- and government-led initiatives, to support the concept and practice of climate-smart farming.

The module creates opportunities for adding value to products grown on climate-friendly farms. For example, a company that already buys Rainforest Alliance Certified™ coffee may encourage its suppliers to seek verification to the Climate Module and incentivize this through favorable contracts or additional premiums.

By implementing mitigation and adaptation practices, farmers increase their resilience, build and restore carbon stocks, and reduce their vulnerability to extreme weather events and climate variability. In addition, recommended practices such as agroforestry may increase income and farm diversity and stability. Farmers who comply with the requirements of the Climate Module will be able to assess the risks posed by climate change to their farms and communities; analyze their practices to quantify and reduce the GHG emissions generated by growing, harvesting and processing activities; and increase the levels of carbon stored on their farms through the restoration of degraded lands, reforestation and improved soil conservation. They will also be able to adapt more readily to altered growing seasons and other changing climatic conditions.

### Standards development process: consulting farmers, experts and the public to develop robust and practical climate criteria

Development of SAN Standards follows ISEAL best practices for standards-setting.[6] As such, these were collaborative, widely consultative and participatory processes. The first iterations of the SAN Climate Module, with its suite of criteria and recommended practices, were drafted and tested in Guatemalan coffee farms in 2009–2010, and later applied to coffee, tea and cocoa farms elsewhere in Central America and in East Africa, West Africa and Southeast Asia. The Rainforest Alliance conducted desk- and field-based research, carried out pilot farm audits and multiple public consultations and had input or worked with leading research institutions (i.e., Universidad del Valle, Guatemala, World Agroforestry Centre, International Centre for Tropical Agriculture), other members of the SAN (i.e., Fundación Interamericana de Investigación Tropical, SalvaNATURA), industry partners (i.e., Efico, a multinational coffee and cocoa trader; Anacafé, the Asociación Nacional de Café en Guatemala; and Caribou Coffee) and smallholder farmers. Results from research, testing and analysis informed the identification and development of a consultation draft version of the climate criteria.

In July 2010, the SAN embarked on a 100-day public consultation to gather worldwide input and further the development of the module. During the public consultation, 132 stakeholders participated through an online consultation website and 172 individuals during local workshops, field visits and/or pilot audits conducted in Brazil, Costa Rica, El Salvador, Ghana, Guatemala, Indonesia and Kenya between July and October 2010. The stakeholders represented 42 countries through a total of 1,035 comments, with particularly strong representation from universities, NGOs and companies in coffee, tea or cocoa. These stakeholder groups accounted for 26%, 44% and 10% of comments, respectively.[7]

The SAN's International Standards Committee (ISC)—a group of 12 independent expert advisors from Latin America, Europe, the United States, Africa and Asia—revised the draft criteria after analyzing the most important trends and issues from stakeholder comments. The committee reframed and adapted the draft criteria to ensure their applicability and relevancy to both smallholder farms and large estates. After revision, the ISC approved the SAN Climate Module in November 2010. As the module is designed to be used with the existing SAN Standard, it is applicable for use by all Rainforest Alliance certified farms, as well as those in process of certification to the SAN Standard.

## Criteria to guide mitigation and adaptation

The SAN Climate Module outlines climate-related farm management practices that can be assessed on any scale of farm seeking certification. The 15 criteria for climate mitigation and adaptation of the SAN Climate Module are:

- The farm's social and environmental management system must assess climate risks and vulnerabilities and include plans to adapt to and mitigate climate change.
- The farm must annually record data about its main GHG emissions sources related to, at minimum, nitrogen fertilizer input, pesticide input, fossil fuel use for machinery, methane generated in waste and wastewater treatment and animal husbandry.
- The farm must obtain available information on climate variability and its predicted impacts and adapt farm practices considering that information.
- The farm must map its land use and keep records of land-use changes.
- The farm's climate change adaptation and mitigation practices must be included in its training and education programs.
- The farm must, to the extent possible, choose service providers that incorporate climate-friendly practices in their operations.
- The farm must reduce vulnerability, prevent land degradation or

enhance ecological functions by planting native or adapted species or promoting natural regeneration.

- The farm must maintain or increase its carbon stocks by planting or conserving trees or other woody biomass. The farm must conduct tree inventories every five years.
- The farm must analyze and implement wastewater treatment options that reduce methane emissions from wastewater treatment and recover the generated methane, to the extent possible.
- The farm must adapt to water scarcity by practices such as harvesting and storing rainwater and selecting drought-tolerant crop varieties.
- The farm must implement an emergency preparedness and response plan for extreme weather events to prevent damage to people, animals and property.
- The farm must initiate or actively participate in the community's climate change adaptation and mitigation efforts, including identification of relevant resources.
- The farm must reduce nitrous oxide emissions through the efficient use of nitrogen fertilizers to minimize the loss to air and water.
- The farm must maintain or increase its soil carbon stocks by implementing management practices, such as crop residue recycling, permanent cover crops reducing tillage, and optimizing the soil's water retention and infiltration.
- The farm must implement organic residue management practices that reduce GHG emissions, such as production of organic fertilizer or biomass energy generation.

As many criteria build upon the existing SAN Standard, farmers can be trained on climate-related criteria during capacity-building programs normally used to prepare farms for certification. Compliance with the SAN Climate Module will be assessed during the normal SAN Standard farm auditing process, which is anticipated to cut costs to farms and increase a farm's efficiency in preparing for verification.

## The challenges to wide-scale adoption of climate-smart farming

Based on our experience, challenges to the wide-scale adoption of climate-friendly farming include: accurate, yet cost-effective verification procedures, enabling smallholder adoption of climate-smart farming, financial incentives, and resolving outstanding data gaps or information needs. We expand upon these issues below.

## Verification methodology: ensuring cost-effective verification of compliance while maintaining accuracy

Many farms, particularly smallholder farms, are not uniform monocultures across the landscape, but rather consist of diverse agricultural activities on multiple integrated parcels, with each parcel exhibiting a distinct emissions profile. In mixed-farming (i.e., cattle vs. coffee-based agroforestry vs. pineapple production), how can GHG emissions be reliably and cost-effectively estimated? Harmonizing the module's verification procedures with emerging farm-level GHG accounting tools, such as the Cool Farm Tool (see Hillier *et al.*), may facilitate cost-effective verification, though further testing and adaptation is needed to ensure the accuracy of these tools in myriad farming systems. It is also the case that climate-friendly, climate-smart farming involves more than GHG emissions measurement, as verification will include adaptation-oriented best practices audited through review of farm records, management plans, protocols, and in dialogue with farmers or community members.

## Smallholder adoption: foster compliance without excluding those most vulnerable to climate change

Smallholder farmers often lack the human, technical and financial resources to implement the practices that would satisfy the SAN Climate Module's criteria. Yet they are often those most at risk of severe climate changes and in greatest need of increasing their adaptive capacity (e.g., Easterling *et al.* 2007; Verchot *et al.* 2004; Simões *et al.* 2010). The climate criteria have been constructed in a way that decouples economic resources from compliance as much as feasible, however investment (labor, financial or both) will be always be required to achieve compliance. In recognition of this, the Rainforest Alliance and SAN aim to roll out training and guidance materials, as well as conduct capacity building and technical assistance for extension agents and farmers to ease the transition for smallholder farmers towards adoption of the module.

## Market incentives: ensuring returns on farmer investment

Climate-smart practices can reduce operating costs and improve a farmer's bottom line, however, a financial structure to incentivize these practices remains to be determined. Aside from a voluntary premium paid by food and beverage companies, farmers implementing the Climate Module could potentially benefit from emerging programs for payment for environmental services systems (see chapters by de Pinto *et al.*, and Wunder and Börner). These could be at the national or international level and should be programmatic approaches that reward specific practices. At present, however, all of these financial incentive structures are speculative. Cost–benefit analyses should be conducted to demonstrate efficiency gains (cost-savings) via module implementation.

### Data gaps and information needs

Sufficient research exists to agree upon the core climate-smart criteria, however additional studies are needed to better quantify farm management practices for specific local contexts and farming types that reduce GHG or help producers prepare for the challenges that a changing climate will bring. A recent Rainforest Alliance literature review of tropical agriculture mitigation and adaptation practices sheds light on some of these data gaps and information needs:

- Research to establish crop rotation best practices to increase soil carbon storage for country- and crop-contexts; current studies are conflicting (IFPRI 2010).
- Improved understanding of $CH_4$ dynamics within agroforestry systems, to better identify priority on-farm interventions to minimize $CH_4$ emissions (Flynn and Smith 2010).
- Further studies to determine the efficacy of enhanced-efficiency nitrogen fertilizers in tropical agricultural systems (e.g. Delgado and Mosier 1996; Halvorson *et al.* 2008; Bronson *et al.* 1992, see also chapter by Norse *et al.*).
- Accurate quantification of GHG emissions from certain farm processing, waste management and processing wastewater management (Rainforest Alliance, in press).

## Priorities and recommendations for action

The systems and tools are emerging to facilitate mitigation and adaptation in agriculture, yet much remains to be done to foster the adoption and successful implementation of these systems, first via pilot projects, and later, at scale. The Rainforest Alliance suggests the following as priority actions to advance voluntary sustainable agriculture certification as a means to encourage adaptation and mitigation:

- *Encourage early adoption and raise awareness*: Demonstrate and quantify, through pilot verifications in different farm types and geographic regions and continuous monitoring and research, the value of climate-friendly farming practices to farmers and local communities.
- *Build capacity amongst local actors*: Develop and disseminate locally-appropriate guidance and training materials and integrate these tools into existing sustainable agriculture training and curriculum.
- *Test financial assumptions*: Analyze the economic impacts that the implementation of climate-smart agricultural practices have on

farmers, considering productivity/yield/efficiencies, added-value through market transactions, and possible benefits from inclusion under incipient payment for environmental services schemes.

- *Work together, better.* Improve coordination and communication amongst agriculture-climate research institutions, government agencies and donor institutions and project-level initiatives to create synergies and accelerate progress.
- *Stimulate the market.* Clarify what kind of claims companies selling climate-friendly products can make, and coordinate the carbon storage and GHG emissions data generated at the farm level with global brands' efforts to quantify their GHG impacts across supply chains.
- *Seek complementarity:* Encourage eligibility of climate-friendly farming for climate finance under REDD+ mechanisms, and search for complementarities between traditional private sector financing of sustainable agriculture (i.e., transactions realized via established commodities markets and supply chains) and new climate finance opportunities.

## Conclusion

Adding voluntary climate-smart criteria to international agricultural certification standards, as the SAN and Rainforest Alliance have done, encourages farm-level adaptation and mitigation and offers the potential to do so at scale.

To build on accomplishments to date, the effectiveness and practicality of the module for farmers must be demonstrated and the economic, marketing and climate benefits explored and verified. This will require enlarging capacity-building programs and building on the ongoing initiatives to train farmers in meeting sustainable agriculture standards.

## Notes

1  See: www.adapcc.org/ and www.fairtrade.net/ for more information.
2  See: www.4c-coffeeassociation.org/en/ for more information.
3  See: www.utzcertified.org/ for more information.
4  Updates on the 4C Association and UTZ Certified initiatives were shared at an ISEAL and GIZ co-convened "Carbon Accounting Workshop," held on 27–28 October 2010 in Bonn, Germany.
5  The SAN functions as the body to set standards and policies that form the basis of farm audits that may result in use of the Rainforest Alliance Certified™ certification seal on farm products. SAN members are: Conservación y Desarrollo (C&D), Ecuador; Fundación Interamericana de Investigación Tropical (FIIT), Guatemala; Fundación Natura, Colombia; ICADE, Honduras; IMAFLORA, Brazil; Nature Conservation Foundation, India; Pronatura-Sur, Mexico; Salva-Natura, El Salvador; and the Rainforest Alliance.
6  See: www.isealalliance.org/ for more information.
7  The public consultation report is available online at: http://sanstandards.org/

userfiles/file/SAN%20Public%20Consultation%20Report%20Climate%20
Module%20February%202011.pdf.

# References

Bronson, K.F., Mosier, A.R. and Bishnoi, S.R. (1992) Nitrous-oxide emissions in irrigated corn as affected by nitrification inhibitors, *Soil Science Society of America Journal*, vol. 56, pp. 161–165.

Café Direct and Deutsche Gesellschaft für Technische Zusammenarbeit (GTZ) (2010) *How can small-scale coffee and tea producers adapt to climate change?* [Online]. Available at: www.adapcc.org/download/Final-report_Adapcc_17032010.pdf (Accessed: 15 November 2010).

Delgado, J.A. and Mosier, A.R. (1996) Mitigation alternatives to decrease nitrous oxides emissions and urea-nitrogen loss and their effect on methane flux, *Journal of Environmental Quality*, vol. 25, pp. 1105–1111.

Easterling, W.E., Aggarwal, P.K., Batima, P., Brander, K.M., Erda, L., Howden, S.M., Kirilenko, A., Morton, J., Soussana, J.F., Schmidhuber, J. and Tubiello, F.N. (2007) *Food, Fibre and Forest Products*. pp. 273–313, in Parry, M.L., Canziani, O.F., Palutikof, J.P., van der Linden, P.J., and Hanson, C.E. (eds). Climate Change 2007: 'Impacts, adaptation and vulnerability', Contribution of Working Group II to the Fourth Assessment Report of the Intergovernmental Panel on Climate Change. Cambridge University Press, Cambridge, UK.

Flynn, H.C. and Smith, P. (2010) *Greenhouse gas budgets of crop production – current and likely future trends*, 1st edn [Online]. Available at: www.sustainablecropnutri-tion.com/ifa/Home-Page/LIBRARY/Publication-database.html/Greenhouse-Gas-Budgets-of-Crop-Production-Current-and-Likely-Future-Trends.html (Accessed: 20 September 2010). International Fertilizer Industry Association, Paris, France.

Halvorson, A.D., Del Grosso, S.J., and Reule, C.A. (2008) Nitrogen, tillage, and crop rotation effects on nitrous oxide emissions from irrigated cropping systems, *Journal of Environmental Quality*, vol. 37, pp. 1337–1344.

IFPRI (2010) *An econometric investigation of impacts of sustainable land management practices on soil carbon and yield risk* [Online]. Available at: www.ifpri.org/publica-tion/econometric-investigation-impacts-sustainable-land-management-practices-soil-carbon-and- (Accessed: 29 December 2010).

IFOAM (2009a) *The contribution of organic agriculture to climate change adaptation in Africa* [Online]. Available at: www.ifoam.org/growing_organic/1_arguments_for_oa/environmental_benefits/pdfs/IFOAM-CC-Adaptation-Web.pdf (Accessed: 8 February 2011).

IFOAM (2009b) *The contribution of organic agriculture to climate change mitigation* [Online]. Available at: www.ifoam.org/growing_organic/1_arguments_for_oa/environmental_benefits/pdfs/IFOAM-CC-Mitigation-Web.pdf (Accessed: 8 February 2011).

Rainforest Alliance, in press. *Potential for Agriculture Mitigation and Adaptation: Opportunities for Climate-Friendly Practices in the Tropics, A Review of Literature*.

Sustainable Agriculture Network (2011) *SAN Climate Module: Criteria for Mitigation and Adaptation to Climate Change* [Online]. Available at: http://sanstandards.org/userfiles/file/SAN%20Climate%20Module%20February%202011.pdf (Accessed: 15 February 2011).

Simões, A.F., Kligerman, D.C., La Rovere, E.L., Maroun, M.R., Barata, M., and Obermaier, M. (2010) Enhancing adaptive capacity to climate change: The Case of smallholder farmers in the Brazilian semi-arid region, *Environmental Science & Policy*, vol. 13, no. 8, pp. 801–808.

Verchot, L.V., Van Noordwijk., M., Kandji, S., Tomich, T., Ong, C., Albrecht, A., Mackensen, J., Bantilan, C., Anupama, K.V., and Palm, C. (2004) Climate change: linking adaptation and mitigation through agroforestry, *Mitigation and Adaptation Strategies for Global Change*, vol. 12 no. 5, pp. 901–918.

# 10

# LESSONS LEARNED FROM REDD FOR AGRICULTURE

*Christine Negra and Eva Wollenberg*

## Introduction

This chapter examines the lessons that can be drawn from experience with Reducing Emissions from Deforestation and forest Degradation (REDD) to inform policies and programs for agricultural climate change mitigation.[1] As an international mechanism to avoid increased emissions in the forest sector, REDD's history yields insights about the international policy process and the elements that could facilitate successful approaches to agricultural mitigation.

## *REDD*

REDD+[2] is a framework developed under the United Nations Framework Convention on Climate Change (UNFCCC) to use financial mechanisms and compensation to retain sequestered carbon in forests. Places with high rates of deforestation such as Indonesia, Brazil and the Democratic Republic of Congo would be eligible for REDD+ funds based on demonstration of additional forest conservation ("avoided deforestation") or management that enhances carbon stocks. Developed nations would provide REDD+ funds to offset their emissions. REDD-type pilot activities have been under way for more than 10 years, and current momentum and investment in REDD suggests that UNFCCC approval is likely in the near term.[3]

## *Agriculture*

Agricultural mitigation is likely to require a broader range of financial measures and incentives than REDD+ due to the diversity of existing practices, larger number of mitigation options and the focus on food production. Existing international and national policies provide scope for mitigation in agriculture (see 'policy windows' below), but few make provisions that currently directly support mitigation measures. Only New

Zealand has enacted a policy to reduce agricultural emissions starting in 2015. A Subsidiary Body for Scientific and Technological Advice (SBSTA) work program for agriculture to explore options under the UNFCCC has been proposed, but not yet approved.

Policy windows for agricultural climate change mitigation through the Kyoto Protocol and the United Nations Framework Convention on Climate Change (UNFCCC):

- Kyoto Protocol. This permits new eligible activities such as management of croplands, grasslands and wetlands as compliance-grade offset credits. Annex I Kyoto Protocol Parties have to account for all non-$CO_2$ greenhouse gas (GHG) emissions from agriculture.
- Clean Development Mechanism (CDM). The CDM has approved several small-scale methodologies for agriculture.
- Reduced Emissions from Deforestation and Forest Degradation (REDD+). Agriculture is mentioned as a driver of deforestation and mitigation strategies could become eligible for finance.
- Nationally Appropriate Mitigation Action (NAMA). NAMAs establish country commitments to reducing GHG emissions, including emissions from agriculture. Developing countries can seek finance for these activities.
- UNFCCC negotiations. Agriculture was mentioned under the UNFCCC cooperative sectoral approaches and sector-specific actions in the lead-up to the climate change conference in Cancún in 2010, and a request has been made to the SBSTA for a work program on agriculture.

Policy windows for agricultural climate change mitigation through national policies:

- National cap-and-trade bills. There is a regional program in Alberta, Canada for soil tillage. New Zealand's Emissions Trading Scheme will include agriculture in 2015.
- Low carbon (or emissions) development strategies. These are national efforts to address mitigation across multiple sectors. Some include agriculture, e.g., Guyana.
- National climate action frameworks or plans. These are national initiatives to coordinate policy on climate change. Some include agriculture, e.g., Brazil, Kenya.
- National mitigation standards and carbon crediting. Australia's National Carbon Offset Standard (NCOS) and proposed Carbon Farming Initiative is an example.
- Intergovernmental collaboration. The Netherlands and Viet Nam are spearheading ministerial level meetings.

## Lessons from REDD+

Based on experience in REDD, we suggest that advancing agricultural climate change mitigation will require parallel advancement in six areas: policy; implementation mechanisms and governance; monitoring, reporting and verification (MRV); finance; capacity strengthening; and co-benefits and safeguards. Table 10.1 provides a summary of the lessons from REDD+ for agriculture in each of these areas.

### *Policy making: create political space*

Support for REDD coalesced when, in 2005, the Coalition for Rainforest Nations (CfRN) reframed 'avoided deforestation' as an economic development strategy for developing countries rather than as an emissions reduction requirement. The Stern Review (2006) then established the rationale and urgency for avoided deforestation and demonstrated its likely cost-effectiveness and rapid impacts (Stern 2006). REDD then developed rapidly once the UNFCCC's Bali Action Plan (2007) introduced the concept and established an intensive two-year process of planning (Tropical Forest Group 2007).

To gain consensus and move forward, agricultural mitigation similarly needs to demonstrate that it can support economic interests, particularly for economic development and food security in developing nations. A road map similar to the Bali Action Plan would provide momentum for a period of dedicated development. A review period should be informed by rigorous analysis of mitigation potential and impacts of proposed mechanisms, as well as on-the-ground projects that demonstrate feasibility. This would allow policy makers to consider technical advances and financial arrangements, building greater confidence.

While REDD+ has developed quickly, civil society engagement has been poor. The development of REDD+ is perceived as top down with highly centralized national processes. To ensure relevant programs and successful implementation, agricultural mitigation should improve on this experience by supporting meaningful participation; free, prior and informed consent; and clear legal rights to carbon.

### *Implementation mechanisms and governance: demonstrate feasibility*

REDD+ drew heavily on experiences with forest conservation projects such as the Noel Kempff Mercado Project in Bolivia (Brown *et al.* 2000). Agricultural mitigation should likewise draw on project experiences in sustainable agriculture, agroforestry, payments for environmental services and the Clean Development Mechanism (CDM). Agriculture can also benefit from standards developed in the voluntary and compliance markets that have been used to define the rules for carbon accounting and project

Table 10.1 Lessons from REDD+ for agricultural mitigation

| Element of REDD+ | Lessons for agriculture |
|---|---|
| Policy making | • Economic incentives for developing countries are essential.<br>• High-level political engagement needs to be maintained.<br>• A designated period of preparation and a phased approach to capacity strengthening will contribute to broad agreement and technical and financial confidence.<br>• Lessons from the field level should be linked back to policy processes.<br>• Political negotiations should focus on larger strategic policies. Technical details are better addressed by experts in relevant fields.<br>• Political participation should be inclusive and transparent. |
| Implementation mechanisms and governance | • Mechanisms should build on existing programs, policies and projects.<br>• Mechanisms and governance measures will be necessary at multiple scales to achieve clear land rights, transparency and accountability.<br>• Technical information should be made accessible to decision-makers early on. |
| Monitoring, reporting and verification (MRV) | • MRV should be simple, streamlined and cost-effective.<br>• A global MRV framework that is accessible and affordable to developing countries is a priority.<br>• Integration of agriculture and forestry would help to address agricultural expansion and leakage.<br>• Balance is needed between precision of measurement and cost.<br>• Independent and reliable standards and verification are necessary.<br>• The concept of additionality should be re-examined. |
| Finance | • Early donor support and leadership is critical for demonstrating feasibility and building readiness.<br>• Coordination of finance among donors and investors is a priority.<br>• Finance should be mainstreamed and integrated with sustainable development investments.<br>• Distribution mechanisms need more attention. |
| Capacity strengthening | • Capacity strengthening at the national level helps build confidence and readiness.<br>• Readiness programs help shape national mitigation programs.<br>• The implementation of readiness plans takes time and will reflect country circumstances. |
| Co-benefits | • Standards and safeguards promote environmental and poverty alleviation aims, but must be independently and robustly implemented.<br>• Early provision for structured participation and attention to free prior and informed consent principles and procedures are priorities. |

design for REDD+.[4] Bringing experts and negotiators together to define operational mechanisms can help to create a shared vision and broader political ownership (Streck *et al.* 2009).

REDD pilot projects also helped to demonstrate feasibility and identify solutions such as creating buffers against leakage risks. A broad foundation of pilot projects is needed in agriculture that can demonstrate verifiable emissions reductions and co-benefits, cost-effectiveness, a critical mass of supply, adequate incentives for farmers and compatibility with national development objectives such as food security. Pilots will also need to cover a range of activities—from livestock and fisheries to irrigation, energy use, land restoration and agroforestry. Outcomes for all three major greenhouse gases (carbon dioxide ($CO_2$), nitrous oxide ($N_2O$) and methane ($CH_4$)) and their interactions will need to be examined.

### *Monitoring, reporting and verification (MRV): keep it simple*

Preparation for REDD+ has created a strong technical foundation for monitoring, reporting and verification (MRV) for forest-related emissions, much of which is relevant to agriculture. The Intergovernmental Panel on Climate Change (IPCC)'s greenhouse gas (GHG) accounting guidelines (IPCC 2006) provide technical guidance useful for agriculture, but to be practiced more widely will require more user-friendly language and formats. Efforts such as the GOFC-GOLD (2009) book and project to create practical guidelines for REDD will be useful for agriculture.

As with REDD, a major obstacle for agriculture will be the lack of capacity in many countries to compile data and estimate uncertainties in line with the IPCC guidelines. Streamlined project approaches, credible verification and rapid capacity strengthening can help to avoid similar obstacles in agriculture. Consensus is needed on the best ways to combine measurement and modeling to provide high-quality, cost-effective GHG estimation and accounting for different types of carbon credits. Estimation at large scales and focusing on management practices as a proxy for emissions are currently being tested in agricultural projects (see chapter by Seebauer *et al.*). There appears to be growing recognition that a combination of models and some on-the-ground measurements can yield robust MRV if applied at large enough scales (see chapter by Olander *et al.*).

Demonstrating additionality for REDD+ has proven to be subjective, costly and a disincentive for projects that already successfully provide mitigation services, suggesting that documenting the positive impacts of agricultural practices on climate change mitigation may be more important than always demonstrating additionality.

*Finance: catalyze early action*

The relatively small number of pilot projects working to produce emissions reductions and sequestration in farm settings is not yet sufficient to mobilize significant investment. Delayed development of domestic trading frameworks, as well as low confidence in the establishment of credible national MRV systems, has contributed to a low value for terrestrial offset credits in the current regulated and voluntary offset credit markets (Terrestrial Carbon Group Project 2008). (See also the chapter by Baroudy and Hooda).

As with REDD+, early financing is needed for readiness and capacity strengthening such as infrastructure planning and development, implementing pilots and synthesizing findings, and developing regionally relevant MRV tools and methods. Finance and incentives should be sought that will support low-emissions agricultural development without significant expansion into forests.

In contrast to forestry, agriculture may rely on more finance through government support and corporate supply chains than offset credits. From a public sector perspective, increasing recognition that national and global security is tied to a resilient and productive agricultural system provides real motivation for public investments that promote farm practices that can advance adaptation, food security and poverty reduction while achieving national mitigation commitments. From the perspective of agribusinesses, there is strong interest in identifying investments that can be shown to stabilize or enhance food production while contributing to corporate or sectoral mitigation targets.

Experience with REDD+ also has demonstrated the need for coordination at the country level among donors (bilateral, multilateral and private) and among domestic government agencies.

*Capacity strengthening: prepare quickly and coordinate*

Capacity strengthening under REDD+ has focused on the national level. Two major multilateral efforts have built confidence and readiness for REDD+: the Forest Carbon Partnership Facility (FCPF 2010), coordinated by the World Bank, and UN-REDD, both established in 2008. These programs have provided funding and technical expertise to help countries shape national REDD+ programs.

Slow transitions from planning to implementation for REDD+ programs, even with these multilateral programs' assistance, suggest that aggressive mobilization targets are unrealistic. A step-wise, "learning-by-doing" approach can encourage early mitigation actions by beginning with pilot projects followed by larger-scale projects that use public funding and simple methodologies and progressing to rigorous quantification of emissions reductions and utilization of incentives from market mechanisms.

Gaps in capacity for agricultural mitigation are most notable regarding knowledge about agricultural mitigation practices, participation in offset markets, MRV and governance. A global research agenda on agricultural practices that combine increased productivity and resilience with reduced net emissions is needed.

### Co-benefits and safeguards: establish and apply principles

Co-benefits refer to the positive environmental and social impacts associated with climate change mitigation, while safeguards seek to limit negative impacts. REDD-type projects have demonstrated the need for generating tangible co-benefits in the form of income and land rights for forest owners and forest-based communities (Cotula and Mayers 2009). They have also shown that distributing benefits through investments in community development, rather than as payments to individuals, may lead to more long-term benefits. Requiring co-benefits under REDD+ has come under criticism by some for detracting attention from the main aim of mitigation, while others consider the inclusion of co-benefits essential for generating stakeholder support and desirable social outcomes.

Standards and certification systems, such as the Climate, Community and Biodiversity Standards, establish guidelines against which the co-benefits and safeguards of REDD+ projects can be assessed. REDD has also generated statements of principles for pro-poor outcomes that can inform agricultural mitigation (Peskett 2008).

Safeguards in agricultural mitigation will need to consider not only improving livelihoods and pro-poor outcomes, but also adaptation to climate change and environmental impacts such as enhanced biodiversity, soil health, weather regulation, and water quality and availability.

## Conclusions

To create the policy space and operational feasibility necessary for agricultural mitigation, simultaneous progress is now needed along four concurrent tracks (see Figure 10.1). These are:

### 1 Develop a shared vision

A shared vision for achieving agricultural mitigation that reflects stakeholders' priorities and identifies the major drivers of agricultural emissions should be developed by:

- developing a common language among technical experts, policy makers and practitioners that allows concerns to be more fluently addressed, and enables the clear framing of policy options

119

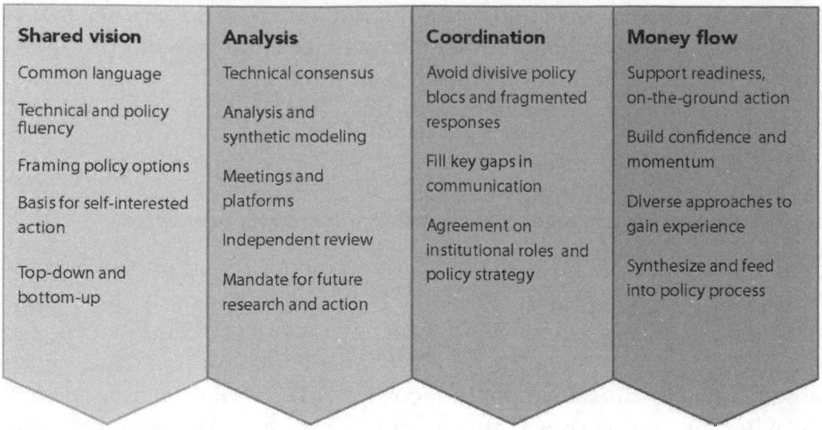

| Shared vision | Analysis | Coordination | Money flow |
|---|---|---|---|
| Common language | Technical consensus | Avoid divisive policy blocs and fragmented responses | Support readiness, on-the-ground action |
| Technical and policy fluency | Analysis and synthetic modeling | | Build confidence and momentum |
| Framing policy options | Meetings and platforms | Fill key gaps in communication | |
| Basis for self-interested action | Independent review | Agreement on institutional roles and policy strategy | Diverse approaches to gain experience |
| Top-down and bottom-up | Mandate for future research and action | | Synthesize and feed into policy process |

*Figure 10.1* Lessons from REDD+ for agricultural mitigation

- acknowledging deadlocks, making the case for self-interested action at national and sectoral scales, and integrating top-down design with bottom-up field experience
- encouraging both formal and informal stakeholder engagement, through a range of events that bring diverse perspectives together, as well as efforts by respected leaders and thinkers.

## 2 Analyze high priority mitigation options and impacts

Policy and implementation options for agricultural mitigation should be informed through:

- focused efforts to promote consensus on technical issues by multilateral agencies, research consortia and other communities of practice through synthetic modeling and analysis, as well as meetings and other platforms
- an authoritative, independent review that puts agricultural mitigation in a global context; rigorously outlines the potential for mitigation options and impacts; identifies the necessary policy and financing strategies; and sets out a mandate for further research and action.

## 3 Coordinate efforts

Countries, agri-business and trade groups, farmers' associations, indigenous communities and multilateral agencies must work together to avoid the creation of divisive policy blocs and fragmented technical and institutional responses. Coordination efforts should:

- be grounded in a comprehensive understanding of the drivers, actors, policies and institutional arrangements currently influencing global agriculture
- identify and fill key communication gaps
- clarify institutional roles and responsibilities, and establish agreement on an overall strategy for developing an agricultural mitigation policy.

### 4 Encourage money to flow

Money from donor agencies, foundations and industry is essential to support preparation for agricultural mitigation, development of infrastructure and project implementation at the local level. Increased investment will build confidence and momentum around agricultural mitigation, and will mobilize technical activity and institutional engagement. Key elements of a strategy to increase the flow of funds include:

- leadership by 'anchor' donors, who invest through bilateral agreements and multilateral programs
- experimentation with supply chain projects, payment for environmental services initiatives and other market incentives, development of mechanisms for sharing results, synthesizing findings and feeding them back into policy processes.

## Notes

1 This chapter is based on a report prepared for the UK Department for International Development by Negra and Wollenberg (2011), "Lessons from REDD+ for Agriculture." Data for the study were drawn from interviews with 32 leaders in REDD and a review of the literature.
2 REDD became a point of focused negotiation in 2005, when the UNFCCC decided to consider 'Avoided Deforestation' as a mechanism for climate change mitigation. Avoided Deforestation evolved to become Reduced Emissions from Deforestation and forest Degradation (REDD), and finally REDD+, which includes the role of forest conservation, sustainable management of forests and enhancement of forest carbon stocks.
3 In December 2010, the Cancún Agreements produced by COP16 called for preparation for REDD+, including requesting developing countries to prepare national strategies, develop reference levels and create monitoring systems. The Cancún Agreements also called for a SBSTA work program on REDD+ to address drivers of deforestation and methodologies, as well as exploration of REDD+ financing options.
4 Mitigation strategies can include increased efficiency of inputs and products through supply chains as well as a broad array of on-farm practices. Co-benefits can include increased yield, improved livelihoods, restoration of degraded lands, and clarity of land tenure.
5 Standards include both registries and certification-type standards such as the Climate, Community and Biodiversity Standard (www.climate-standards.org). To date, the only registry with a global scope is the Verified Carbon Standard (www.v-c-s.org).

6 The 2010 Cancún Agreements included a request to developing countries to develop information systems for tracking how well safeguards are included in REDD+ implementation. Key actors supporting better REDD governance include Forest Law Enforcement, Governance and Trade (FLEGT); Chatham House; Global Witness; Forest People's Programme; World Resources Institute; Imazon; Rights and Resources Initiative; the Forests Dialogue and CIFOR (see chapters by these organizations in Saunders and Reeve 2010).

# References

Brown, S., Burnham, M., Delaney, M., Powell, M., Vaca, R. and Moreno, A. (2000) 'Issues and Challenges for Forest-based Carbon-offset Projects: A Case Study of the Noel Kempff Climate Action Project in Bolivia', *Mitigation and Adaptation Strategies for Global Change*, vol. 5, pp. 99–121.

Cotula, L. and Mayers, J. (2009) 'Tenure in REDD: Start-point or afterthought?' Natural Resource Issues No. 15. International Institute for Environment and Development, London.

FCPF (2010) 'Synthesis Report: REDD+ Financing and Activities Survey', World Bank Forest Carbon Partnership Facility.

GOFC-GOLD (2009) 'A sourcebook of methods and procedures for monitoring and reporting anthropogenic greenhouse gas emissions and removals caused by deforestation, gains and losses of carbon stocks in forests remaining forests, and forestation'. GOFC-GOLD Project Office, Alberta, Canada.

Intergovernmental Panel on Climate Change (2006) *2006 IPCC Guidelines for National Greenhouse Gas Inventories, Volume 4: AFOLU* [Online]. Available at: www.ipcc-nggip.iges.or.jp/public/2006gl/index.html.

Negra, C. and Wollenberg, E. (2011) *Lessons from REDD+ for Agriculture*, CCAFS Report no. 4 [Online]. Available at: www.ccafs.cgiar.org. The CGIAR Research Program on Climate Change, Agriculture and Food Security (CCAFS), Copenhagen, Denmark,

Peskett, L., Huberman, D., Bowen-Jones, E., Edwards, G. and Brown, J. (2008) Making REDD work for the poor [Online]. Available at: www.iucn.org/about/work/programmes/economics/?2052/Making-REDD-Work-for-the-Poor. Poverty Environment Partnership (PEP)

Saunders, J. and Reeve, R. (2010) 'Monitoring Governance for Implementation of REDD+', paper presented at the Monitoring Governance Safeguards in REDD+ Expert workshop, London, UK.

Stern, N. (2006) 'Stern Review: the economics of climate change'. Cambridge University Press, Cambridge, UK.

Streck, C.L., Gomez-Echeverri, L., Gutman, P., Loisel, C. and Werksman, J. (2009) *REDD+ Institutional Options Assessment: Developing an Efficient, Effective, and Equitable Institutional Framework for REDD+ under the UNFCCC* [Online]. Available at: www.redd-oar.org/. Meridian Institute.

Terrestrial Carbon Group Project (2008) *How to Include Terrestrial Carbon in Developing Nations in the Overall Climate Change Solution* [Online]. Available at: www.terrestrialcarbon.org.

Tropical Forest Group (2007) 'A History of Climate Change and Tropical Forest Negotiations', Tropical Forest Group Articles.

# 11

# SUSTAINABLE LAND MANAGEMENT AND CARBON FINANCE

## The experience of the BioCarbon Fund

*Ellysar Baroudy and Neeta Hooda*

### The BioCarbon Fund

Established in 2004, the BioCarbon Fund supports projects that sequester or conserve greenhouse gases (GHGs) in forests, agricultural and other ecosystems, mainly in developing countries. It is one of two carbon funds at the World Bank dedicated to land use (the other is the Forest Carbon Partnership Facility dedicated to large-scale Reducing Emissions from Deforestation and forest Degradation (REDD+)); 10 remaining carbon funds and facilities do not deal with land issues. The Fund purchases carbon credits resulting from net reductions in GHG emission. The reductions are additional, i.e., would not have happened anyway relative to a baseline scenario, and, to the extent possible, produce environmental and social benefits. In addition, the Fund supports the development of the carbon market, with the elaboration of carbon accounting methodologies, tools to help project developers access the market and information dissemination.

To date, the BioCarbon Fund has committed US$90 million to over 20 Afforestation and Reforestation (A/R) projects, five projects that reduce emissions from deforestation and forest degradation (REDD+) and four sustainable agriculture land management pilots. The majority of funds are dedicated to projects under the Clean Development Mechanism (CDM), one of the mechanisms under the United Nations Framework Convention on Climate Change (UNFCCC) that allows developed countries to reduce their GHG footprint through enabling sustainable development in developing countries. The CDM, however, allows only for A/R projects in the land-use sector. Avoided deforestation or agriculture land management projects currently cannot be accounted for under the CDM. The latter are being developed by the BioCarbon Fund in the voluntary carbon market.

This chapter will focus on the more recent experience of the BioCarbon Fund with carbon from sustainable land management (SLM).

## Carbon finance and sustainable land management

The BioCarbon Fund started working on agriculture in 2007, when it set up its second tranche of funding and included resources for sustainable land management. Two elements prompted the BioCarbon Fund to consider expanding to SLM: (1) the natural progression to deal with all land-use issues in the fund; and (2) the evolving debate in the international climate negotiations on agriculture. The BioCarbon Fund resources dedicated to sustainable land management support the development of an institutional framework around the net reduction of GHGs from soil, including the development of an accounting methodology, and will test the possibility for the purchase of emission reductions from such projects. The selection of projects is under way but the fund will select a variety of project types (e.g., smallholder and medium-sized farmers; varying technologies). Some lessons that were learned with respect to carbon finance through earlier experiences of the BioCarbon Fund can be applied in this sphere. These are described below for finance, the carbon market and carbon accounting.

## Carbon finance

Three lessons learned from the current state of finance of agricultural carbon are the need for strong project entities, adequate financing, and overcoming the risk associated with testing a new market (see also discussion in Shames *et al.* and Havemann).

A key element in any carbon project is to have a dedicated project champion (also referred to as the project entity or developer; see discussion in chapter by Tennigkeit *et al.* of the role of the project developer). This will be the lead person who typically manages the project, and interacts with the auditors and BioCarbon Fund. A dedicated champion is especially important for complex projects, or ones involving many farmers and plays a significant intermediary role in such cases. In a smallholder agriculture project for example, the farmers will typically work with the project entity who aggregates their activities (in such case the project entity can also be referred to as a project aggregator), with whom the Fund interacts. It would be impossible to undertake cost-effective carbon transactions between individual farmers and the Fund.

For any carbon project to succeed, adequate financing needs to be secured. For a carbon project, two financing streams are needed: (1) the upfront investments needed to undertake the project, and (2) the financing needed to develop the carbon asset. The BioCarbon Fund was not set up to allow for upfront investments in the project, but can help with costs associated with the development of the carbon asset, as well as later to purchase carbon credits. Since the remuneration of credits is based on performance, revenues only flow once a project is established and implemented. The amount of revenues depends on a project's size and timely emission reductions delivery. Raising resources for the initial financing of the project has been a challenge for most carbon projects and has frequently been the main impediment for progress. To date, projects in the fund have financed initial investments through either public grants or, in the case of non-governmental organizations (NGOs), their own funding.

In A/R projects, both the private sector and grants have been primary sources of carbon finance. Private investment has accounted for 50% of investments in the BioCarbon Fund portfolio, including short- and long-term loans from commercial banks. This is also expected to be the case in sustainable land management projects over time.

In order to help the agriculture carbon sector as a whole, the BioCarbon Fund has also dedicated resources to cover the cost of (1) developing a new methodology and the double approval process involved in getting the methodology approved in a recognized standard; (2) assessing the carbon baseline and project preparation; and (3) investing in capacity building and training for project developers and entities. Some resources allocated were not expected to be needed immediately, but were deemed necessary to invest in as the methodology development progressed. This included stakeholder meetings with experts in soil science.

The risks to buyers and sellers at such an early stage of the agricultural carbon market are significant. Most carbon transactions in the voluntary market have taken place in developed countries, mainly the United States and Canada, where agricultural emissions data are abundant, farms are often large and financial risks are lower than in developing nations. Very few transactions have been recorded from developing country projects.

For sellers (usually a project developer) in developing countries, entering these untested waters requires dedicating resources to a new area of expertise within their teams and for the transaction costs associated with developing the carbon asset. Project preparation costs including developing a Project Design Document (PDD), the initial step in the carbon-asset documentation process, is time intensive and costly. Project developers often have limited capacity to establish projects, which is not unusual in such a new field. In A/R projects, the cost of project development can range from US$100,000 to $400,000; this can include delays incurred by lack of existing methodologies and where a methodology has to be developed at the same time as the project. SLM project development costs have been at the lower end of this range. See chapter by Tennigkeit *et al.* for a detailed breakdown of project development and other costs.

For buyers, their willingness to purchase agricultural carbon credits has been limited to date. The BioCarbon Fund tested the market with co-purchasers from outside the fund, but some buyers preferred to wait for developments in the sector; others did not want to invest in the carbon methodologies, a fundamental element for transactions to be possible. Carbon methodologies can cost US$100,000–150,000 to develop; they are ultimately a public good so private companies can be reluctant to invest in these. Because of the pioneering nature of SLM, no buyers were willing to make advance payments for the carbon as risks were deemed too high. This should change as the market develops in the coming years.

If lessons learned are applied from the more advanced nature of other sectors in the carbon market such as A/R and REDD, it is expected that funding and financing bottlenecks will decrease in time as methodologies become available to use and carbon transactions increase.

## Sustainable land management—carbon gaps

In addition to the fundamental investments and project finance, a number of factors are needed in developing carbon markets and carbon accounting systems to open up space for rural communities to access potential financing.

### Regulatory and voluntary markets

Carbon credits normally can be sold into the regulated market (where buyers have an obligation through regulation to reduce their GHG footprint) or into the voluntary market (where buyers are not forced to reduce their footprints but do so for reasons such as corporate social responsibility). The regulated markets are most significant in terms of volumes and values of transactions (Kossoy and Ambrosi 2010). These regulations are set, for example, by the UNFCCC Kyoto Protocol and the European Union's Emission Trading Scheme (EU-ETS). However, both of these have not been as favorable to credits from developing countries that have originated from forestry or SLM projects, as from other sectors. In the land-use sector, the CDM only allows credits to be generated from A/R projects. A few of these A/R carbon methodologies allow for accounting of the soil carbon pool, but most account only for above- and below-ground biomass. Credits generated from these projects are deemed to be "temporary" credits because the gains over time in carbon dioxide ($CO_2$) sequestration could be overturned by fire, disease, etc. Credits from other sectors (industrial gas, waste, energy efficiency, etc.) are deemed permanent. Governments are allowed to buy credits from the forestry sector in the CDM, but these are capped at 1% of their 1990 emissions. Although European companies in the EU-ETS can use CDM credits to offset their emissions, the EU-ETS bans completely the use of forestry credits from the CDM because they are not permanent credits, and therefore not fungible in their system. The regulations governing the CDM and the EU-ETS disadvantage many developing countries where better land-use management could be incentivized through carbon revenues.

The UNFCCC negotiations are evolving to include other land use and forest activities, beyond A/R. Most advanced is reducing deforestation and forest degradation (REDD) under the Long-term Cooperative Action track in the negotiations. Another track in the negotiations, the Kyoto Protocol, contains draft text that suggests more eligible activities in land use, including cropland management, and also suggests ways of addressing temporary credits. The latter would mean that a forestry credit could be counted as a permanent credit rather than a temporary one, if an insurance scheme or similar were in place in case of reversal of sequestration.

Meanwhile, the voluntary carbon market has continued to grow mainly with interests of companies with green agendas. Developments in sustainable land management and REDD+ are happening more quickly in this market, and there has been a recent increase in carbon accounting methodologies for REDD+ (Hamilton *et al.* 2010a, b). The voluntary standards have a different approach when it comes to permanence of carbon, adopting reserves or a buffer that sets aside emission reductions as a form of 'insurance' (see chapter by Swickard and Nihart on the Verified Carbon Standard). In practice, for agriculture land management for example, setting aside up to 60% of emission reductions, the top limit of some buffers, is a significant disadvantage to project developers. This means that if that buffer limit is set, project developers are able to gain revenues from only 40% of the credits generated by their projects. Overall, this voluntary market remains limited with few methodologies and activities being undertaken. It would be helpful for regulators, whether in the compliance or voluntary markets, to assess how to make rules and regulations friendlier to sustainable land management because of its importance for rural livelihoods in developing countries.

### Carbon accounting

The BioCarbon Fund has gained much experience in the development of methodologies, playing a key role in current A/R and REDD+ methodologies. It is also developing a methodology for SLM. As work on methodologies has evolved over time, it is clear that there is a need for simpler accounting methods if land use is to become part of the climate solution and if there is to be a wide-scale contribution to sustainable development in degraded areas.

The BioCarbon Fund is piloting a new approach for estimating changes in the soil carbon pool when sustainable land management practices that enhance soil carbon sequestration are adopted on farmlands. The methodology uses an approach where instead of making direct soil carbon measurements, the emphasis is on robust monitoring of adopted agricultural practices which are known to have impacts on soil carbon (positive or negative), and then using the monitored parameters as inputs into models for estimating the soil carbon change (see chapter by Seebauer *et al.*).

Such an approach has to be strongly backed up by long-term research results that document the impacts of practices on soil carbon and can work if there is confidence in the science supporting the results. This implies a need for scaled-up, multi-dimensional scientific experiments that show the impacts of activities on soil carbon and test new activities with the potential to sequester carbon. Such independent experiments need to be designed in a way that the results can be applicable to projects being implemented at the national or sub-national scales, but the investments for the research should not be the responsibility of the project entity as this places too much of a burden on the rural farmer. There is an urgent need for the scientific community to collaboratively invest in research experiments for testing activities likely to influence soil carbon, document the results and translate results, for example, for use as defaults that can be used by project developers and small-scale farm holders to get benefits.

Classification of land use into agro-ecological zones, preparation of baseline soil carbon maps and addressing information gaps such as climatic data at the sub-national level are other areas that would facilitate the use of new accounting approaches. Often lack of sufficient information at the project level has hindered the application of models.

Many land uses that have significant GHG reduction potential are not yet tested using GHG accounting methodologies, including reducing methane emissions from rice paddies, and grassland and pastureland management. Pilot activities and development of new methodologies will need to be explored. As more land-use activities are brought under the umbrella of the regulated mechanisms, there will be a need to test how these can co-exist, for example A/R and REDD+ or the linkage between REDD+ and SLM.

## What is needed to help SLM carbon take off?

There are inherent difficulties in detecting and providing proof of soil C change at short time intervals (see also chapters by Milori *et al.* and Seebauer *et al.*). Until now we have seen largely approaches and modalities developed for the non-land use sector adapted for application to the land-use sector. This was the case in the CDM and Verified Carbon Standard (VCS) where differences of the land-use sector have been captured in the adapted modalities and in some ways the land-use sector projects have been subjected to a higher level of scrutiny. Given the inherently high spatial variability of land use, the desirable accuracy of measurements of carbon pools and soil C change in particular needs to be revisited by the global community to foster realistic, yet conservative approaches to estimations. Incentivizing adoption of practices by the farming community globally will require a big leap towards simplification of procedural and accounting complexities entrenched in current standards.

Regional organizations such as rural development banks in countries need to be sensitized about agriculture and carbon programs. Strengthening national rural finance and microfinance institutions can play a big role in bridging this gap towards upfront financing. An example is the network of rural and community banks in Ghana. Nair and Fissha (2010) assess the role and impact this banking network has had in the rural economy and the inroads that have been made so far. There is a need to explore, nurture and support such financial institutions further. Corporations have an interest in fostering social responsibility projects that can demonstrate yields from fields where sustainable practices have been applied.

Carbon finance projects in the sustainable land management sector needs to be more holistic. Whilst carbon finance has the potential to increase the financial returns of certain farming practices, the primary reason behind adoption of practices from the farmers' perspective are increased yields or income. If better management practices can provide increased yields, add value along the production chain and simultaneously result in carbon gains, then the practices are likely to be taken up more readily. The key question is whether there will be incentives so that farmers will adopt sustainable land management practices in their own best interests. Some models of this nature are already beginning to emerge in a few African countries (see chapter by Lager and Nyberg).

Currently there is a mismatch between the complexity of procedures and the implementation capacity at the grassroots. A two-pronged approach that bridges this gap by enhancing capacities of beneficiaries and project implementers on the one hand and simplifies procedures on the other is required.

The size of sustainable land management and A/R pilots in the BioCarbon Fund currently ranges from a few hundred to a few thousand hectares. With an increase in project size, transaction costs can be reduced and a higher share of carbon revenues can be shared with the beneficiaries. However the current capacity of project entities and the difficulties in leveraging finances for upfront financing pose barriers to reaching economies of scale. Valuation of other environmental services such as biodiversity, soil conservation, improved water quality and water retention capacities need to be incentivized so the participating communities are compensated for primary and secondary benefits emanating from the project.

Sustainable land management practices and their use for GHG mitigation in most countries currently are not embedded in national legislation. Activities are undertaken at the initiative of project entities and thus remain localized efforts. It would be strategic if the use of carbon finance could be anchored within the agricultural policy discussions at the national or sub-national level.

Carbon projects based on sustainable land management will require clear land tenure and emission reduction rights. Investors will not take the risk of working in environments where ownership of any asset, whether agricultural produce, or carbon are not clear. From the BioCarbon Fund A/R experience, carbon finance has helped to increase land tenure security in project areas. It is expected that this should also be the case in sustainable land management projects.

A strong institutional and management framework is important for the success of these projects. Investing and sustaining local capacities can ensure permanence of carbon initiatives, and successful projects rely on equitable benefit-sharing schemes that improve local livelihoods. Technical expertise is helpful within a project team, but this can be outsourced; what is critical and essential is solid management capacity.

## Conclusion

The BioCarbon Fund is furthering work on sustainable land management as this should lead to a "triple-win" situation: (1) increasing productivity and improving farmers' livelihoods by generating additional incomes; (2) contributing to the climate change solution through mitigation of GHGs; and (3) contributing to the adaptation agenda by helping rural communities become more climate resilient. These three traits have been apparent in all BioCarbon Fund land-use projects to date; taking this to scale and facilitating the adoption of these projects is one of the priorities of the Fund going forward.

## References

Nair, A. and Fissha, A. (2010) 'Rural banking: The case of the rural and community banks in Ghana', *Agriculture and Rural Development Discussion Paper No. 48*. World Bank, Washington, DC.

Kossoy, A. and Ambrosi, P. (2010). 'State and trends of the carbon market 2010'. The World Bank, May 2010.

Hamilton, K., Chokkalingam, U. and Bendana, M. (2010a) 'State of the forest carbon markets 2009', *The Ecosystem Marketplace.*

Hamilton, K., Peters-Stanley, M. and Marcello, T. (2010b) 'Building bridges: State of the voluntary carbon markets 2010', *The Ecosystem Marketplace, Bloomberg.*

For additional resources: www.biocarbonfund.org (click on: Useful LULUCF Resources).

# 12

# FINANCING MITIGATION IN SMALLHOLDER AGRICULTURAL SYSTEMS

## Issues and opportunities

*Tanja Havemann*

This chapter describes obstacles to financing mitigation in smallholder agricultural systems, and provides recommendations to overcome these, with an emphasis on overlaps with carbon finance.

### Carbon finance in the context of smallholder AFOLU practices

Smallholders can generate carbon credits through energy and land-use (LU) practices. This chapter focuses on LU practices, summarized in Table 12.1.

Net increases in mitigation attributable to improved practices, compared to usual practices, are used to estimate carbon credit volumes. Credits must be quantified and independently verified according to a chosen carbon credit standard and a methodology under that standard.

Various barriers have prevented sales of carbon credits from being a significant addition to smallholder incomes. One barrier has been a lack of accepted standards with methodologies to quantify mitigation from agricultural practices, illustrated by Table 12.1. A number of new methodologies are currently being developed, for example the Kenya Agricultural Carbon Project (KACP) is developing a Verified Carbon Standard (VCS) methodology to credit increases in both above- and below-ground carbon stocks.

General barriers to mitigation projects involving smallholders include the:

- small size of benefit per smallholder: projects require significant spatial scale to be economically viable, given the transaction costs of monitoring mitigation achieved;
- informational complexities faced by the smallholder during the carbon credit registration and issuance processes;
- sizeable upfront cost of project development;

Table 12.1 Mitigation options for agriculture[1]

| | Carbon credits | Applicable standard | | Carbon credits | Applicable standard |
|---|---|---|---|---|---|
| **Crop rotations and farming systems design** | | | **Nutrient and manure management** | | |
| Improve crop varieties | X | N/A | Improve nitrogen (N) use efficiency | X | N/A |
| Feature perennials in crop rotations | L | CDM A/R, VCS, Plan Vivo | Adjust fertilizer application to crop needs | X | N/A |
| Use cover crops to avoid bare fallows | X | N/A | Use slow-release fertilizers | X | N/A |
| Enhance plant and animal productivity and efficacy | X | N/A | Apply N when crop uptake is guaranteed | X | N/A |
| Adopt farming practices with reduced reliance on external inputs | X | N/A | Place N into soil to enhance accessibility | X | N/A |
| **Livestock management, pasture and fodder supply improvement** | | | Avoid any surplus N application | X | N/A |
| Reduce lifetime emissions | X | N/A | Manage tillage and residues conservatively | X | N/A |
| Breed dairy for lifetime efficiency | X | N/A | Reduce unnecessary tillage using minimum and no-till strategies | X | N/A |
| Breed and manage to increase productivity | X | N/A | **Maintaining fertile soils and restoring degraded land** | | |
| Plant deep-rooting species in primary production | L | CDM A/R, VCS, Plan Vivo | Revegetate | L | CDM A/R, VCS, Plan Vivo |
| Introduce legumes into grasslands | L | CDM A/R, VCS, Plan Vivo | Improve fertility by nutrient amendment | X | N/A |
| Prevent methane emissions from manure heaps and tanks | E | CDM, VCS, Gold Standard | Apply substrates such as compost and manure | X | N/A |

| Practice | | Standard | Practice | | |
|---|---|---|---|---|---|
| Utilize biogas as a resource | E | CDM, VCS, Gold Standard | Halt soil erosion and carbon mineralization by soil conservation techniques, e.g., reduced tillage, contour farming, strip cropping and terracing | X | N/A |
| Compost manure | E | CDM, VCS, Gold Standard | Retain crop residues as covers | X | N/A |
| | | | Conserve water | X | N/A |
| | | | Sequester carbon by increasing soil organic matter content | X | N/A |

**Abbreviations**
CDM A/R: Clean Development Mechanism, Afforestation/Reforestation methodologies
VCS: Verified Carbon Standard

*Legend*
X = Not yet possible to generate carbon credits.
L = Generates LU credits.
E = Generates energy credits.

*Note*
The table only refers to standards applicable in developing countries.

- uncertainty of cash flow ex-ante; projects using a voluntary carbon standard can be difficult to value, as Voluntary Carbon Markets (VCMs) are relatively untransparent and illiquid.

Additional barriers faced by LU projects include:

- lumpy cash flows; credits can typically only be sold after a minimum of five years of operation,[2] whereas significant upfront costs are incurred. Credit volumes are relatively small in the first issuance periods;
- demand is primarily from VCMs;
- engagement in a carbon credit project may reduce the already limited smallholders' land management options, which may be a disincentive.

## Characteristics of smallholder agriculture finance in developing countries

Many barriers to smallholder carbon credit mitigation projects are general issues associated with investing in smallholder agriculture, rather than carbon finance per se.

### Markets and investments domestically driven and focused

Most agricultural investments in developing countries are funded domestically (Ritchie 2010). Sources of capital can be classified as informal (personal loans from family members or informal lenders) or formal. Formal sources include trade credit and commercial lending. Non-domestic sources include loans associated with international agriculture companies, banks, donors and non-governmental organizations (NGOs).

Formal financial agents can engage with smallholders in a variety of ways, broadly categorized into short- and long-term credit, provision of risk mitigation instruments, and equity. Table 12.2 summarizes formal financing types that are not associated with an aggregating institution, such as contract farming. The cost of providing capital varies depending on factors including duration and location. Monies can be distributed through a company, individual or cooperative.

Access to capital tends to reflect smallholder category; these are classified according to production for:

- subsistence, as opposed to sale
- sale in domestic markets as opposed to for export
- export, as part of an aggregating institution, or on a more individual basis, for example, through a local trader.

Access is also influenced by location and local infrastructure (physical, social, institutional).

It may be easiest for smallholders producing for export, particularly those part of an aggregating institution, to access formal sources of capital.

*Table 12.2* General agricultural finance categories

| Financing type | What is it? | Example relevant to a smallholder | Example of requirements to access, if available |
|---|---|---|---|
| Short-term credit | Loans with a maturity ≤ 12 months | Small loans to individual based on personal profile | Group backing/personal reputation. May require collateral. |
| Long-term credit | Loans with a maturity of > 12 months | Larger bank loans | More sizeable collateral including future production, land. Bank account. |
| Risk mitigation instruments | Insurance products, savings | Micro-insurance, weather-based insurance | Reliable local weather data, mobile phone. |
| Equity | Ownership in a commercial company | Equity in a producer organization | Bank account, legally incorporated entity. |

*Investments reflect stakeholders' risk and return profiles*

Smallholders that are highly reliant on the land for their livelihood may strongly favor short-term returns. They may also engage in many different income-generating activities, on and off the land, to diversify their incomes and minimize risk. Risk related to uncertain land tenure and land availability may increase preferences for quicker returns.

'Sustainable agriculture' investments,[3] that maintain long-term productivity, may require significant upfront costs, potentially accompanied by a reduction in short-to-medium-term income. Smallholders must therefore be helped to overcome initial upfront costs, and the short-to-medium-term penalties associated with implementing a longer-term improved land management practice.

These improved practices may in some cases require a price premium on the product to be viable; however it may be difficult to claim a premium for 'climate-friendly' agricultural products for smallholders who only supply their local markets, where customers typically are not in a position to pay a 'sustainability' premium for products.

*There may be additional risks associated with production for export*

Although farmers may receive higher prices for export crops, particularly for certified produce, risks may also increase. For example, in committing to deliver a certain type and volume of produce, the farmer may divert productive capacity away from directly feeding their family, and reduce income diversity (of products and seasonally). An example of this risk is the experience of DrumNet in Kenya, where a company buying produce from smallholders suddenly stopped doing so because of lack of compliance with new European import requirements. This forced producers to sell their product below expected price and they returned to their subsistence crops (Ashraf *et al.* 2010).

Carbon credits could be considered a type of export produce, and could lessen or exacerbate existing risks. Risks could be exacerbated if benefits associated with mitigation are only provided to the smallholder after credits are issued and if the smallholder has to bear upfront costs. The effect on smallholders depends on the basis on which benefits associated with mitigation are issued, e.g., payments made for changing practices versus for the production of specific carbon credit volumes. The form and timing of benefits influence smallholders' appetite for engagement.

## Overlaps between financing for agricultural products and mitigation

Profitability of projects is a factor of production volumes, expected product prices and the size and timing of production costs. While there is

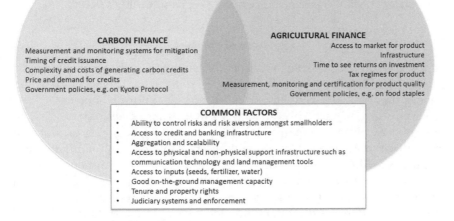

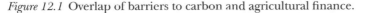

*Figure 12.1* Overlap of barriers to carbon and agricultural finance.

some difference between specific barriers associated with improved and increased investment in smallholder agriculture and carbon credit projects, significant overlap exists (Figure 12.1). Selected overlapping factors are described below.

### *Ability to manage risks*

The ability to control risks helps to smooth income. Smallholders may be sensitive to agricultural risks due to lack of income or food source diversity. They face risks related to production, overdependence on a few crops, price volatility, and changes to regulatory frameworks

Many smallholders rely on semi-formal arrangements to manage financial risk, e.g., social networks, informal loans, contract farming and sharing liability amongst a group. These methods are limited in their ability to transfer and diversify risk and are context sensitive. More formal risk mitigation methods, e.g., insurance and guarantees are less readily available in developing countries. Combining micro-insurance with access to credit has demonstrated some successes, however this is still relatively new.

Risk within a project should be transferred to the entity best able to control it. There are risks specifically associated with a carbon credit project (registration, monitoring, reporting and verification), in addition to the general implementation risks. The project developer is often best equipped to manage carbon credit project risks.

Links to adaptation could also be explored. For example, a program to develop increased smallholder resilience could support a third-party insurer to provide affordable insurance against non-controllable risk related to climate change. Access to low-cost insurance could encourage the small-holder to invest in mitigation and at the same time increase their resilience to climate change; an example of this is the Peruvian El Niño Southern Oscillation (ENSO) insurance program (Skees and Collier 2010).

### Access to inputs

Access to credit and banking infrastructure can help smallholders reduce risk, manage cash flows and invest in increased productivity. Table 12.3 describes some of the commercial credit sources—it excludes informal sources.

Various actors have roles to play in improving smallholder access to resources. Fertilizer companies are, for example, repackaging products to make them more accessible to smallholders. Mobile telephone banking is helping smallholders to access micro-insurance. Government policies could help improve access to credit, e.g., by developing credit bureaus and property registries, information technology infrastructure to support monitoring and gather weather data, financial education and development of risk transfer mechanisms and guarantees. Mainstreaming recognition of other sources of collateral, including warehouse receipts, accounts receivable and standing crops is also necessary. Contracts for mitigation could be a form of collateral.

### Aggregation

Aggregation can improve access to credit and risk mitigation products and can increase bargaining power. Commercially viable models exist for engaging smallholders, but require aggregation to reduce barriers (see also Shames et al.).

The Chiansi irrigation project in Zambia developed by InfraCo provides an example of how benefits can be provided to smallholders through aggregation and 'patient capital. 'Patient capital' is '... *long-term, subordinated capital invested at sub-commercial cost, which is used to fund the one-off start-up costs and part of the cost of the very long-life assets*' (Palmer et al. 2010). This helped to overcome investment barriers, including high upfront costs, long payback periods and a perception of high risk by investors.

Business models to engage smallholders through different aggregation models are summarized in Table 12.4 (Cotula and Vermeulen 2010); many variations exist. 'Out-grower scheme' and 'contract farming' are used interchangeably and describe a situation where a smallholder has a contractual relationship with a purchaser of agricultural goods.

Formal credit attached to the supply chain has been the dominant source of working capital for smallholders (Doran et al. 2009). However, this has been hampered by lack of organization, transparent pricing and fragmentation (van Empel 2010).

Table 12.3 Overview of credit sources

| Provider | Description | Issues |
|---|---|---|
| Local bank | Bank with significant operations in the country. | Many are not active in rural areas, e.g., local banks in sub-Saharan Africa allocate on average less than 8% of their lending to agriculture.* |
| Agricultural bank | Specialized bank lending money to farmers, typically over long time periods and at low interest rates. Often government subsidized. | Few successful agricultural banks exist and many are government controlled. They may only invest in projects of a certain size or producing a particular product. |
| Microfinance institution | Organization providing small, short-term loans to individuals. | Traditionally not active in rural areas as they require scale. |
| Public sector credit | Credit, and other resources are provided at below-market rates by the government. | Hostage to government policies and management infrastructure. Long-term sustainability may be an issue. |
| Rural financial cooperatives | Refers to several models, including Village Savings and Loans Associations (VSLA). | Success depends on terms and conditions of loan provision and how funds are distributed. A poorly diversified customer base may increase risk; governance and management may be poor and the facility vulnerable to liquidity shortages. |
| Trade credit | Financing based on the product value chain. | Limited flexibility and funds are tied to the product; may be open to misuse by the credit provider. |
| Socially Responsible Investor (SRI) | Investors seeking social and environmental outcomes as well as financial returns. | Investors typically require a minimum return and scale, and tend to invest in the generation of a specific product. Subsidized investments may not be sustainable. |
| Carbon financier | Investment based on expected carbon credit revenues. | Requires a minimum scale, i.e., tons of carbon mitigated. |

Note
* Doran et al. 2009.

Table 12.4 Overview of aggregation models

| Model | Sub-model | Description |
|---|---|---|
| *Driven by smallholders* | | |
| Cooperatives and farmer-controlled institutions | Associations, trusts, enterprises, cooperatives, farmer-owned companies | Formalized groups of smallholders with legal standing. Different structures exist depending on the institution's purpose. |
| Contract farming | N/A | Smallholders group together to lease land to a third party. |
| *Driven by third-party such as an agri-business company or exporter* | | |
| Contract farming | High centralized | Institution that buys from a large number of smallholders and imposes demands on produce quantity and quality. |
| | Nucleus estate | Institution that buys through a centralized model, combined with a nucleus estate managed by the institution. |
| | Multipartite | Joint venture between a third party and a local entity representing smallholders and in contractual relationships with them. |
| | Informal | Verbal purchase agreements, usually completed on a seasonal basis. |
| | Intermediary | Institution that has a contract with an intermediary who signs up individual smallholders. |
| Tenant farming or share cropping | N/A | Contracting of smallholders to manage land owned or leased by a third party. |

Despite aggregation model diversity, some common ingredients for long-term success exist, including:

- clear participation criteria for the farmers, e.g., minimum landholding, basic understanding of business planning and farm management skills;
- transparent terms of reference for the product, e.g., quality requirements, pricing arrangements and services provided;
- trust between parties in the model, e.g., agreed codes of conduct;
- registration and record keeping.

### *Best practices and capacity*

Many smallholders lack access to extension services that could help them improve yields without increasing costs. This may be paid for by the farmer, an NGO or the government or provided by the aggregating entity. It is likely that both carbon credit project developers and agricultural investors will have to invest in improved extension services.

### *Tenure, property rights and enforcement*

Various access and user rights may be attached to the land; these may be allocated to parties other than the landowner. Improper consideration of tenure issues may exacerbate inequality within the community and lead to conflict, compromising returns. Enforcement of rights including proper arbitration processes for dispute resolution processes (DRPs) may also be lacking. DRPs must be transparent, accessible and must not be too lengthy. For example, Indonesia is considering establishing 'Green Benches' to tackle disputes arising as a result of carbon-related investments.

## Conclusions

Existing carbon finance approaches, with their complex procedures, unpredictable and often long payback periods are exacerbated by existing agricultural finance barriers. Financing barriers faced by smallholders are more fundamental to address than barriers specifically related to carbon credit development. If properly designed, carbon finance could result in win–win situations for smallholders and reduce some of these hurdles by providing revenue diversification tied to sustainable practices and by encouraging aggregation. The forms of carbon finance most appropriate to smallholders will likely be those which result in long-term productivity increases while at the same time increasing mitigation.

Recommendations to various groups for developing win–win approaches that address smallholder financing (and carbon finance) barriers are given below:

### Governments, multilaterals and donor agencies

- Develop sectoral approaches, e.g., Nationally Appropriate Mitigation Actions (NAMAs), encouraging smallholders to adopt improved practices by providing trade credit and adopting legislation requiring banks to lend to small and medium-sized enterprises (SMEs) that promote climate-friendly agriculture, subsidizing certification.
- Support infrastructure development that improves smallholders' access to financial inputs (e.g., credit bureaus).
- Support controlled productivity gains, e.g., access to input through broad co-investment subsidies.
- Evaluate government policies, e.g., taxation to encourage increased investment in improved smallholder agriculture.
- Re-examine trade rules that act as barriers to agricultural producers in developing countries and 'climate-friendly' labeling.
- Develop public–private partnerships catalyzing improved investment in agriculture, e.g., 'patient capital' and credit guarantee facilities. An example of this is the US$10m guarantee facilities developed by Alliance for a Green Revolution in Africa (AGRA) and its partners.
- Support pilots that promote improved agricultural practices.

### Companies, including investors and banks

- Test the application of carbon as an additional revenue stream in existing smallholder production systems.
- Develop new products that can be used by smallholders to overcome barriers, e.g., M-PESA, the Kenyan mobile phone-based money transfer system.
- Test the use of new forms of collateral, e.g., carbon credit purchase contracts.
- Help leverage and invest 'patient capital'.

### Non-governmental organizations and research organizations

- Explore opportunities for adding a mitigation element to existing private and public extensions services (e.g., piggybacking on PepsiCo's Indian distribution system).
- Develop and test instruments to help farmers overcome barriers, e.g., risk mitigation instruments, securitization of future carbon revenues.
- Facilitate aggregation, e.g., support the building of farmers' organizations.

To leverage the greatest impact, 'fit-for-purpose' carbon finance must be designed with an emphasis on overcoming the fundamental, traditional barriers faced by smallholders and to financing sustainable agriculture in general, rather than on generating discrete units of mitigation.

## Acknowledgement

Thank you to the following for their input: John Booth (Conservation Finance International), Tracey Campbell, Lizzie Chambers (Beyond Carbon), Professor Calestous Juma (Harvard Kennedy School), Dr. Namanga Ngongi (AGRA Alliance). The opinions expressed in this paper are solely those of the author.

## Notes

1 Adapted from Smith *et al.* (2007).
2 An exception to this is the Plan Vivo standard, which allows for quicker crediting, but suffers from relatively poor demand.
3 'Sustainable agriculture' is defined according to the US Congress 'Farm Bill' (Food, Agriculture, Conservation, and Trade Act of 1990).

## References

Ashraf, N., Giné, X. and Karlan, D. (2010) 'Finding missing markets (and a disturbing epilogue): Evidence from an export crop adoption and marketing intervention in Kenya,' in *Innovations in Rural and Agricultural Finance*, Kloeppinger-Todd, R. and Sharma, M. (eds). Focus 18, IFPRI and World Bank.

Cotula, L. and Vermeulen, S. (2010) 'Making the most of agricultural investment: A survey of business models that provide opportunities for smallholders'. IIED and FAO.

Doran, A., McFadyen, N. and Vogel, R.C. (2009) 'The Missing Middle in Agricultural Finance: Relieving the capital constraint on smallholder groups and other agricultural SMEs'. OXFAM Research Report.

Palmer, K., Parry, R., MacSporren, P., Derksen, H., Avery, R. and Cartwright, P. (2010) *Chiansi irrigation: Patient capital in action*, InfraCo Briefing Paper, [Online]. Available at: www.keithpalmer.org/pdfs/Chiansi_Irrigation_Briefing_Paper.pdf.

Ritchie, A. (2010) 'Community-Based Financial Organisations: Access to Finance for the Poorest,' Brief 3 in *Innovations in Rural and Agricultural Finance*, Kloeppinger-Todd, R. and Sharma, M. (eds). Focus 18, IFPRI and World Bank.

Skees, J.R. and Collier, B. (2010) 'New Approaches for Index Insurance: ENSO Insurance in Peru', Brief 11 in *Innovations in Rural and Agricultural Finance*, Kloeppinger-Todd, R. and Sharma, M. (eds). Focus 18, IFPRI and World Bank.

Smith, P., Martino, D., Cai, Z., Gwary, D., Janzen, H., Kumar, P., McCarl, B., Ogle, S., O'Mara, F., Rice, C., Scholes, B. and Sirotenko, O. (2007) 'Agriculture in Climate Change 2007: Mitigation,' in Metz, B., Davidson, O.R, Bosch, P.R., Dave, R. and Meyer, L.A. (eds), *Contribution of Working Group III to the Fourth Assessment Report of the Intergovernmental Panel on Climate Change*. Cambridge University Press, Cambridge, UK and New York, NY.

van Empel, G. (2010) 'Rural Banking in Africa: The Rabobank Approach,' Brief 4 in *Innovations in Rural and Agricultural Finance*, Kloeppinger-Todd, R. and Sharma, M. (eds). Focus 18, IFPRI and World Bank.

# 13

# ECONOMICS OF AGRICULTURAL CARBON SEQUESTRATION IN SUB-SAHARAN AFRICA

*Timm Tennigkeit, Fredrich Kahrl, Johannes Wölcke, and Ken Newcombe*

This paper examines the economics and institutional dynamics of an agricultural carbon sequestration program in sub-Saharan Africa (SSA). To maintain simplicity and transparency, as well as due to a lack of data, we do not attempt to capture the diversity among and within countries in SSA, but instead use a range of consensus-based estimates and transparent assumptions to build intuition about the magnitude and importance of different biophysical and economic variables.

We emphasize the importance of the non-mitigation benefits. Agricultural carbon sequestration projects in SSA have important benefits that cut across major policy goals: soil productivity, agricultural productivity and food security; rural development; and water use and climate adaptation. How these goals are prioritized has important consequences for the roles that project developers and institutions should play in agricultural sequestration programs. Below, we use four representative sequestration packages—packages of practices to be adopted by farmers—and a transparent accounting framework to begin to quantify these consequences.

## Sequestration packages

We consider four representative sequestration packages:

- Package 1. *No external inputs.* Sequesters carbon through changes in agronomic practices, tillage and residue management, and nutrient management, without seed or fertilizer inputs;
- Package 2. *Medium external inputs.* Package 1, with seed inputs;
- Package 3. *High external inputs.* Package 1, with seed and fertilizer inputs;

- Package 4. *Agroforestry*. Includes practices from Package 1 but sequesters significant additional carbon in above- and below-ground biomass through agroforestry, with required seed inputs but little or no fertilizer inputs.

In the first three cases, we assume that project farmers are principally maize growers and grow an average of 1.5 maize crops per year. This average is intended to reflect both variability in planting frequency (e.g., in some years farmers can or do not grow two crops) and crops (i.e., the second crop may not be maize, but we assume inputs and revenues are roughly similar). In the case of agroforestry we assume that farmers fallow land and grow, on average, one crop per year.[1] We assume that sequestration packages would require some additional seed and potentially fertilizer inputs, but would have low or no fixed costs. The main additional required input would be household labor.[2]

Crop response is a key variable in sequestration projects. Agricultural sequestration projects are assumed to improve soil fertility and increase average crop yields on the average plot within an average project. Crop response indicates the change in crop yield from a base year ($t_0$, the initial year of the project) that results from the adoption of new agricultural practices. In our four representative packages, some increases in yield are possible without increases in the use of improved seeds or synthetic fertilizer, but yield improvements are significantly higher with seed and fertilizer inputs.

Estimates for the values of key variables in these four packages are given in Table 13.1. Carbon sequestration rates are estimated on an annual (per yr) basis, while crop response and inputs are estimated on a seasonal ('per sn') basis.

We assume that in Package 4 (Agroforestry), no additional nitrogen-based fertilizer inputs are needed. Mafongoya *et al.* (2004) report that leguminous crops commonly used in improved fallows, such as *Crotalaria grahamiana*, *Tephrosia vogelli*, pigeon pea, and *Calliandra calothyrsus*, can generate more soil nitrogen than added in Project 3 (35, 50, 57, and 31 kg N ha⁄yr, respectively).[3] To the extent that other macronutrients (phosphorus (P) and potassium (K)) are limited, inorganic fertilizers might still be required in Project 4.

Table 13.2 contains median estimates for price, yield, and household income and land ownership that are needed to complement the project-specific values in Table 13.1.

Again, these four packages are not meant to capture the full spectrum of conditions, but are rather intended to be representative. We turn now to the two key actors in those projects: project developers and farmers.

145

Table 13.1 Carbon sequestration rates, crop response and inputs for four representative packages

| | Package 1: No external inputs | Package 2: Medium external inputs (seeds only) | Package 3: High external inputs (seeds and fertilizer) | Package 4: Agroforestry |
|---|---|---|---|---|
| Carbon sequestration rate | 0.5 t $CO_2$/ha/yr | 1 t $CO_2$/ha/yr | 1.5 t $CO_2$/ha/yr | 4 t $CO_2$/ha/yr |
| Crop response | 150 kg/ha/sn (225 kg/ha/yr) | 1,000 kg/ha/sn (1,500 kg/ha/yr) | 2,000 kg/ha/sn (3,000 kg/ha/yr) | 1,500 kg/ha/sn (1,500 kg/ha/yr) |
| Seed inputs | | 25 kg/ha/sn (38 kg/ha/yr) | 25 kg/ha/sn (38 kg/ha/yr) | 28 kg/ha/sn (30 kg/ha/yr) |
| Fertilizer inputs | | | 30 kg N/ha/sn (45 kg/ha/yr) | |
| Additional labor inputs | 20 days/ha/sn (30 days/ha/yr) | 30 days/ha/sn (45 days/ha/yr) | 40 days/ha/yr (60 days/ha/yr) | 50 days/ha/sn (50 days/ha/yr) |

*Notes*
"Sn" refers to season.

Data sources are as follows:

Package 1: Carbon sequestration rate is based on a range of estimates. Crop response without inputs is based on a mid-range value from the estimated range (30–300 kg/ha) for maize in Lal (2004). Additional labor inputs are authors' estimates.

Package 2: Carbon sequestration rate is based on a range of estimates. Crop response, seed inputs, and additional labor inputs are authors' estimates.

Package 3: Carbon sequestration rate is based on a range of estimates. Crop response, seed inputs, fertilizer inputs, and additional labor inputs are authors' estimates.

Package 4: Carbon sequestration rate is based on the estimate for soil carbon in agroforestry project, and assuming another 2 t$CO_2$/ha/yr can be sequestered in above- and below-ground biomass. Crop response assumes that agroforestry leads to a significant increase in maize yields (e.g., as per Amadalo et al. 2003), but lower than that in Project 3 because of competition between crops and woody plants. Maize seed inputs are authors' estimates; agroforestry seed estimates are based on Amadalo et al. (2003), and assumed to be used once every four seasons. Additional labor inputs assume that, on top of the additional 30 days per year in Project 2, farmers would need to spend an additional 20 days per year on average planting and maintaining trees.

*Table 13.2* Values for key variables in the four packages

|  | *Median values* |
| --- | --- |
| Carbon price | $10/t CO$_2$e |
| Maize seed delivered cost | $0.8/kg |
| Agroforestry seed delivered cost | $2/kg |
| Fertilizer delivered cost | $0.4/kg N |
| Labor cost | $1.50/day |
| Maize farm gate price | $150/ton |
| Current maize yield | 1.0 t/ha |
| Land ownership[4] | 1.5 ha/farmer |

Data sources: Maize seed delivered costs are authors' estimates. Agroforestry seed delivered costs are based on a mid-range value from Amadalo *et al.* (2003). N fertilizer delivered costs are based on Nkonya *et al.* (2005). Labor costs are authors' estimates. Maize farm gate prices are based on a mid-range value from Woelcke *et al.* (2006), Djurfeldt *et al.* (2005), and 1991–2006 FAOSTAT data on producer prices in 21 sub-Saharan African countries. Current maize yield is a conservatively low estimate based on 1980–2006 historical data for East African countries from FAOSTAT (2010).

## Project developers

Project developers might be non-profit or for-profit organizations; local, regional, or international; do their own aggregation or contract other organizations to do it; and might be integrated into local research and extension institutions or work parallel to them. As a diversification strategy, it may be advantageous to work across countries, or for logistical reasons it may be more cost-effective to cluster projects within one country. Project developers have five basic roles in agricultural sequestration projects:

- designing projects, conducting the baseline survey, and preparing the documentation necessary to generate credits;
- providing or organizing extension services;
- aggregating farmers and cost-effectively scaling up these services;
- pooling and managing risk; and
- directly or indirectly quantifying the amount of sequestered carbon.

In addition, project developers might play a more direct role in the buying and selling of credits, depending on how markets are structured.

As currently conceived, project developers would be given long-term (e.g., 20-year) contracts to sell their carbon credits at a pre-negotiated price. For instance, for a 200,000 ha project with an average carbon sequestration rate of 1 t CO$_2$/ha and a carbon price of US$10/t CO$_2$, project developers might receive annual payments of US$2,000,000, for a total contract value of US$40 million over 20 years.

Project developers have three main cost categories:

- *fixed (and sunk) costs* of setting up the project;
- *operational costs* of overheads, identifying farmers, conducting initial soil carbon measurements, aggregating farmers and providing extension services, monitoring, verification, and enforcing contracts; and
- an *insurance pool* that covers project developers against the risk of contract default, measurement or prediction error, sampling error, climate disturbances, or natural disaster.

We calculate rule-of-thumb estimates for project set-up costs, annual operating costs, and an annual insurance pool in Table 13.3. Project set-up costs are roughly scale invariant, which gives project developers an incentive to establish larger projects to dilute these costs. Operational costs will scale with the physical size (i.e., number of hectares and farmers) of the project. Insurance costs, alternatively, will scale with the number of credits, which may not be linearly tied to the physical size of the project. We assume that project developers will require a profit margin of 15%.

Our estimate of annual operating costs includes wages for management and extension staff; monitoring, verification and enforcement costs, and the project developers' profit requirements. Staff wages are based on an extension model that assumes that farmers are self-organized into groups of 30–40 households, and extension workers work with a limited number of these farmer groups. For a project of 200,000 ha, an average farm size of 1.5 ha per household, and 40 households per farmer group, the project would need to provide extension to 3,333 farmer groups. Assuming that each extension worker can manage 30 farmer groups, the project would need to hire 111 extension workers. There are likely to be economies of scale between extension and monitoring, reporting, and verification (MRV), and we assume that MRV costs are low (US$0.50) on a per-hectare basis (Table 13.4).

To protect themselves against the risks mentioned above, project developers would have to set aside a certain number of their credits in an insur-

*Table 13.3* Cost estimates for agricultural sequestration projects

|  | Cost |
|---|---|
| Fixed costs for project set-up | $300,000 |
| Annual operating costs | $2.39/ha/yr |
| Annual insurance pool | 10% × annual project revenues |
| Project developer profit | 15% × annual project revenues |

*Note*
All values are authors' estimates. Fixed costs do not include financial costs, which are potentially more flexible and which we discuss later.

*Table 13.4* Detailed breakdown of annual operating costs for a 200,000 ha project

|  | Cost breakdown ($) | Total annual cost ($) | Annual cost per ha ($) |
|---|---|---|---|
| Extension workers | 111 × 2,500/year | 277,500 | 1.39 |
| Management staff | 2 × 25,000/year | 50,000 | 0.25 |
| Project director | 1 × 50,000/year | 50,000 | 0.25 |
| Total staff costs |  | 377,500 | 1.89 |
| MRV costs |  | 100,000 | 0.50 |
| Total operating costs |  | 477,500 | 2.39 |

*Note*
All values are authors' estimates.

ance pool to cover a potential shortfall in credits generated during a verification period.

In Table 13.3, the expected number of credits required to cover the shortfall in carbon credits would be 10% of total credits (e.g., 20,000 t $CO_2$ for a 200,000 t $CO_2$ per year project), which would require a 10% reduction in the prices and payments given to farmers to maintain the same level of profitability for project developers. Carbon credits held by project developers as insurance thus represent a real cost in the form of a larger discount in the carbon price given to farmers.

Because operating costs are proportional to project area and insurance costs are proportional to project revenues, insurance costs will increase with project revenues (linearly here), while operating costs decline as carbon prices increase. This discrepancy in cost scaling means that, at higher carbon prices, a larger share of project revenues is available for farmers.

The set-up costs in the above cost structure imply that project developers would require a minimum scale to achieve profitability. For a 20-year project with US$300,000 in fixed costs, at an average carbon sequestration rate of 1 t $CO_2$/ha and a carbon price of US$10/t $CO_2$, the required scale would likely be around 200,000 ha, or roughly 133,000 households.[5] At 200,000 ha per project, an average sequestration rate of 1–2 t $CO_2$/ha/yr and a total cropland sequestration potential of 73 Mt $CO_2$/yr, the total number of cropland sequestration projects in SSA could reach between 200 and 350.

As Table 13.5 illustrates, if project set-up funds are borrowed, project developers would need to spend the first few years paying off principle and interest on the loan. Over the entire 20-year project, project developers would require 37–39% of project revenues to cover their costs, which would leave US$6.09– US$6.31/t $CO_2$ (or equivalently per ha) for farmers.

*Table 13.5* Revenues and costs for a typical package (package 2)

| Project characteristics | | Revenues and costs | |
|---|---|---|---|
| Negotiated carbon price | $10/t CO$_2$ | Annual revenues | $2,000,000 |
| Average sequestration rate | 1 t CO$_2$/ha | Project set-up costs (debt financed) | $300,000 |
| Total project area | 200,000 ha | Annual loan repayment | $42,713 |
| Project duration | 20 years | Annual operating costs | $239,000 |
| Annual operating costs | $2.39/ha | Annual insurance costs | $200,000 |
| Insurance pool | 10% | Annual profit | $300,000 |
| Expected rate of return | 15% | Residual (first 10 years) | $1,218,287 (61% of revenues) |
| | | Price for farmers (first 10 years) | $6.09/ha |

## Farmers

Agricultural carbon sequestration projects provide incentives, access, and expertise that help to overcome farmers' adoption barriers. If sequestration projects increase soil fertility and crop yields, farmers are being incentivized to change their behavior in ways that are in their medium-to-longer-term interest. Farmers need high initial cash or in-kind payments, but might be willing to accept lower payments later once the benefits of higher yields have taken effect. In this section we focus on a static snapshot of farmers' revenues and cost from agricultural sequestration projects, and take up the issue of timing in the next section. When farmers will adopt is a difficult question; therefore our focus here is assessing at what levels of revenues and costs it will become profitable for them to adopt.

Assuming, momentarily, that increases in carbon stocks, yields, and input costs are constant over time, to maintain profitability in each period revenues from net improvements in maize yields and generating carbon would need to exceed the increase in cost associated with those practices.

(Maize Price × Yield Improvements) + Carbon Payment > Cost Increase

Table 13.6 provides a static (annualized) cost accounting for both project developers and farmers (farmer accounts are in US$/ha), assuming that projects are developed based only on one package and that farmers are

Table 13.6 Annual revenues and costs to project developers and farmers for all packages

| | Package 1 ($) | Package 2 ($) | Package 3 ($) | Package 4 ($) |
|---|---|---|---|---|
| **Project developers** | | | | |
| Total annual project revenues | 1,000,000 | 2,000,000 | 3,000,000 | 8,000,000 |
| Annual principal and interest | 42,713 | 42,713 | 42,713 | 42,713 |
| Annual operating costs | 478,000 | 478,000 | 478,000 | 478,000 |
| Annual return | 150,000 | 300,000 | 450,000 | 1,200,000 |
| Annual insurance | 100,000 | 200,000 | 300,000 | 800,000 |
| Annual net revenues | 229,287 | 979,287 | 1,729,287 | 5,479,287 |
| Price to farmers ($/t $CO_2$) | 2.29 | 4.90 | 5.76 | 6.85 |
| **Farmers (all values in $/ha)** | | | | |
| Annual carbon payments | 1.15 | 4.90 | 8.65 | 27.40 |
| Annual revenues from yield improvements | 34 | 225 | 450 | 225 |
| Total additional revenues | 35 | 230 | 459 | 252 |
| Seed costs | 0 | 29 | 29 | 23 |
| Fertilizer costs | 0 | 0 | 60 | 0 |
| Additional labor costs | 45 | 68 | 90 | 75 |
| Total additional costs | 45 | 68 | 150 | 75 |
| Net revenues | –10 | 162 | 309 | 177 |

paid the residual between total project revenues and total costs to project developers. Implicitly, this approach assumes 100% adoption, which, while unrealistic, provides a more consistent representation of project developer and farmer costs. Adoption and disadoption rates will have a range of impacts on project costs. All of the estimates in Table 13.6 are based on the values in Tables 1, 2, 3, and 4.

Table 13.6 illustrates several important points. First, even with relatively low crop response, maize yields are a much larger source of income than carbon payments. Across all four packages, revenues from increased maize yields are 9 to 53 times higher than the income from carbon payments. Second, under our assumptions here, net revenues are positive for farmers in all but the lowest case (Package 1); without the co-benefit of higher maize yields (i.e., with carbon payments alone), however, net revenues are negative in all cases. Third, all four packages could, in principle, be economically viable for farmers. Although net revenues from Package 1 are negative in Table 13.6, for risk-average households with surplus labor Package 1 might still be net positive. In more absolute terms, the attractiveness of carbon payments to farmers will depend on the size of their land holdings.

## Timing and payment options

Timing is central to the economics of agricultural sequestration projects. Three of the four sequestration packages described above require a significant upfront investment in inputs, whereas the yield benefits from these inputs would take time to more fully materialize.

For the sequestration packages, we assume that the following market and agronomic conditions hold true:

- Carbon revenues are paid to project developers at a constant rate. Thus, while carbon sequestration is likely to be concave as a function of time, carbon revenues are paid at default values that reflect an annualized average of expected cumulative sequestration over the contract period.
- Baseline maize yields are declining gradually by 0.1% per year.
- Crop response to new practice packages is positive and concave in time. For Packages 1 and 4, higher maize yields do not materialize until period 2. For Packages 2 and 3, increases in maize yields occur in period 1 (Figure 13.1).[6]

Project developer's and farmer's costs are assumed to be as described in Tables 1, 2, 3, and 4. The simple profit condition for farmers states that discounted revenues should be greater than discounted costs, but we add the condition that revenues are determined by net yields (i.e., the difference between yield improvements and business as usual (BAU) yield decreases).

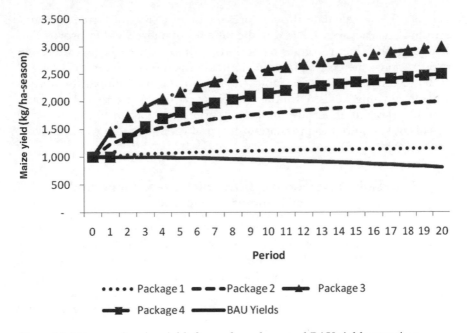

*Figure 13.1* Seasonal maize yields for each package and BAU yields over time.

Note
See endnote 6 for a description of how these curves were generated.

$$\sum_{t=1}^{n} \beta^{t}[P_{MZ} \times (Y_{IMP} - Y_{BAU}) + P_{C}S] > \sum_{t=1}^{n} \beta^{t}[P_{F}F + wL]$$

where $P_{MZ}$ is the price of maize (US$/ton), $Y_{IMP}$ is improved yields (t/ha), $Y_{BAU}$ is baseline yields (t/ha), $P_{C}$ is the price of carbon (US$/t $CO_2$), S is the carbon sequestration rate, $P_{F}$ is the price of fertilizer (US$/kg nutrient), F is fertilizer use (kg nutrient), w is the wage rate (US$/day), L is the additional labor inputs (days), and $\beta^{t}$ is a standard discount factor.

Table 13.7 shows farmers' net revenues per hectare from each of the four packages under the assumptions detailed above, and using a discount

*Table 13.7* Net revenues for all sequestration packages

| Period | Package 1 ($) | Package 2 ($) | Package 3 ($) | Package 4 ($) |
|---|---|---|---|---|
| 1 | −44 | −42 | −21 | −77 |
| 2 | −35 | 19 | 70 | 5 |
| 3 | −30 | 41 | 113 | 36 |
| ... | ... | ... | ... | ... |
| 20 | 33 | 205 | 399 | 206 |
| NPV | (91) | 140 | 397 | 107 |

rate of 30% to calculate the net present value (NPV) of farmer revenue streams. For Packages 2, 3 and 4, the benefits of maize yield improvements lead to positive net revenues by period 2, whereas in Package 1 the lower crop response means that farmers do not break even until period 12.

Positive net revenues in Table 13.7 are driven predominantly by revenues from rising yields. With the notable exception of Package 4 (agroforestry), carbon revenues are unlikely to bridge the gap in timing between initial costs required to purchase inputs and revenues from yield improvements (Table 13.8).

## Broader institutional, distributional, economic, and environmental impacts

Agricultural carbon sequestration projects on the scale that would be required to sequester 40 Mt C/yr in SSA would have impacts beyond project boundaries, both positive and negative. Although there are a number of potential impacts, we focus on four here:

- displacement of resources from existing institutions, programs, and projects;
- longer-term distributional consequences;
- shorter- and longer-term repercussions for factor markets; and
- large-scale, unanticipated greenhouse gas (GHG) leakage.

Significant financial and human resources will be required to carry out larger-scale agricultural sequestration projects in SSA. This infusion of resources should be additional to Official Development Assistance (ODA) as frequently expressed by developing countries. Investments in agricultural sequestration should be a complement rather than a substitute for agricultural sector investment in SSA.

If agricultural carbon sequestration would be adopted on a larger scale, i.e., 20–30% of land and households in rural SSA, the direct and indirect impacts on land and labor markets would be significant.[7] Because of the global complexity of land, labor, and agricultural price effects and the lack of high resolution data, attempting to model more precisely these impacts is unlikely to provide a sufficient basis for designing agricultural sequestration programs. Rather, we advocate the use of indicators and a cautious scaling up of projects over the next 5 to 10 years that allows for some degree of adjustment.

The risk of leakage associated with soil carbon sequestration is expected to be low because increasing agricultural productivity is expected to significantly reduce emissions from deforestation. In the long term, tenure ambiguities that still hamper agricultural investments in SSA (Deininger and Jin 2006) have to be addressed and a national GHG inventory system should be established to enable sectoral land-use mitigation activities.

*Table 13.8* Constant annual carbon payments to farmers vs. the cost of seed and fertilizer

| Package | Total annual project carbon revenues ($/yr) | Total annual project cost of seed/fertilizer ($/yr) | Annual carbon payments to farmers ($/ha) | Annual farmer seed/fertilizer costs ($/ha) |
|---|---|---|---|---|
| Package 1 (no external input) | 1.0 million | 0.0 million | 1.15 | 0.00 |
| Package 2 (seed input only) | 2.0 million | 6.1 million | 4.90 | 30.40 |
| Package 3 (seed and fertilizer input) | 3.0 million | 8.5 million | 8.65 | 42.40 |
| Package 4 (agroforestry) | 8.0 million | 6.0 million | 27.40 | 30.00 |

*Note*
For agroforestry, again, we are assuming no use of inorganic fertilizers is required. If soils are P or K limited, agroforestry would face greater difficulties in financing initial costs with carbon revenues.

## Conclusions

- From the perspective of long-term soil fertility enhancement and food security, the question is not whether to undertake a larger-scale agricultural carbon sequestration program in SSA but rather how.
- Without actual experience and data, desk studies provide only a limited basis for understanding the opportunities and challenges. Early projects such as the Kenya Agricultural Carbon Project (KACP) provide valuable opportunities to learn.
- The amount of carbon revenues available for farmers will depend on the costs of deploying extension services and a monitoring, verification, and enforcement infrastructure. For a 200,000 hectare project with average sequestration rates of 1 ton $CO_2$ per hectare per year and a price of US$10 per ton $CO_2$, we estimate that project developers will need roughly 40% of carbon revenues to cover their costs, with the remainder (around US$6 per hectare per year) available for farmers.
- Annual low-carbon payments (e.g., US$6 per hectare per year) are unlikely to be sufficient to cover all initial costs for adopting "packages" of sequestration practices and technologies that include commercial seed and fertilizer inputs.
- Cash revenues or avoided expenditures resulting from higher yields are significantly larger than carbon payments but carbon finance can make a significant contribution to remove respective adoption barriers.
- With some amount of external finance, sequestration projects can reduce these barriers. For instance, with an initial loan project developers could provide farmers with an initial infusion of resources for seeds and fertilizer.

  Even if the public costs (e.g., underwriting project developers' risks) for the former are substantial, it may still be more cost-effective than financing a large-scale soil fertility enhancement program.

## Acknowledgement

The initial investigation for this chapter was financed by the BioCarbon Fund of the World Bank in early 2009. Views expressed in this article are the sole responsibility of the authors, not the associated institutions.

## Notes

1 This might assume, for instance, that land is fallowed once every four years, but when scaled by our factor of 0.75 to reflect variability (i.e., assuming 1.5 crops in a normally 2 crop per year cycle) agroforestry would have just over 1 crop per year, which we round to 1.

2 Some practices, such as reduced tillage, could indeed be labor saving, but an average package of improved management practices on an average plot is likely

to require additional labor. Improved practices are assumed to be labor productivity enhancing (i.e., $\Delta MP_L > 1$).

3 These values only account for direct soil N additions and should thus be viewed as conservative estimates.

4 Although the small size of farms in SSA is commonly discussed, assessments of average farm size across the region are difficult to make. Land holdings vary significantly across SSA, from land-poor countries like Rwanda (mean 0.71, 2000) to land-rich countries like Zambia (mean 2.76, 2000), with significant variation in land holdings within countries as well (UNCTAD 2006).

5 At zero economic profit, discounted net revenues minus fixed set-up costs are equal to zero, or $\beta^t[P_c C_{seq} T - \alpha P_c C_{Seq} T] - FC = 0$, where $\beta^t$ is a discount factor, $P_c$ is the price of carbon (in $US\$/tCO_2$), $C_{Seq}$ is the cumulative amount of carbon sequestration (in $tCO_2/ha$) over the contract period, T is the land area in the project (in ha), and $\alpha$ is the percentage of non-set-up costs in total revenues, which includes both a reasonable rate of return for project developers and the percentage of revenues required to encourage farmer adoption. The project size (i.e., amount of land T) required to solve this equation is $T = FC/[\beta^t P_c C_{seq}(1 - \alpha)]$. The above calculation uses a discount rate of 10% and a conservative $\alpha$ value of 0.95.

6 For crop response functions we use a logarithmic function of the form a + bln(t), where a is the initial yield and b is a fitted coefficient. The curves in Figure 13.1 are intended to be heuristic; the b coefficients are not empirically derived.

7 For more on land and labor dynamics of payments for environmental services, see Zilberman et al. (2008) and Lipper and Cavatassi (2004).

# References

Amadalo, B., Jama, B., Niang, A., Noordin, Q., Nyasimi, M., Place, F., Franzel, S. and Beniest, J. (2003) 'Improved fallows for western Kenya: an extension guideline'. Nairobi: World Agroforestry Centre.

Deininger, K. and Jin, S. (2006) 'Tenure security and land related investment: Evidence from Ethiopia', *European Economic Review*, 50(5), 1245–1277.

Djurfeldt, G., Holmen, H., Jirstroml, M. and Larsson, R. (2005) *The African Food Crisis: Lessons from the Asian Green Revolution.* CABI Publishing, Wallingford, UK.

FAO (2010). FAOSTAT [Online]. Available at: http://faostat.fao.org/.

Lal, R. (2004) Soil carbon sequestration to mitigate climate change, *Geoderma*, vol. 123, pp. 1–22.

Lipper, L. and Cavatassi, R. (2004) Land-use change, carbon sequestration, and poverty alleviation, *Environmental Management*, vol. 33, pp. S374–S387.

Mafongoya, P., Giller, K.E., Odee, D., Gathumbi, S., Ndufa, S.K. and Sitompul, S.M. (2004) 'Benefiting from $N_2$-fixation and managing rhizobia', in Van Noordwijk, M., Cadisch, G. and Ong, C.K. (eds) *Below-Ground Interactions in Tropical Agroecosystems: Concepts and Models with Multiple Plant Components.* CABI, Wallingford, UK.

Nkonya, E., Pender, J., Kaizzi, C., Edward, K. and Mugarura, S. (2005) 'Policy options for increasing crop productivity and reducing soil nutrient depletion and poverty in Uganda', *IFPRI EPT Discussion Paper 134*.

United Nations Conference on Trade and Development (UNCTAD) (2006) *The Least Developed Countries Report: Developing Productive Capacities.* UN, New York, NY and Geneva, Switzerland.

Wölcke, J., Berger, T. and Park, S. (2006) 'Sustainable land management and technology adoption in eastern Uganda', in Pender, J., Place, F. and Ehui, S. (eds) *Strategies for Sustainable Land Management in the East African Highlands*, IFPRI, Washington, DC.

Zilberman, D., Lipper, L. and McCarthy, N. (2008) When could payments for environmental services benefit the poor, *Environment and Development Economics*, vol. 13 pp. 255–278.

# 14

# THE POTENTIAL FOR MICROFINANCE AS A CHANNEL FOR CARBON PAYMENTS

*Fred Werneck*

This chapter addresses how microfinance institutions (MFIs) could support carbon finance for agricultural mitigation projects. It provides an introduction to microfinance, lists challenges faced by the sector and presents some opportunities for collaboration between MFIs and agricultural mitigation project developers. By showing potential areas of cooperation, this chapter aims to identify priorities and provide recommendations for project developers, as well as policy makers. The chapter draws primarily on the personal experience of the author.

## Brief introduction to microfinance

Microfinance is generally defined as the provision of financial services at small amounts for productive purposes.[1] This reduced transaction volume is the attraction and the key challenge of microfinance.

The attraction is that small amounts per transaction allow MFIs to serve low-income individuals who do not typically have access to banking services. The development of the microfinance sector has expanded the financial market frontier, providing opportunities to businesses and entrepreneurs at the bottom of the economic pyramid.

The main challenge is to minimize transaction costs, making them commensurable with transaction size. This is where traditional banks struggle. Microfinance activities in most countries are based on clients in the informal sector, who lack collateral or credit history, so MFIs have developed specific technologies to streamline their services. Key aspects of the credit analysis process and client outreach strategy[2] are not compatible with a traditional bank's structure.

## How microfinance institutions work

The main products MFIs offer to clients are micro-loan related, and the business model is developed around the provision of such small loans. The MFI may also provide other services such as micro-savings and different types of micro-insurance. Besides the social arguments for providing these additional services, there are economic benefits to expanding the product range such as reduction of the cost of capital, increase in the margin earned per client and client retention.

To simplify the analysis of the business model, let's assume an MFI has only one product: micro-loans. The MFI uses equity[3] and debt as sources of funding and has no assets other than its loan portfolio to low-income clients. The interest received on the micro-loan portfolio constitutes its financial revenues. The MFI pays interest on the debt, which is classified as financial expenses. The difference between the two accounts ("financial margin") is used to pay for expenses such as personnel and administration, ideally leaving the MFI with a profit. This profit is the remuneration of the equity.

To illustrate the mechanism described above, Table 14.1 below shows the figures for three leading MFIs and the sector average, as calculated by MIX Market (www.mixmarket.org)—a microfinance information provider, which consolidates data from over 1,100 MFIs worldwide.

*Financial revenues* depend on market conditions (e.g., interest rate caps imposed by the regulator, competition) and are limited by the repayment

*Table 14.1* Sample of MFI financial details

| 2009 data | Sample of top players | | | MIX Market |
|---|---|---|---|---|
| MFI | Grameen Bank | SKS | MiBanco | benchmark |
| Country | Bangladesh | India | Peru | (various) |
| Total assets (USDm) | 1411.4 | 897.9 | 1278.7 | 7.0 |
| Yield on gross portfolio (real) | 13.6% | 15.6% | 26.2% | 21.4% |
| Financial revenues/assets | 16.3% | 27.2% | 26.2% | 23.1% |
| Financial expenses/assets | −8.1% | −8.3% | −5.4% | −4.8% |
| Financial margin | 8.2% | 18.9% | 20.8% | 18.3% |
| Provisions/assets | −1.3% | −1.5% | −6.3% | −1.3% |
| Personnel expenses/assets | −4.4% | −6.1% | −6.6% | −7.6% |
| Administrative expenses/assets | −2.1% | −3.7% | −4.3% | −6.0% |
| Taxes and other expenses (net)/assets | 0.0% | −2.7% | −0.8% | −1.9% |
| Total expenses/assets | −7.7% | −13.9% | −18.0% | −16.8% |
| Return on assets | 0.4% | 5.0% | 2.8% | 1.5% |
| Net margin | 2.6% | 18.2% | 10.7% | 6.5% |
| Return on equity | 5.7% | 21.6% | 32.6% | 7.4% |

*Source:* MIX Market.

capacity of the clients. Installment amounts are based on the loan amount, interest rate charged, loan length and the number of repayments.

*Financial expenses* represent the cost of third-party funding to the MFI (debt), which can be as large as six to eight times the capital provided by the owners.

*Provisions* refer to likelihood of defaulting clients, reflecting the management's estimate of how much "bad debt" has been provided. Provisions tend to work as a balancing factor for unsustainable growth and inappropriate products.

*Personnel expenses* are a key factor to the business. It is critical to maximize the financial profits per loan officer.[4] The viability of the business relies on an efficient way of deploying employees and engaging clients.

### Individual vs. group loans

One answer to the challenge of small transaction amounts is the provision of group loans.[5] This practice requires more than putting clients together to provide one larger "aggregated" loan. Group loans have proven successful when they rely on peer pressure for appropriate fund use and repayment, risk diversification at client level and statistical reduction of transaction defaults.

Group loans are typically made to groups of around five clients, each performing different business activities, requiring loans of similar amounts and durations, and who know each other well enough to perform a self-assessment of the combined credit risk. To enforce the group aspect, MFIs often create "rituals," which are performed at every meeting as a team-building exercise. Group loans, however, require clients' willingness to share personal financial information with other group members.

Individual loans are based on the same principal as those applied by traditional banks. The MFIs perform a simplified credit analysis based on an estimate of the applicants' business financials and personal references. The typical lack of enforceable collateral, track record and formal accounting, however, led to the development of a more flexible approach compared to traditional banks. Based on scarce evidence, loan officers have to quickly assess the client's business outlook and willingness to repay the micro-loan. Such assessment requires specific training and skills. The analysis per individual client is much more detailed than that on a group loan.

Given the differences between these two models, it is not easy to switch between individual and group loans in a given area. Constraints include extensive loan officer training requirements, changes in corporate culture, inadequate information systems and process design. These constraints are described more fully in the section titled 'Challenges in adding new products to an MFI' below.

### *Urban vs. rural outreach*

Serving clients in rural locations differs from urban microfinance models in many aspects, including:

* lower client densities result in higher transaction costs;
* credit risks are typically associated with agricultural activities, even if the client is not a farmer;[6]
* agricultural activities are inherently risky due to weather variability, risk of pests and diseases and fluctuation in labor availability; and
* it is harder to recruit loan officers in rural areas, as education levels tend to be lower and working conditions more demanding.

As a result, MFIs have tended to concentrate on urban and peri-urban areas. However, competition in, and saturation of, many urban markets are increasingly driving MFIs into rural areas.

Products offered in rural areas tend to differ from those in urban areas. A concentration on working capital micro-loans for trading and services (i.e., short-cycle businesses with continuous cash inflows) in urban areas allows MFIs to work with regular repayments and flexible schedules. In rural areas MFIs tend to offer micro-loans with repayments matching crop seasons—it is common to observe microfinance products with a grace period or no coupons.[7]

The purpose of the micro-loans is also different. Besides financing purchase of seeds, fertilizers and other agricultural inputs (including cows and tools), rural clients also use the loan to smooth household income between crops.

## Credit risk and analysis at the base of the economic pyramid

The success of a client's application for a micro-loan is determined by the assessment of the client's repayment capacity. As part of the credit analysis, the loan officer assesses the impact of the micro-loan on the clients' living costs and on potential income generation.

Some MFIs are attempting to implement pre-approval processes for certain products, such as solar lanterns and efficient cook stoves. However, clients may not always use the products in the anticipated way (e.g., continue to use kerosene lamps for certain activities even after the purchase of solar lanterns). Thus the product may increase living standards but not reduce costs to the extent expected.

Potential increases in income are harder to assess. Credit analysis typically doesn't consider potential additional income as future repayment capacity. The excess cash available to pay for the installments is calculated based on the current business outlook, which in many cases is not sufficient to meet the client's requirements. When the clients find the amount inadequate, they tend to seek additional sources of funding, such as another MFI or an informal moneylender.

Multiple loans have become the norm in many microfinance markets. It is now accepted, even in countries that have a functioning credit bureau[8] (e.g., Peru), that clients also have micro-loans with other MFIs.

## Current status of the market

With total assets over USD 7 billion,[9] the microfinance sector is still maturing.

In urban areas, the aggressive growth promoted by managers, shareholders and creditors resulted in pockets of client over-indebtness, market saturation and intensive competition for market share. Leading MFIs reached significant scale, attracting media and political attention. In contrast, in most rural areas MFI growth still represents an expansion of the financial frontier (i.e., market growth), as a result of the existing dearth of financial services.

For 2011, growth in the number of clients remains a high priority, but MFIs are also increasingly seeking ways to expand their business activities with existing clients (e.g., through broader product range). Interesting mobile phone solutions, micro-insurance, micro-savings and even carbon-related products are therefore attractive bolt-on opportunities for many MFIs.

## Microfinance and agricultural carbon projects

Synergies between MFIs and agricultural mitigation projects exist; the question is how to make such collaboration attractive for all parties involved.

MFIs are financial intermediaries, and therefore not in a position to undertake project implementation. There are two key areas where MFIs can support carbon project developers: by providing funding to the project or by using their infrastructure to distribute the project benefits.

### *Micro-loans for agricultural mitigation practices*

Design of a new microfinance product starts with an analysis of a number of characteristics of the project, such as:

- smallholder farmers' interest in the project or product;
- financing requirements of the smallholder farmer to implement the project;
- MFI's interest in the product (i.e., micro-loan) that will support such a project; and
- smallholder farmers' capacity to repay the micro-loan.

Smallholder farmers' interest in a project depends on (1) the initial investment required and (2) time before tangible results are observable. A clear reward within a reasonable time frame is key to maintain smallholder farmers' engagement, especially when the project deprives them of short-term income (e.g., by temporarily reducing the availability of productive land).

Timing is also critical to the MFI, since risk is affected by contract length (see discussion of contracts in De Pinto *et al.*). A typical microfinance loan ranges from 6 to 24 months, so the project should be able to generate sufficient benefits to allow the micro-loan repayment within this period. It is also important to note that after the loan has been repaid, the MFIs will stop monitoring the client, unless a clear incentive is provided (e.g., a new micro-loan).

Synergies are most apparent when projects:

- require upfront investments (to be financed by the MFI);
- reduce living costs or generate additional income (i.e., increase repayment capacity);
- produce short-term results (to keep the smallholder farmers engaged); and
- have a relatively short payback period.

Table 14.2 summarizes those characteristics for some agricultural mitigation activities eligible for carbon credits.

### Disbursement of carbon-related benefits

In addition to micro-loans, a potential area of collaboration is disbursement of carbon-related benefits to smallholder farmers. Assuming the carbon project developer is responsible for contracting smallholder farmers, securing financing and calculating individual benefits, there are still incentives for MFIs to engage in benefit distribution, such as:

- increase in client's income, supporting the repayment of existing micro-loans;
- potential for micro-savings, directly (carbon benefit through a savings account) or indirectly;
- synergies with the MFI business, e.g., sharing client monitoring costs; and
- marketing, if the MFI can be associated with a reputable partner.

It is important to remember that the volume of carbon-related benefits per client can be very small, and may not cover the costs of servicing smallholder farmers that are not existing MFI clients.

An alternative for such disbursements is the use of mobile phone payment services, e.g., M-Pesa in Kenya. However, the availability of such

Table 14.2 Characteristics of smallholder farm mitigation activities eligible for carbon credits

| Mitigation activity | Financing requirement | | Repayment sources | | | Timing for results/revenues | |
| | Upfront investment?* | Continuous investment?* | Living costs reduction?** | Additional non-carbon income?** | Tangible short-term results? | Expected payback |
|---|---|---|---|---|---|---|---|
| Afforestation and reforestation activities | Yes | Limited | Potentially | Potentially | No | Few years |
| Avoided deforestation | No | No | No | No | No | No |
| Utilize biogas (e.g., from manure) | Yes | Limited | Yes | No | Yes | 6–12 months |
| Compost manure | Limited | Limited | Yes | Potentially | Yes | <6 months |

Notes
* for the project activity, does not consider carbon credit-related costs.
** for the smallholder farmer.

services is still limited and the mitigation project developer may see benefits in having a physical presence in the field.

## Challenges in adding new products to an MFI

Introduction of a new product or service to the MFI's portfolio tends to be a challenge. As in any other business, the MFI management must be convinced that the new product is (1) profitable, (2) attractive to the clients, (3) appropriate considering the skills and expertise of employees, and (4) that it will not conflict with the MFI's existing products. MFIs are also likely to require support (technical or financial) to adjust their credit analysis processes, monitoring, employee incentive schemes, accounting records, reporting, processes, etc.

MFIs are lean organizations, constantly fighting to increase number of clients and profits. Most of them are young companies with a singular focus on growth, but lacking adequate infrastructure to support it. Typical limitations are observed in management information systems (MIS), branch outreach, financial and sales training, corporate governance and management skills. In summary, MFIs are often not well-structured businesses that have the capacity to readily broaden the product range.

### Existing product portfolio

The evaluation of a new product must consider, amongst other factors, asset and liability management, client repayment capacity, transaction and monitoring costs, and market conditions. New products must be aligned with the MFI's existing procedures and not compete with existing products for the client's repayment capacity. Agricultural mitigation projects are more likely to attain MFI support if they promote an increase in the clients' income or a relevant reduction in living costs, therefore justifying an additional micro-loan to clients.

### Management information systems (MIS)

With some exceptions, MFIs do not have state-of-the-art MIS. Due to high license and IT personnel costs, combined with rapid company growth, the IT platform is often outdated and unable to provide appropriate support. To reduce costs and enforce standardization, the MIS tends to have limited flexibility. It is therefore not easy to add a new product to the MIS, in terms of accounting, reporting, management tools, internal controls, etc.

In a typical MFI, the loan officers would collect client data by completing forms. Data generally consists of the client's basic personal information (e.g., name, age, gender, address, ID number), a description of the client's business and main assets. Upon return to the MFI branch, the loan officers, or typists, add the data into the MIS, which records the micro-loan application for approval and monitoring.

### *Loan officers*

Loan officers tend to be young professionals with limited and diverse qualifications. Most working hours are spent in the field prospecting for new clients and monitoring existing ones. They receive some training when they are hired and, in some cases, have access to continuous training. A typical routine includes an early meeting at the MFI branch, followed by monitoring or collections in the morning and prospecting for new clients in the afternoon. It is often difficult to provide them with additional training for new products, due to availability at branch and backgrounds.

Loan officers are usually the only contact between the MFI and the client. They use public transport, bicycles or scooters to visit clients—usually a number of clients on a single trip. Cash is often transported in plastic bags or backpacks. Therefore, loan officers are more appropriately considered a source of information, rather than a potential distributor of agricultural inputs and tools. This is an important point, as mitigation project developers tend to overestimate the potential contribution of an MFI in the project.

### *Cash disbursement process*

It is important to consider the MFI's processes before suggesting a new product. To prevent fraud and/or guarantee the loan officer security, some MFIs only disburse loans and accept repayments at a branch. Although the loan officers still promote loans and monitor the client activities in the field, this practice forces the clients to visit the MFI branch regularly. Logistic costs can make the proposal less attractive to the client, but these visits can potentially increase the range of products offered by the MFI and facilitate cross selling.

### *Financing*

A new product also requires additional funding. MFIs rely on a number of funding sources, from public and private investors to donors and micro-saving activities.[10] A typical structure to restrict the use of funds to pre-defined activities (e.g., the new product) includes the provision of earmarked loans to the MFIs. However, earmarked loans require internal control mechanisms to trace the funds and report on their use. Such mechanisms tend to be costly for nascent MFIs, creating additional accounting expenses and demanding management time. In the event of MFI liquidation, earmarked loans receive the same treatment as other loans. In summary, a mitigation project developer must consider its ability to analyze in detail if the MFI is a suitable partner.

## Conclusions and recommendations

In general terms, MFIs already help to transition smallholder farmers towards mitigation and adaptation activities by improving financial understanding and literacy, and indirectly by increasing access to information. Even though MFIs may not directly support carbon credit projects, they are helping to develop some of the local capacity required to implement future mitigation projects.

A closer collaboration between the microfinance sector and agricultural mitigation projects depends on the ability of the mitigation project developer to provide the right incentives to the MFIs. This chapter briefly described a basic microfinance business model and listed a number of challenges, but also some key aspects of microfinance product design. Understanding the business principles of an MFI is a critical step to start a profitable collaboration.

This chapter also provided two potential ways to close the gap between MFIs and mitigation projects developers:

*Micro-loans for agricultural mitigation projects:* A critical element of an MFI is its capacity to assess credit risk—the client's ability and willingness to repay the micro-loan. This means that micro-loans with repayment capacity associated with carbon credit revenues tend to be less attractive to an MFI, since loan officers (and probably most MFI managers) are not in a position to assess the likelihood of such additional revenue. Mitigation project developers must structure the agreement in a way that limits the MFI's exposure to carbon credit risk.

*Disbursement of carbon-related benefits:* MFIs are financial intermediaries, not providers of delivery services. To make a collaboration proposal attractive, agricultural mitigation projects must provide clear incentives to the MFIs.

The potential for microfinance as a channel for carbon payments depends solely on the structuring of the mitigation project. Developers have to find ways to make the collaboration proposal appealing to smallholder farmers (via financing, support and benefits); MFIs (through viable business opportunities and protection against carbon delivery risk); and the project investors.

## Notes

1 Loans to purchase consumer goods are usually not considered a microfinance activity.
2 Given the small transaction volume, the profit per client is very small. MFIs require a large client base to benefit from economies of scale and become profitable.

3 Equity refers to the owner's funds in the company, while debt refers to funds borrowed by the company—typically from third parties.

4 Loan Officer is a typical job title of the employee responsible for most for the client-facing activities: finding clients, analyzing credit risk, placing micro-loans and collecting repayments.

5 A group of clients signs a loan contract stating that each member of the group is liable for the aggregated loan amount. If one client cannot repay the loan, the other group members have to cover him/her otherwise all members are considered default clients.

6 Although not every client is fully subject to agricultural seasonality (e.g., egg and milk producers, or farmers that benefit from remittances from family members working in urban areas), the rural economy as a whole (i.e., local markets, civil construction, labor availability, etc.) oscillates according to the crop income.

7 'Grace period' refers to the portion of a loan when no payment is due or accrued; and 'no coupons' refers to bullet loans, i.e., when no installment is due until the final settlement of the contract.

8 Credit bureau is an institution that maintains credit data and provides credit information.

9 As reported by MIX Market, based on 1,109 MFIs, by the end of 2009.

10 Some MFIs are allowed to offer deposit and savings accounts to their clients. The money received from the clients can constitute an important funding source due to its low cost and potential scale.

# 15

# PAYMENTS FOR ENVIRONMENTAL SERVICES TO MITIGATE CLIMATE CHANGE

## Agriculture and forestry compared

*Sven Wunder and Jan Börner*

### Land uses and payments for environmental services

Payments for environmental services (PES) schemes have received much attention as potentially cost-effective and equitable means to yield environmental benefits, nested in forestry (Pagiola *et al.* 2002; Landell-Mills and Porras 2002) and agriculture (FAO, 2007; Ribaudo *et al.* 2010). PES schemes are voluntary and conditional (cash or in-kind) transfers from at least one buyer to minimum one seller, aimed at increasing environmental service (ES) provision, relative to a given baseline (Wunder, 2007). For mitigation purposes, PES have the advantage of being performance-focused, thus easing links to carbon markets and specific mitigation targets.

In this chapter, our particular interest is in land use, land-use change and forestry (LULUCF)-oriented PES schemes. Two fundamentally different approaches exist:

1 *forest-based PES*—conserving, managing and restoring standing forests, afforestation/reforestation (A/R), reducing agricultural expansion into forests, retiring cropland for restoring natural vegetation (typically, forests), versus:
2 *agricultural PES*—changing the use of agricultural products, technology and management practice (e.g., organic production, no-tillage, or no-burn techniques), to reduce negative/increase positive environmental externalities.[1]

No global-coverage PES surveys exist that would allow us to document empirically the quantitative distribution between (1) and (2) types of implemented PES for climate change mitigation.[2] However, the general pattern is that publicly financed agri-environmental policies for both agricultural practices and land retirement dominate PES schemes in developed countries (Baylis *et al.* 2008; Claassen *et al.* 2008), while forest-based PES dominate in developing countries—especially so in terms of areas and landholders enrolled. For instance, the Chinese sloping land conversion program targets land retirement and reforestation on about 15 million ha, while Costa Rica's national PES program focuses mostly on forest conservation (270,000 ha). In comparison, as a Food and Agriculture Organization (FAO) assessment concluded, "relatively few PES programs have targeted farmers and agricultural lands in developing countries" (FAO 2007:9)—and yet many observers see a huge potential for increasing the scope for agricultural PES (Niles *et al.* 2002; FAO 2007).

This chapter will scrutinize four hypotheses as to why agriculture might so far have lagged behind forest-based in terms of PES implementation. For each of these, we will provide some tentative indicators for their validity, although only further research will be able to deliver more conclusive evidence.

## Mitigation effectiveness

---

*Hypothesis 1:* Agricultural change tends to biophysically produce less environmental services per land unit than forest-based changes.

---

ES buyers will tend to prefer options that deliver more services. In a report to the UNFCCC[3] Secretariat, Verchot (2007) identified agroforestry and grassland management, each with a global annual mitigation potential of over 2,000 Mt $CO_2e$/yr, as prime agricultural mitigation options. In absolute terms, the mitigation potential of changes in agricultural practices is thus impressive. However, according to Sohngen and Sedjo (2006), this is still almost three times lower than the annual mitigation potential of reducing deforestation to almost zero. The Intergovernmental Panel on Climate Change (IPCC)'s Fourth Assessment Report gave estimates for the per-hectare mitigation potential in agriculture and land retirement (Table 15.1). A quick look reveals that agricultural management options provide returns in the 0–3 t $CO_2$/ha/year range—much lower than use-restricting forestry and soil restoration. Restoration, requiring the retirement of agricultural land, yields annually 73.33 t $CO_2$/ha on organic soils; this is more than 20 times the carbon mitigation from the best agricultural option. On heavily degraded soils, carbon-accumulation rates after land retirement can be much lower. Afforestation and reforestation sequester 1–35 t $CO_2$/ha/year, depending on tree species and site characteristics. Related soil

*Table 15.1* Average annual $CO_2$ mitigation potential for selected LULUCF options

| Activity | Practice | Dry climate | all [t $CO_2$/ha/year] | Moist climate |
|---|---|---|---|---|
| Land retirement | | 1.61 | | 3.04 |
| Manure/biosolids | Management/application | 1.54 | | 2.79 |
| Croplands | Water management | 1.14 | | 1.14 |
| | Tillage and residue management | 0.33 | | 0.7 |
| | Agroforestry | 0.33 | | 0.7 |
| | Agronomy (e.g., improved varieties) | 0.29 | | 0.88 |
| | Nutrient management | 0.26 | | 0.55 |
| Grasslands | Grazing, fertilization, fire | 0.11 | | 0.81 |
| Restoration | Organic soils (e.g., wetlands)[4] | 73.33 | | 73.33 |
| | Degraded lands | 3.45 | | 3.45 |
| Afforestation | | | 1–35 | |

*Source:* Metz *et al.* (2007).

carbon changes can be positive (1–1.5 t $CO_2$/ha/year) or negative (up to –2.2 t $CO_2$/ha/year) (Metz *et al.* 2007). A caveat is that agroforestry, with a low value in Table 15.1, in other studies was shown to accumulate up to 18 t $CO_2$/ha/year in above-ground vegetation (Mutuo *et al.* 2005).

Although emission reductions from avoided deforestation and forest degradation (REDD) are gains estimated for a single time period and thus not directly comparable to the annual accumulation figures in Table 15.1, high estimate ranges of 350–900 t $CO_2$/ha of prevented forest loss put both land retirement and agricultural change mitigation options into further perspective. Even if replaced by highly efficient energy crops, such as sugarcane, 'neutralization' of emissions from preceding tropical rainforest conversion would take decades (Gibbs *et al.* 2008). Obviously, this requires that REDD interventions are able to target spatially the forests threatened by deforestation: if payments have to be made for 10 hectares, of which only 1 hectare is truly threatened, then the impressive biophysical efficiency would decline correspondingly to one-tenth (Wunder *et al.* 2008).[5]

In conclusion, forest- and tree-based options such as avoided deforestation, reforestation, and land retirement-cum-regeneration tend to clearly have the upper hand for carbon services delivered in per-hectare terms, as long as these interventions are well designed and targeted. Standing forests represent a huge carbon stock, with which agricultural change cannot easily compete.

## Competitiveness of land-use options

*Hypothesis 2:* Opportunity costs and adoption barriers in agricultural PES tend to be higher than in forest-based PES, thus effectively limiting their acceptance among landowners.

Competitiveness is not only about service quantities delivered, but also about costs and complexities. Opportunity costs (profits lost from foregoing first-best land-use options) tend to dominate service providers' costs, followed by transaction costs (see 'Transaction costs' below). A key argument for REDD is that per-hectare opportunity costs of protecting formerly threatened forests are low, because most converted uses provide low returns (Stern 2006). When considered in terms of cost per ton of emission reduction, forest-based options often also come out favorably, because of the high quantities of carbon contained in trees (see 'Mitigation effectiveness' above; Swallow *et al.* 2007).

Conversely, improved cropping techniques ideally pay for their adoption through yield and income increases; they are win–win options whose profits out-compete forestry (Koohafkan and Stewart, 2008; Critchley, 2009). Yet, farmers are often too risk-averse to adopt complex technologies requiring upfront investments, maintenance, and training (Mercer 2004). In many developing-country contexts, technological complexity puts off farmers facing binding constraints of capital, labor and know-how; lacking supply for new required inputs; or markets for their new outputs. It may be much simpler for land stewards to delimit a forest area as a 'no-go zone', or set aside a marginal production area for natural regeneration: these use-restricting solutions face usually positive but numerically small opportunity cost, and little or no investments are required, making them low-risk to adopt.

In conclusion, mitigation opportunity costs in forestry and agriculture can span a wide range, but we suspect that typically one of the following scenarios predominates:

a    agriculture's per-hectare opportunity costs are lower than forestry's, but not enough so to compensate the significantly lower per-hectare carbon quantities (see 'Mitigation effectiveness' above);

b    agriculture's opportunity costs are negative (i.e., expected profits exceed the former best option), but high technological complexity, input requirements, and risks imply that they are not adopted by farmers on a large scale;

c    per-hectare opportunity costs of agricultural change are higher than for REDD forest mitigation options—especially when including upfront investment costs.

173

## Transaction costs

---

*Hypothesis 3:* Transaction costs of agricultural PES tend to exceed those in forest-based PES.

---

Transaction costs are part of mitigation costs and occur both on the service provider and buyer side. For service buyers, transaction costs can be defined as all costs that are not payments proper, and can be divided into start-up and running costs (Wunder *et al.* 2008). Transaction costs per carbon unit tend to decline with increasing carbon volume because of the importance of fixed costs, which in the buyer selection represents a key entry barrier for small-scale PES, such as many smallholder carbon schemes (Cacho *et al.* 2005). See also discussion by Shames *et al.* and de Pinto *et al.*

Some factors clearly favor our hypothesis. As discussed above, carbon volumes in agriculture may be significantly lower on a per-hectare basis. In prime agricultural areas, population density is often higher than in forest margins, so that agricultural PES implementers have to deal with a higher number of payment recipients for any fixed-quantity area enrolled. However, as a contrary factor, land-tenure conditions are usually more secure in consolidated agricultural areas than in forest frontiers. If frontier land-tenure regularization has to precede PES, start-up transaction costs could become prohibitively expensive.

Empirical data on PES transaction costs are scarce, partly because most schemes are still fairly young, but one global case-study review (Wunder *et al.* 2008: 844–8) found many schemes to incur high start-up costs for establishing land-use linkages, reference scenarios and payment negotiation. Start-up costs varied between 76 and 4800 US$/ha. However, recurrent annual transaction costs tended to be at least one order of magnitude below start-up costs. No systematic variation between agricultural and forest-based transaction costs were found in that sample.

While the impacts of transaction costs are context-specific, and thus tend to be mixed for our two PES approaches, a more clear-cut case showing the high transaction costs of agriculture can be made specifically for monitoring costs. Monitoring active land management, i.e., making sure a farmer is continuously using no-tillage farming, terracing, or mulching practices (and calculating how much carbon this provides) will normally have to be monitored through costly direct field visits. Even very sophisticated remote-sensing techniques will typically fail to distinguish, for instance, between annual crops cultivated using alternative land preparation technologies. If use-modification increases the presence of trees in land-use systems, such as in agroforestry, low-cost remote sensing-based monitoring can partially be employed, combined with some ground-truthing. Nonetheless, field

174

verification needs may still be far above the requirements for monitoring activity-restricting REDD programs, where much verification can be accomplished through remote sensing.

It thus seems that our transaction cost hypothesis (3) is less powerful than the biophysical service provision potential (1) and their competitiveness (opportunity costs and adoption barriers) (2) ones in explaining why forest-based PES schemes outnumber agricultural initiatives. Apart from the local context, transaction costs are very much influenced by project design and carbon buyers' monitoring needs, so that we cannot generalize. However, monitoring costs may become a serious bottleneck whenever the adoption of alternative land-use options is costly to observe with accuracy.

## Spillover effects

*Hypothesis 4:* Negative environmental spillover effects tend to reduce PES effectiveness more in agricultural than in forest-based interventions.

PES interventions affect farmers' supply and demand decisions, which could create environmental spillover effects in three ways. First, the change promoted through PES could itself produce unintended environmental externalities, e.g., when carbon forestry projects introduce fast-growing monoculture plantations that negatively affect biodiversity and groundwater reserves (Jackson *et al.* 2005) (i.e., *losing other services*). In order to concentrate on the mitigation picture, we will not discuss this factor more here, but just note that fully omitting this aspect would make some mitigation options look too promising vis-à-vis their full environmental externality implications. Second, agricultural PES, packaged with technical assistance and marketing efforts, could make promoted uses so profitable that they are expanded into previously unused land, such as natural forests, independent of initial PES subsidies (*overshooting adoption scale*). And third, by affecting output and factor markets (labor, capital), PES interventions can cause price effects and factor movements that shift spatial pressures and trigger unintended land-use change in areas not targeted by the PES scheme (*leakage*).

As for the overshooting risk, imagine that, contrary to Hypothesis 2, agricultural innovations are successfully introduced on a large scale: the innovations are long-term profitable, and an initial PES overcomes traditional adoption barriers, e.g., through a package containing a subsidized credit, technical and marketing assistance. Conservationists often argue that this would create positive spillover effects on the environment: using an intensified cropping technique, farmers would need less land to match current incomes and thus decide to clear less land (Davidson *et al.* 2008).

175

The underlying assumption is that farmers would only produce just what they need, and then prioritize leisure once that target is reached. More often than not, however, farmers actively *expand* any new and more profitable methods into new production areas to raise their incomes, including by converting forests, especially where demand for the product is open-ended (Angelsen and Kaimowitz 2001). Agricultural PES might thus easily become victims of their own profitability success, at the cost of reduced environmental efficiency.

As for leakage, forest-based schemes can 'push' economic activities into non-target areas, in particular when they are of use-restricting nature (set-aside protection). In the Noel Kempff project in Bolivia, a national park was extended in 1997 to stop deforestation and logging. The stop-logging component was estimated to cause 2–42% leakage—i.e., at worst logging demand would compensate 42% of the shortfall by moving extraction elsewhere, including abroad (Sohngen and Brown 2004). These large ranges illustrate the uncertainties involved in modeling and quantifying leakage effects. Expected leakage amounts in REDD would depend on a series of parameters of flexibility in output and production-factor markets: generally the more flexible the economy, and the smaller the REDD intervention is, the more it will succeed in substituting production in space, and thus likely raise leakage (Wunder 2008).

In conclusion, the extent to which the two carbon spillover pathways become relevant strongly depends on local contexts. Agricultural PES cannot generally be regarded as environmentally riskier than forest-based options. Experiences in developed country settings, such as the EQIP in the US, show that large-scale agricultural schemes can be implemented without major undesired side-effects (Baylis *et al.* 2008). Dealing realistically with negative spillover effects is thus also a matter of careful scheme design and of the proper carbon discounting of quasi-inevitable (or too costly to address) impacts, thus taking relevant local particularities and responses into account.

## Conclusions and discussion

This chapter has provided a general background on PES as a tool to enhance environmental service provision, including climate change mitigation, in agriculture and forestry. Recent global assessments have emphasized the technical potential for both mitigation types. Empirically, the global picture is differentiated strongly between North and South. In Northern developed countries, most PES schemes are agri-environmental, aiming either to change agricultural practices or to take land out of agriculture—and often back into forests. However, in developing countries the majority of PES schemes—and especially the largest among them—have been forest-based rather than agricultural. A main focus in their forest transition stage is on avoiding deforestation at the forest margin, as the prime globally relevant environmentally sensitive process happening in those countries. In turn, in developing countries, the corresponding main vegetation-cover and land-use and processes for PES to leverage are changes in agricultural practices and agricultural land retirement with subsequent (natural or aided) conversion to more carbon-rich vegetation covers.

While we cannot here conclude on all the reasons for the dominance of forest-based over agricultural PES in developing countries, we discussed four hypothetical substantive explanations, related to biophysical quantity of service provision, opportunity costs and adoption barriers, transaction costs, and spillover effects, respectively. We assessed their likely explanatory power. We found that, as a root cause, carbon mitigation quantities are often much higher on a per-hectare basis in forest-based than in agricultural PES schemes: the biophysical service potential per hectare of tree-based interventions is clearly highest, both for carbon retention and sequestration (Hypothesis 1). As a non-trivial caveat, this verdict requires the forest-based schemes to achieve true additionality, which is often harder to do (and demonstrate) than in agricultural interventions.

On the opportunity cost and adoption side (Hypothesis 2), the two types of schemes have widely different rankings. Probably in a minority of cases, when agricultural schemes are cheaper to implement, they often fail to make up for the less carbon they provide, compared to forest-based options (e.g., Börner and Wunder, this volume, with examples from the Amazon). And when they are competitive, to the extent of becoming privately profitable in their own right, e.g., with improved cropping techniques mitigating both carbon and yielding higher farmer incomes, then severe adoption barriers (risk aversion, liquidity and know-how constraints, and diverse market imperfections) often make them de facto unrealistic alternatives.

For our two last hypotheses, on transaction costs and spillover effects, the overall evidence is ambiguous, with many scenario-specific differences. Transaction cost proxies, such as population densities and the degree of tenure security, may work in opposite directions, depending on where and how schemes are implemented (Hypothesis 3). That said, quite a few promising agricultural interventions are expensive to monitor for land-use compliance and service provision, because significant on-the-ground presence is needed. Finally, for negative environmental spillover effects (Hypothesis 4), heterogeneity also rules. Forest-based approaches (e.g., REDD) are generally more likely to produce leakage effects, especially when the interventions are substitutive enough to affect output and factor markets. However, use-modifying PES in agriculture (and even forestry) can sometimes become victims of their own success with farmers by overshooting adoption into environmentally sensitive areas (Angelsen and Kaimowitz 2001).

Hence, our discussion suggests that there are some hard facts to explain why PES implementation has de facto been biased towards forests and use-restricting interventions, especially with respect to the quantity of mitigation services provided. Our assessment of global PES examples is far from exhaustive, and specific institutional and historical factors (global climate negotiations and mitigation crediting rules, environmentally more active forest-sector lobbying, etc.) should not be overlooked. However, at the margin our findings should at least caution against limitless optimism as to how much agricultural PES-based mitigation can be achieved, given some solid obstacles, and at this stage still limited funding flows: there is a substantial rationale explaining the forest-based PES implementation bias we have seen until now in developing countries.

What would this imply for the on-the-ground prospects of implementing agricultural PES schemes? Certainly, there does still seem to be much room to experiment. Some recent pro-carbon and biodiversity interventions, financed by the Global Environmental Facility, have apparently with some success used PES- conditional incentives in helping farmers to overcome adoption hurdles, e.g., in a three-country program for silvipastoral adoption (Pagiola *et al.* 2004), but it is still too early to say if they have been effective in maintaining adoption in the medium term. Agricultural PES will be relevant and viable in many countries where forest resources and tree growth potential are not a competitive alternative—or for other environmental services, such as watershed services on degraded lands. Agricultural PES could also be boosted to the extent that LULUCF-based mitigation options globally are being upgraded and better funded—so that focus becomes less on forest–agriculture competition than on higher competitiveness vis-à-vis third, non-LULUCF-based mitigation options. But that does not nullify agriculture's current global-level limitations, as analyzed in this chapter, which agricultural mitigation initiatives would have to overcome.

## Notes

1 Agroforestry is a prime borderline case; other land management changes (e.g., grassland restoration) may fall between the two categories.
2 Landell-Mills and Porras (2002) survey 287 global PES cases, but focus a priori on forest-based ones. Porras *et al.* (2008) review 50 cases of both forest-based and agricultural PES, but only for watershed services.
3 United Nations Framework Convention on Climate Change.
4 Some of these potential gains are offset by increased methane emissions from restored organic soils.
5 In this section, for simplicity we only deal with carbon sequestration, ignoring the mitigation potential of $CH_4$ or $N_2O$.

## References

Angelsen, A. and Kaimowitz, D. (2001) *Agricultural technologies and tropical deforestation*. CIFOR, CABI Publishing New York, Oxon.
Baylis, K., Peplow, S., Rausser, G. and Simon, L. (2008) Agri-environmental policies

in the EU and United States: A comparison, *Ecological Economics*, vol. 65, pp. 753–764.

Cacho, O., Marshall, G.R. and Milne, M. (2005) Transaction and abatement costs of carbon-sink projects in developing countries, *Environment and Development Economics*, vol. 10, pp. 597–614.

Claassen, R., Cattaneo, A. and Johansson, R. (2008) Cost-effective design of agri-environmental payment programs: U.S. experience in theory and practice, *Ecological Economics*, vol. 65, pp. 737–752.

Critchley, W. (2009) 'Soil and water management techniques in rainfed agriculture: state of the art and prospects for the future', in *Background Note for the World Bank's Water Anchor*. World Bank, Washington, DC.

Davidson, E. A., De Abreu Sá, T.D., Reis Carvalho, C.J., De Oliveira Figueiredo, R., Kato, M. d. S.A., Kato, O.R. and Ishida, F.Y. (2008) An integrated greenhouse gas assessment of an alternative to slash-and-burn agriculture in eastern Amazonia, *Global Change Biology*, vol. 14, pp. 998–1007.

FAO (2007) *The State of Food and Agriculture: Paying Farmers for Environmental Services*. Food and Agriculture Organization of the United Nations, Rome.

Gibbs, H.K., Johnston, M., Foley, J.A., Holloway, T., Monfreda, C., Ramankutty, N. and Zaks, D. (2008) Carbon payback times for crop-based biofuel expansion in the tropics: the effects of changing yield and technology, *Environmental Research Letters*, vol. 3, 034001.

Jackson, R.B., Jobbagy, E.G., Avissar, R., Roy, S.B., Barrett, D.J., Cook, C.W., Farley, K.A., le Maitre, D.C., McCarl, B.A. and Murray, B.C. (2005) Trading Water for Carbon with Biological Carbon Sequestration, *Science*, vol. 310, pp. 1944–1947.

Landell-Mills, N. and Porras, I.T. (2002) *Silver bullet or fool's gold? A global review of markets for forest environmental services and their impact on the poor, Instruments for Sustainable Private Sector Forestry*. IIED, London.

Koohafkan, P. and Stewart, B.A. (2008) *Water and cereals in drylands*. Earthscan, London and Sterling, VA.

Mercer, D.E. (2004) Adoption of agroforestry innovations in the tropics: A review, *Agroforestry Systems*, vol. 61–62, pp. 311–328.

Metz, B., Davidson, O.R., Bosch, P.R., Dave, R. and Meyer, L.A. (eds) (2007) *Contribution of Working Group III to the Fourth Assessment Report of the Intergovernmental Panel on Climate Change, 2007*. Cambridge University Press, Cambridge, United Kingdom and New York, NY, USA.

Mutuo, P., Cadisch, G., Albrecht, A., Palm, C. and Verchot, L. (2005) Potential of agroforestry for carbon sequestration and mitigation of greenhouse gas emissions from soils in the tropics, *Nutrient Cycling in Agroecosystems*, vol. 71, pp. 43–54.

Niles, J.O., Brown, S., Pretty, J., Ball, A.S. and Fay, J. (2002) Potential carbon mitigation and income in developing countries from changes in use and management of agricultural and forest lands, *Philosophical Transactions of the Royal Society of London. Series A: Mathematical, Physical and Engineering Sciences*, vol. 360, pp. 1621–1639.

Pagiola, S., Bishop, J. and Landell-Mills, N. (eds) (2002) *Selling forest environmental services. Market-based mechanisms for conservation and development*. Earthscan, London and Sterling, VA.

Pagiola, S., Agostini, P., Gobbi, J., de Haan, C., Ibrahim, M., Murgueitio, E., Ramírez, E., Rosales, M. and Ruiz, P.R. (2004) 'Paying for biodiversity conservation services in agricultural landscapes', in *Environment Department Paper #96*. World Bank, Washington, DC.

Porras, I., Grieg-Gran, M. and Neves, N. (2008) *All that glitters: A review of payments for watershed services in developing countries*. International Institute for Environment and Development.

Ribaudo, M., Greene, C., Hansen, L. and Hellerstein, D. (2010) Ecosystem services from agriculture: Steps for expanding markets, *Ecological Economics*, vol. 69, no. 11, pp. 2085–92.

Sohngen, B. and Sedjo, R. (2006) Carbon Sequestration in Global Forests Under Different Carbon Price Regimes, *The Energy Journal*, vol. 27, pp. 109–126.

Sohngen, B. and Brown, S. (2004) Measuring leakage from carbon projects in open economies: a stop timber harvesting project in Bolivia as a case study, *Canadian Journal of Forestry Research*, vol. 34, no. 4, pp. 829–39.

Stern, N. (2006) *Stern Review: the economics of climate change*. Cambridge University Press, Cambridge, UK.

Swallow, B., van Noordwijk, M., Dewi, S., Murdiyarso, D., White, D., Gockowski, J., Hyman, G., Budidarsono, S., Robiglio, V., Meadu, V., Ekadinata, A., Agus, F., Hairiah, K., Mbile, P., Sonwa, D. and Weise, D. (2007) *Opportunities for Avoided Deforestation with Sustainable Benefits. An Interim Report by the ASB Partnership for the Tropical Forest Margins*. Nairobi, Kenya.

Verchot, L. (2007) *Opportunities for Climate Change Mitigation in Agriculture and Investment Requirements to take Advantage of these Opportunities*. World Agroforestry Center, Nairobi, Kenya.

Wunder, S. (2007) The efficiency of payments for environmental services in tropical conservation, *Conservation Biology*, vol. 21, pp. 48–58.

Wunder, S., Engel, S. and Pagiola, S. (2008) Taking stock: A comparative analysis of payments for environmental services programs in developed and developing countries, *Ecological Economics*, vol. 65, pp. 834–852.

# 16

# AVOIDED DEFORESTATION ON SMALLHOLDER FARMS IN THE BRAZILIAN AMAZON

*Osvaldo Stella Martins, Paulo Moutinho,*
*Erika de Paula Pedro Pinto, Ricardo Rettmann,*
*Simone Mazer, Maria Lucimar Souza, Ane Alencar,*
*Isabel Castro, and Galdino Xavier*

## Introduction

In the Brazilian Amazon, deforestation and its associated greenhouse gas (CHG) emissions will only be effectively lowered if a new rural development model is adopted, allowing forest conservation to contribute to local economic growth. This article presents an initiative called "REDD for Amazon Smallholders" which is under development to address this challenge. The initiative involves 350 smallholdings located in an area characterized by the expansion of the agricultural frontier in the Brazilian Amazon (Transamazon Highway region—BR230—in the southwestern region of the State of Pará). Projects that help contain deforestation and reduce frontier expansion can play an important role in climate change mitigation (Carvalho *et al.* 2004, p. 172), especially if at the same time they create opportunities for compensating those who are able to maintain forest carbon stocks.

While land-use change accounts for 15–18% of annual carbon emissions (9.9 billion tons of carbon) globally (Le Quéré *et al.* 2009, p. 832), in Brazil it accounts for the majority of the country's emissions (Figure 16.1) and is the primary reason why Brazil has been the fourth largest carbon dioxide ($CO_2$) emitter in the world (MCT, 2010). Since 1960, approximately 17% of the Amazon rainforest (68 million hectares, an area equivalent to the size of France or almost twice the State of Maranhão), has been converted to other land-use activities (INPE 2009). Brazil's emissions of $CO_2e$ from land-use change, forestry and agriculture accounted for almost 80% of its total emissions in 2005 (MCT 2010, vol. 1, p. 151).

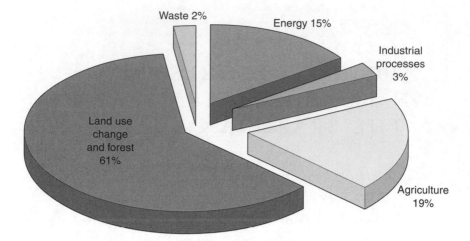

*Figure 16.1* Brazilian emissions by sector, 2005 ($CO_2$e).

Studies predict that in the Amazon, the temperature may increase 5–8°C by 2100 and reduction in rainfall could reach 20% (Marengo *et al.* 2007). Variation in climate and rainfall patterns may change significantly the agricultural vocation of the region. In the tropics, it is expected that crop yields will decrease with even slight increases in local temperature of 1–2°C (Magrin *et al.* 2007). This scenario represents a threat in terms of food security, since smallholders and subsistence family producers may be particularly vulnerable to climate change impacts and their adaptation options may be more limited (Magrin *et al.* 2007 p. 602).

In addition, further deforestation, logging and forest fires associated with more frequent and intense El Niño events could significantly increase carbon emissions, further contributing to global climate change and related impacts (Figure 16.2).

Because of their historical involvement with the rural communities of the Transamazon region, the Amazon Environmental Research Institute (IPAM) and the Living, Producing and Preserving Foundation (FVPP) decided to concentrate their efforts on finding alternatives to compensate the smallholders for the environmental services they provided through the adoption of sustainable land-use activities that reduced pressure on standing forest. These efforts led to the development of the so-called "REDD for Amazon Smallholders" initiative. Presented to the Amazon Fund, this is the first Reducing Emissions from Deforestation and forest Degradation (REDD) project in Brazil involving family producers with the goal of seeking compensation for those who promote the reduction of emissions coming from deforestation and burning, as well as to provide incentives for large-scale and long-term change in regional land-use patterns, which have been historically responsible for high deforestation rates in the

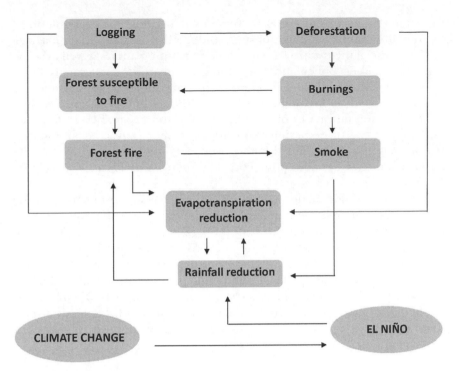

*Figure 16.2* Relationship between land use and climate change in the Brazilian Amazon.

region. The project's main challenge is to simultaneously address natural resource conservation, improve productivity and increase income generation for disadvantaged populations, promote sustainable forest management, and strengthen grassroots organizations, while ensuring food security for smallholder agriculture.

It is expected that by the end of a five-year period the project will result in an effective change in the local economic model, by associating improvements in the generation of income and quality of life of rural communities with the maintenance of the standing forests and their environmental services.

## Background

The design of this project results from the experience gained under the Proambiente program, which addressed the concerns related to small-scale family producers (Figure 16.3) in the Brazilian Amazon, intending to promote sustainable production systems by adopting integrated management of natural resources in smallholdings. In 2004, the Proambiente project was officially adopted by the federal government as a response to the large demand from social movements that called for recognition of the multifunctional character of small and family-based production associated with social inclusion and environmental conservation, as well as social control of public policy and the valuing of environmental services provided by smallholders, since it generates benefits for society as a whole.

It is important to highlight that IPAM, FVPP and other partners technically supported all phases of the trajectory of the Proambiente Program, especially one of its experimental centers, in the Transamazon Highway region (covering around 350 family producers' properties located in the municipalities of Senador José Porfírio, Anapu and Pacaja—see Figure 16.3).

The field studies carried out by IPAM and partners were used as the basis for the principle of integrated management of smallholdings and the need to adopt production diversification (Ferreira Neto 2008; Medeiros 2007). Another major contribution of IPAM was to strengthen the discussion on the importance of valuing ecosystem services generated in the properties from the adoption of sustainable production practices (Lima *et al.* 2007; IPAM 2011). Unfortunately, Proambiente as a Brazilian federal public policy was not effectively implemented due to the lack of a legal framework in Brazil that recognizes environmental services, their valuation or payment, among other reasons.

*Figure 16.3* Family producers working in their property in the municipality of Pacajá.

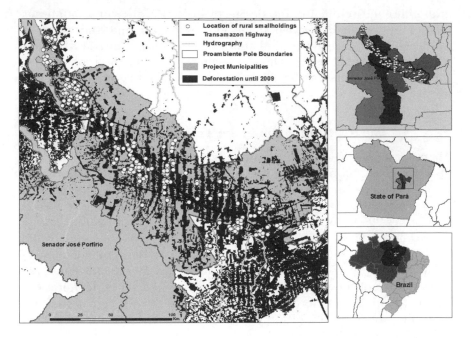

*Figure 16.4* Project area with the 350 smallholdings located along the Transamazon Highway.

## An opportunity to promote REDD

The potential of this REDD initiative represents an unprecedented opportunity for an economically efficient, socially fair and politically feasible strategy. It could promote environmental conservation and sustainable use of forest resources on a significant scale, while improving the quality of life of disadvantaged populations that live in tropical forests. After a wide public consultation in the Transamazon region, IPAM and FVPP submitted the REDD for Amazon Smallholders project to the Amazon Fund (Brazilian government initiative to raise funds for projects that potentially promote forest conservation and sustainable use of the Amazon biome, while reducing emissions from deforestation and forest degradation). The Amazon Fund is a private fund managed by the Brazilian Development Bank (BNDES) according to guidelines and criteria set by a steering committee that includes representatives from federal and state governments, non-governmental organizations (NGOs), social movements, indigenous people, scientists and industries (Stella Martins *et al.* 2010)

Although the project is still subject to approval by the Amazon Fund, IPAM and FVPP have already motivated local stakeholders and governments to discuss possible solutions to deal with the high rates of deforestation and incidence of fires in the region at a larger scale. This discussion has raised awareness and strengthened partnerships with local public

authorities and resulted in the design of the first Consortium of Municipalities for Sustainable Development of Transamazon and Xingu regions,
involving the municipalities of Altamira, Anapu, Brasil Novo, and Senador
José Porfírio (IPAM 2011). This consortium will allow the replication of
the REDD for Amazon Smallholders project, increasing its potential
impacts.

## The REDD program in the Transamazon region

All previous engagement of smallholders' families in the Proambiente
Program created a perfect condition to develop the first REDD program
in the Transamazon region, involving 350 smallholdings distributed in 15
rural communitarian groups, occupying an area of 31,745 ha. The average
farm size is 90.7 ha—ranging from 25 to 250 ha (FVPP 2002). On average,
approximately 55% of each property or 49.9 ha is covered by forest, including legal reserves and permanent protection areas, both protected by Brazilian law (Figure 16.5). These areas are most at risk of deforestation as a
consequence of the growing demand for more land for agricultural production. Until 1970 deforested areas in the Amazon for agricultural purposes did not reach 3% of the total area. Now, it represents more than
10% (Margulis 2003).

## Methodology

To estimate the avoided deforestation in properties that are part of the
project, the regional historical rates of deforestation between 2000 and
2008 were defined to identify the project's baseline based on annual forest

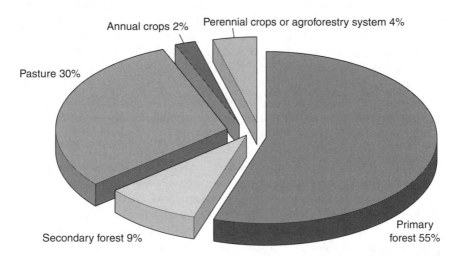

*Figure 16.5* Proportion of each land-use type in relation to the average area
(90.7 ha) of each property.

loss of each property. The payment is made for those who do not clear the forest.

### Historical baseline scenario

To define the deforestation rates and baseline scenarios, Landsat satellite imagery from 1998 to 2008 from the Brazilian National Institute for Space Research (PRODES/INPE) was used.

Applying the historical average rate of deforestation of 4.8% per year in an average property area in the region, results in a potential annual forest loss of 2.39 ha per property. Each hectare of forest covered by the project roughly corresponds to 426 tons of $CO_2$e, resulting in 1,107 tons of $CO_2$e emissions per property annually due to deforestation. Based on this, 5,029.70 tons of $CO_2$e per property are expected to be emitted into the atmosphere over the next five years. These emissions can be avoided by providing incentives to family producers to change their land-use model. As 350 families are involved in the project, a total of about 1.8 million tons of $CO_2$ emissions are expected to be avoided in a five-year period.

### Carbon stock

Models and studies were already developed to estimate the biomass and carbon stock distribution in the Amazon Basin. Estimates suggested 60 to more than 200 tons of carbon exist per hectare, across different land use, including forest (Saatchi et al. 2007; Soares-Filho et al. 2006; Fearnside et al. 1997). A model of the region was developed using the average amount of carbon per hectare, based on the studies of Saatchi et al. (2007). This model indicates an average carbon stock of 126 tons per hectare in the project area and different stocks for other land uses.

### Transition investments

To change land-use historical patterns, which are based on slash-and-burn agriculture and extensive pasture, investments will be focused on activities that could increase the productivity and profitability of the areas already cleared with adoption of new technologies and development of forest management plans. In addition, the project must address land tenure regularization-related issues and strengthening of grassroots organizations, among others.

### Opportunity cost

In this project, opportunity cost is the value of the productive activities that communities forego when they conserve forest instead of converting it. Without attractive economic alternatives to stop the slash-and-burn agri-

187

culture cycle and the need for clearing new areas of native forest, the trend for the smallholders of the Transamazon Pole to maintain the activities of pasture openings and crop plantations of low productivity will continue. This project aims to make the preservation of forest ecosystems economically more attractive to the smallholders. The analysis of the opportunity cost was based on the average income of slash-and-burn agriculture and extensive pasture and the historical annual average deforestation rate (4.8%). Based on it, each family will receive, monthly, an average of US$87 for avoiding deforestation in those areas that annually would be cleared considering a business-as-usual (BAU) scenario.

Thus, it is expected that, at the end of the five-year project, smallholders must achieve (1) an effective change in the local economic model based on low-impact farming systems, (2) improvements in income generation, (3) increase of forest conservation and environmental services provision.

## Challenges related to the project and lessons learned

All the accumulated experience of this project in recent years, mainly on the approach of payment for environmental services (PES) which now comes within the context of REDD, has generated lessons that currently allow us to identify more clearly how to deal with key challenges within the context in question. The main lessons are described below:

### 1 Family producers' participation

This project has ensured the participation of the 350 families involved in it at all stages. At the beginning, a capacity-building process was carried out in order to qualify the debate with the smallholders, as well as local technicians, Rural Labor Unions and social movements' representatives. In 2007, IPAM and FVPP trained the communities' leaderships for updating the data on land use of all the properties of the pole. After that, a REDD proposal was designed in a participatory way based on the information surveyed. Smallholders' involvement at all stages of this project has helped them to recognize the importance of the role they can play as environmental agents,[1] as well as strengthening the credibility of IPAM and FVPP in the region and ensuring the success of implementation of proposed actions.

### 2 Social control

The mechanism of social control proposed by the smallholders under the Proambiente program was maintained since it allows the project's participatory monitoring regarding the progress of the actions to which each community has committed itself through the establishment of Community Agreements. Table 16.1 shows the degree of compliance to the commit-

Table 16.1 Commitments made in Community Agreements and compliance by the families involved in it

| Commitments | Number of communities that have signed the commitment | Accomplishment level[2] |
|---|---|---|
| Fire control and management | 15 | high |
| Preservation and recovery of Protected Areas | 12 | medium |
| Reduce use of pesticides | 11 | low |
| Preservation and recovery of the Legal Reserve | 8 | medium |
| Preserve and restore riparian areas | 7 | medium |
| Implement the system called "agriculture without burning" | 7 | medium |
| Use of organic fertilizer | 5 | medium |
| Construction of nursery | 4 | high |
| Planning location of crops to avoid pressure on Protected Areas | 3 | medium |
| Implementation and management of agroforestry systems | 3 | medium |

ments made by smallholders within the Proambiente context, even with the uncertainties of the program, in an evaluation carried out in 2007 based on the Community Agreements signed by those 15 rural community groups of the Transamazon Pole (in 2005/2006).

### 3 Land tenure regularization

One of the project's priorities is to address the land tenure regularization of rural properties to thus contribute toward the smallholders' compliance with environmental law, while giving legal certainty to them and allow their access to financing. In the project's area, about 95% of the owners do not have the land title.

### 4 Lack of legal framework for PES

The lack of a legal framework for payment of environmental services (PES) in Brazil was one of the reasons that made the consolidation of the Proambiente program impossible. One of the challenges of this project is the PES for the smallholders, since the transfer of resources cannot even legally be characterized as payment or compensation for environmental services. Thus, a legal consultancy was hired to identify how to allow appropriate and legal transfers of resources, without allocation of charges, which would make the project unfeasible to be replicated in other areas due to the high costs involved in it.

189

*5 Sustainability: transaction investments*

The so-called "transition investments" to be provided by this project is a key component that will allow changes in current land-use patterns with the adoption of sustainable production systems and forest management activities, while increasing productivity and profitability of the areas already cleared, with investment in technology, training, technical assistance, infrastructure, among others. The strategy is to promote, over a period of five years, effective changes in production patterns and local economic logic, combining income generation with forest conservation, so that the payment for environmental services is no longer a constraint for the producer to continue in this process after project completion.

*6 Incentives and policy required to maintain agricultural boundary*

The experiences accumulated in the region in a joint effort with social movements and smallholders under this project, led by IPAM, in partnership with FVPP, initiated in 2008 a joint process with the local government of the municipalities of this region. The goal has been to contribute to the development of public policies related to controlling and reducing emissions from deforestation, burning and forest fires in order to expand the scale of this project. The project has potential to be replicated to about 10,000 farms in the municipalities of Altamira, Brasil Novo, Pacajá, Anapu and Senador José Porfírio. Such efforts have resulted in the development of the 1st Consortium of Municipalities for Sustainable Development, focused on Reducing Deforestation and Forest Fires.

## Conclusions

The REDD for Amazon Smallholders project has a total cost per ton of $CO_2$ avoided of about US\$8. By comparison, the average price of carbon traded on the European Union Emissions Trading Scheme (EU-ETS, greater global carbon market) was US\$17.70/ton of $CO_2$ in 2010. The project's costs, considering the price of each ton of $CO_2$, demonstrate how it is competitive in terms of carbon mitigation cost, not counting the co-benefits that will be achieved as a result of the project's investments. Besides promoting compensated reduction of deforestation, the project will also lead to a change in the land-use patterns towards a new model of rural development based on low emission for smallholders. The idea is to make economically feasible the adoption of new technologies and cover the costs for technical assistance in order to increase agro-cattle productivity and profitability of already-cleared areas. The project will enhance recognition of the value given to the role played by biodiversity in the ecosystems' balance through the establishment of sustainable production systems, which are more efficient and less vulnerable to pests, while generating income for families and providing multiple environmental services.

The chances of success of this project are very high due to the involvement of families from its first design until its submission to the Amazon Fund, which strengthens its commitment in meeting the goals set under this project and the common interest of seeking effective change in the historic model to use land that has contributed little to improving the quality of life of these families.

In a broader context, this project forms part of a regional strategy to reduce deforestation and forest fires (the creation of the Consortium of Municipalities cited above), which guarantees the methodology replication at larger scale and increase of potential impacts.

## Notes

1 The role of environmental agents was created within the Proambiente program. Each rural community would elect a representative, namely the Environmental Agent, who then became the link between the small producers and the technical team. The agent's mission was mainly to encourage community participation to achieve the project's objectives.
2 **High**—in most groups, the agreement is fulfilled by 50% of families or more. **Medium**—in most groups the agreement is fulfilled by 25% to 50% of families. **Low**—in most groups, the agreement is completed by less than 25% of families.

## References

Carvalho, G., Moutinho, P., Nepstad, D., Mattos, L. and Santilli, M. (2004) An Amazon perspective on the forest–climate connection: opportunity for climate mitigation, conservation and development? *Environment, Development and Sustainability*, vol. 6, pp. 163–174.
Fearnside, P.M. (1997) Greenhouse gases from deforestation in Brazilian Amazonia: net committed emissions, *Climatic Change*, vol. 35, pp. 321–360.
Ferreira Neto, P.S. (2008) Avaliação do Proambiente (Programa de Desenvolvimento Socioambiental da Produção Familiar Rural), Brasília, Ministério do Meio Ambiente.
Fundação Viver, Produzir e Preservar (FVPP) (2002) Diagnóstico Rápido e Participativo, Programa de Desenvolvimento Sustentável da Produção Familiar Rural da Amazônia, Altamira, PA.
Instituto de Pesquisa Ambiental da Amazônia (IPAM) (2011) A Região da Transamazônica Rumo à Economia de Baixo Carbono: Estratégias Integradas para o Desenvolvimento Sustentável, Brasília.
Instituto Nacional de Pesquisas Espaciais – INPE (2009) 'Monitoramento da Floresta Amazônica Brasileira por Satélite', Projeto PRODES.
Le Quéré, C., Raupach, M.R., Canadell, J.G. and Marland, G. (2009) Trends in the sources and sinks of carbon dioxide, *Nature Geoscience*, vol. 2, pp. 831–836.
Lima, E.S. and Kurisaka, A. (2007) Diagnóstico Regional FLOAGRI: Diagnóstico Socioeconômico e Ambiental dos Municípios da Área de Estudo do Projeto FLOAGRI na Rodovia Transamazônica, Instituto de Pesquisa Ambiental da Amzônia, Belém.

Magrin, G., Gay García, C., Cruz Choque, D., Giménez, J.C., Moreno, A.R., Nagy, G.J., Nobre, C. and Villamizar, A. (2007) 'Latin America. Climate Change 2007: Impacts, Adaptation and Vulnerability', in Parry, M.L., Canziani, O.F., Palutikof, J.P., van der Linden, P.J. and Hanson, C.E. (eds) *Contribution of Working Group II to the Fourth Assessment Report of the Intergovernmental Panel on Climate Change.* Cambridge University Press, Cambridge, UK, pp. 581–615.

Marengo, J.A. and Valverde, M.C. (2007) Caracterização do Clima Atual e Definição das Alterações Climáticas para o Território Brasileiro ao longo do Século XXI, *Multiciência,* no. 8, pp. 6–28.

Margulis, S. (2003) 'Causas do Desmatamento na Amazônia Brasileira', 1ª edição, 100 p, Banco Mundial, Brasília.

Medeiros, C.B., Rodrigues, I.A., Buschinelli, C., Mattos, L.M. and Rodrigues, G.S. (2007) 'Avaliação de serviços ambientais gerados por unidades de produção familiar participantes do programa Proambiente no estado do Pará', Jaguariúna, Embrapa Meio Ambiente.

Ministério da Ciência e Tecnologia (MCT) (2010) *Segunda Comunicação Nacional do Brasil à Convenção-Quadro das Nações Unidas sobre Mudança do Clima* [Online], Coodenação-Geral de Mudanças Globais do Clima, Brasília. Available at www.mct.gov.br.

Saatchi, S.S, Houghton, R.A., Alvala, R.C.D.S, Soares, J.V and Yu, Y. (2007) Distribution of aboveground live biomass in the Amazon basin, *Glob. Change Biol.,* vol. 13, pp. 816–837.

Soares-Filho, B., Nepstad, D.C., Curran, L., Cerqueira, G.C., Garcia, R.A., Ramos, C.A., Voll, E., McDonald, A., Lefebvre, P. and Schlesinger, P. (2006) Modelling conservation in the Amazon basin, *Nature,* vol. 440, pp. 520–523.

Stella Martins, O., Moutinho, P., Rettmann, R. and Pinto, E. (2010) 'Brazil', in Springate-Baginski, O. and Wollenberg, E. (eds) *REDD, forest governance and rural livelihoods: the emerging agenda,* Center for International Forestry Research (CIFOR), Bogor, Indonesia.

# 17

# AN AUSTRALIAN LANDSCAPE-BASED APPROACH

## AFOLU mitigation for smallholders

*Penny van Oosterzee, Noel Preece, and Allan Dale*

### Introduction

It is not possible to avoid dangerous climate change without taking into account the natural and agricultural ecosystems of the planet (Trumper *et al.* 2009), a sector second only to the energy sector in its potential to mitigate greenhouse gas (GHG) emissions. Each year, tropical forests draw down 15% of global emissions (Trumper *et al.* 2009), while deforestation and agricultural emissions are responsible for over 20% of global emissions (McKinsey & Company 2009).

In Australia, the Agriculture, Forestry and Other Land Use (AFOLU) sector provides the largest abatement opportunity available, accounting for 25% of the country's total emissions (Australian Government 2010). A recent assessment of GHG abatement potential through change in rural land-use demonstrates for the state of Queensland, the second largest state in Australia, that the overall attainable GHG abatement was 140 Mt $CO_2$e/yr, or 77% of that state's emissions (Eady *et al.* 2009). By 2020 rural land use is projected to offer about 40% of the low-cost emissions reductions opportunity (ClimateWorks Australia 2010). Given the high contribution to total emissions, the AFOLU sector could potentially transform Australia's mitigation effort and influence global mitigation effort (Garnaut 2008). At the same time, the right landscape actions, including improved farming practice, wetland restoration, and forest, grazing and cropland management, could also support landscape adaptation to the impacts of climate change (Verchot *et al.* 2007).

### Regional approaches

One of the challenges to incorporating the AFOLU sector under the Convention on Climate Change and its Kyoto Protocol is the lack of regionally differentiated management approaches, since climate change causes regionally differentiated impacts that require local knowledge to manage effectively (Steffen *et al.* 2009). The response to climate change also requires regionally-specific technical solutions and management approaches, and incurs different transaction costs (Robledo and Blaser 2008). Impacts, such as catchment disturbance, biodiversity loss and pollution can act in unison at a regional scale, suggesting that fragmentary approaches to management should be replaced by integrative strategies that seek to alleviate multiple sources of threat (Vorosmarty 2010). But even potentially climate-friendly activities such as reforestation need to be implemented with some caution since single-species plantations for carbon or for biofuels, for instance, could lead to highly simplified industrial landscapes with stressed hydro-ecological systems and low biodiversity when compared with native mixed plantations or agroforestry systems (Keenan *et al.* 1997; Hartley 2002; Sayer *et al.* 2004; Marcot 2007; Verchot *et al.* 2007).

## Australia's natural resource management framework

Landscape-scale management across regions and institutional frameworks for natural resource management (NRM) already exist in Australia, largely through adaptively managed and collaborative frameworks negotiated between the Federal and State governments. These nationally coordinated and bilaterally agreed arrangements become a firm framework for managing carbon abatement with simultaneous biodiversity, food security and water quality services. In terms of climate change mitigation, the agriculture, forestry and other land-use sectors could be integrated into such frameworks using accepted international methodologies for measuring carbon benefits.

The foundations for natural resource management in Australia are embedded in formally negotiated Commonwealth–State arrangements. The Commonwealth of Australia comprises six states, each with its own legislature and parliament. Each state also has a number of definable agro-ecological regions that form a sensible management scale in biophysical, social and administrative terms. For example, Figure 17.1 displays the location of the Wet Tropics region within the Queensland State context.

Regional approaches to natural resource management (NRM) have been evolving across the globe for several decades, and are appropriate to effective landscape-scale management of natural resources (Dale *et al.* 2008). As the two levels of government cooperate in many areas in Australia, one important area is in the delivery of community-based NRM through strategic investment at a regional scale. The NRM regions generally mirror one or more bioregions of the Interim Biogeographic Regionalisation of Australia, which are large areas (mostly > 10,000 km$^2$) defined by similarity of biophysical attributes (Commonwealth of Australia 2005).

Under these community-based arrangements, regional NRM bodies— regionally constituted groups that undertake strategic planning, deliver

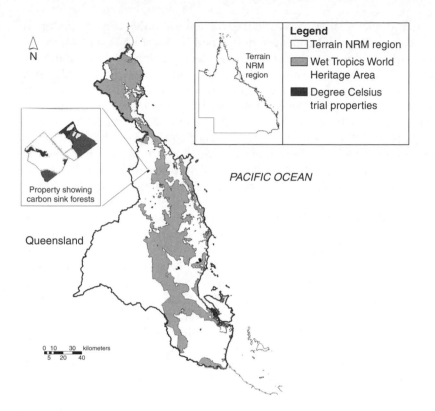

*Figure 17.1* The location of the Wet Tropics region within the state of Queensland.

natural resource programmes and engage the majority of the community's interests—develop regional NRM plans. These plans are jointly accredited by State and Commonwealth governments and comprise scientifically informed, but regionally negotiated targets and priorities. These targets comprise time-bound Resource Condition Targets (e.g., 'halt of the decline of water quality into the Barrier Reef Lagoon by 2020') and associated Management Action Targets (e.g., 'rehabilitate 25 kilometres of region's riparian zone by 2015').

These plans have been built around nationally consistent targets and principles, but are flexible enough to address regional needs and priorities. Once designated by both Commonwealth and State agencies, these plans form the basis for investment in implementation of identified strategic natural resource management actions from Australian, State and Local governments and the community and private sectors.

Regional NRM plans require the engagement of a wide range of sectors as they seek to secure regional consensus regarding critical aspirational and resource condition targets aimed at securing the health of defined natural assets (e.g., soil, water, biodiversity and other assets). NRM bodies

195

are usually steered by Boards which are representative of landholders in their regions. Regional plans are developed by NRM body staff in consultation with landholders. Through consequent Regional Investment Strategies, regional NRM bodies work with a wide range of capable local delivery agents (e.g., landcare groups, industry bodies, local councils, etc.) to motivate and engage all smallholders to improve their natural resource management practices and to take on collaborative local projects. This gives Regional NRM bodies a very wide reach to a region's landholders on issues pertaining to the effective management of natural resources, including those activities that might deliver opportunities in the AFOLU sector. This approach is facilitated by Australia's secure land tenure system and clearly defined property rights embedded in law. It also is able to deal with Australia-wide diversity of property sizes from extensive grazing properties (ranging from hundreds to thousands of square kilometres), intensive farms (that can be smaller than 10 hectares), traditional owner homelands and major and minor conservation holdings. Landholders are not bound by the NRM plans, but elect to become part of the NRM process, either through their own volition or as a result of NRM extension activities. The NRM bodies often provide funding and other resources to landholders to implement actions to fulfil the objectives of the NRM bodies.

This institutional framework can be adapted for managing the landscape-scale impacts of climate change, and for aggregating GHG abatement and sequestration activities including integrating and managing co-benefits. Regional NRM already contributes to sustainable economic development by integrating economic, social and environmental policies across regions through on-ground implementation (Williams *et al.* 2005). A key finding of the Assessment of Australia's Terrestrial Biodiversity 2008 (van Oosterzee 2009) is that the strengthening and consolidation of the regional delivery model for NRM has also assisted delivery of biodiversity outcomes because regions provide a geographical basis for the assessment and reporting of biodiversity trends (Wentworth Group 2009).

Regionalising NRM planning and delivery activities also allows for decentralisation of decisions closer to the community—at the property and local scales (e.g., at a sub-catchment or locality scale). This facilitates and enables more open participatory decision-making. Integration of effort is the key since, as Australia does not have regionalised governments that deal with NRM issues, there is a tendency for local, state and federal government agencies to focus on narrow sets of fragmented objectives and to operate within the confines of their own agendas.

Currently, Regional NRM Bodies are supported to invest in actions that uphold national and state governments' priorities. This, unfortunately, subjects them to short-term, erratic funding cycles based on shifting politics, constraining actions that help build resilient regional landscapes (Robins and Dovers 2007). Consequently, the benefits from aggregating multiple NRM activities of small landholders and trading the resultant credits in carbon markets could be a strong driver of investment in Australian natural resource management and regional NRM groups on a consistent basis.

## The Degree Celsius Wet Tropics Project

A project, developed for the Wet Tropics Region in north-eastern Australia, demonstrates how the NRM activities of Australia's existing regional NRM bodies can be aggregated for both regional and larger-scale delivery of climate mitigation and abatement. The Degree Celsius Wet Tropics Biocarbon Sequestration and Abatement Project aggregates multiple 'small-scale' sub-projects, which are too small to be profitable by themselves in the carbon market. Landholders have elected to join the Wet Tropics project, and agreed to sell their rights to carbon through the project. They have provided information on their activities to the project, and advised on practices they have implemented. They have also enabled establishment of permanent monitoring plots on their properties. In Australia, 86% of agriculture and forestry businesses are small businesses (ClimateWorks Australia 2010).

Three of the major land uses of the Wet Tropics NRM region are shown on Figure 17.1. Much of the accessible land is privately owned, while the more rugged parts of the region are mainly leasehold land, State forest or National Park. Most of the forest in these areas is contained and protected within the Wet Tropics World Heritage Area.

Agriculture is the main land use with, in year 2001 figures, nearly 130,000 ha under cropping and about 47,000 ha under horticulture. Improved pasture for grazing accounted for about 65,000 ha. In the coastal areas the main crops are sugarcane and bananas. Extensive grazing is the main land use in the western part of the region. Forestry as an industry has declined in recent times, though rainforest vegetation covers about 350,000 ha of freehold land (Department of Environment and Resource Management 2009).

The main land uses on the lowlands include small (100–1000 ha) sugar farming and other agriculture (e.g., bananas and tropical fruits), livestock, private forestry and aquaculture. The relative balance between and extent of these crops fluctuates over time depending on commodity prices.

Potential GHG mitigation and abatement project activities in these industries include avoided deforestation, avoided degradation, reforestation using native species and agricultural land management through fertilizer reduction. Using the existing accredited regional NRM plan, with its established priorities, as the framework for the aggregation of carbon sequestration activities for the market, the Wet Tropics project ensures the delivery of complementary biodiversity, sustainable agriculture, water quality and community benefits (Wentworth Group 2009).

The Wet Tropics project currently uses existing methodologies of the Clean Development Mechanism (CDM) of the United Nations Framework

Convention on Climate Change (e.g., UNFCCC 2008) and of the Intergovernmental Panel on Climate Change (IPCC 2006). The nitrous oxide emission calculation guidelines of the IPCC (Volume 4, IPCC 2006), for instance, were used to calculate the abatement effect of reduced use of fertilizers on agricultural land. The National Carbon Accounting Toolbox (NCAT) FullCAM programme (Department of the Environment and Heritage 2005; Richards and Evans 2009)—which was initiated by the Australian government in response to the UNFCCC Reporting Guidelines on Annual Inventories for GHG and for Kyoto Protocol compliance monitoring—is used to model reforestation and avoided deforestation and forest degradation activities.

Specific Regional Ecosystem (RE) mapping data for the state of Queensland (Environmental Protection Agency 2007), which are based on a finer mapping scale, augmented the national data. The RE mapping provides an important tool in both assessing the carbon sequestration of the project area and allocation of zones into potential future land-use categories. High-resolution digital ortho-rectified aerial photography supplemented the RE mapping layer. This enabled categorization of the vegetation into environmental plantings and farm forestry (A/R), avoided deforestation and reduced or avoided logging (REDD). Calculation of the carbon stocks in the project area was based on a pooled value of site-specific NCAT modelling informed by the combination of spatial imagery and historical data. The NCAT modelling resulted in a set of Tier 2 values for the carbon pools of individual sites.

Activities conducted by landholders could potentially earn money for both the regional NRM groups and for the landholders and contribute to reducing deforestation and degradation, increasing reforestation and improving agricultural practices. This would significantly enhance the capacity of regions to secure the aspirational targets in their NRM plans.

Australia is currently in the process of legislating the "Carbon Farm Initiative" (CFI), a scheme to facilitate the sale of verified carbon credits to national and international voluntary markets (Australian Government 2011). The initiative will have clear rules for the recognition of the carbon credits. The CFI is derived essentially from the offsets provisions of Australia's proposed Carbon Pollution Reduction Scheme (CPRS), a national emissions trading scheme, which was shelved in 2010. In 2009, the original CPRS took a narrow view of offsets, essentially following the CDM in only including afforestation and reforestation. After significant public input an amended (and subsequently shelved) scheme provided a national framework to include abatement from agriculture, forestry and land use, which was a world first.

Without a national emissions trading scheme, the CFI can allow trading only within voluntary markets. This poses a number of significant challenges for rural smallholders, in particular the lower prices available in these markets compared to a carbon price established under a national scheme. Compounding this would be the high transaction costs associated with administration, and verification under voluntary schemes. The carbon price and the amount of carbon revenues are the two main factors that will generate meaningful offsets based on AFOLU. Internationally, administrative and transaction costs have acted to stall AFOLU projects, and have resulted in a loss of carbon revenue opportunities of nearly a billion dollars, without any commensurate environmental benefits (World Bank 2010).

The ongoing deforestation rate in the Wet Tropics, for example, over the past 20 years is over 1,700 ha/yr (Department of Environment and Resource Management 2009), equivalent to a release of at least $4.5 \times 10^5$ t $CO_2e$/yr. The aspirational target for the region's biodiversity is to maintain and enhance existing native vegetation. If farmers could be rewarded for the carbon stored in these non-remnant, unprotected forests, they would have an incentive to retain them, reducing the emissions caused by clearing and helping to secure the region's biodiversity targets.

Additionally in the Wet Tropics, forest practices such as logging are degrading biodiverse remnant forests. Degradation, in the form of logging of remnant forests for timber is also an ongoing practice, and based on recent trends in submitting notices of intention to log (unpublished data from Department of Environment and Resource Management database sources, first half 2009), the potential savings in avoided degradation in the Wet Tropics could be over $1.2 \times 10^6$ t $CO_2e$/yr, if landholders who intend to log choose to retain their trees to sell the carbon sequestration benefits.

By contrast, recent data show that reforestation using environmental plantings is occurring at about 40 ha/yr, which is small compared to the area being cleared (>1,700 ha/yr over the past 20 years). Current reforestation, therefore, accounts for less than 0.5% of the potential sequestration benefits of avoided deforestation and degradation, but with financial incentives to plant, the reforestation rates would increase.

Sugarcane plantations cover over 180,000 ha of the Wet Tropics region (McDonald and Weston 2004), and lie in the catchment of the Great Barrier Reef. The average application rate of nitrogenous fertilizers to sugarcane was estimated to be 200 kg/ha (N. Preece, unpublished data from Incitec Pivot). The emissions from these fertilizers averaged $5 \times 10^5$ t $CO_2e$/yr, averaged over 10 years of records, modelled using algorithms developed by IPCC (IPCC 2006). The Wet Tropics carbon project would help to reduce this application rate by 20% or more by encouraging farmers to trade the difference in emissions below legislated regional baselines. Landholders in the region have shown strong interest in measures to reduce their fertilizer use and trade the resultant reduced emissions.

## Conclusions

Lessons from Australia's Wet Tropics project can help define and highlight implementation issues that move beyond land tenure, governance and technical capacity; issues Australia has mostly resolved. In some States, the rights to carbon have not yet been assigned. While the Australian government has accounted for Australia's greenhouse gases regardless of the tenure of the land, it is the right of the States to determine ownership of carbon on private land. In Queensland, for instance, the rights to carbon are assigned through legislation to landholders on freehold land, but not yet on leasehold land. In this light the Queensland government has established a policy that those rights will be assigned to leaseholders when a carbon trading scheme is enacted. Once issues such as this are resolved, then the Australian Wet Tropics model of aggregating smallholdings could be appropriate as it provides the processes and mechanisms to aggregate multiple small carbon sequestration projects and reduces transaction costs, which often are impediments to carbon trading.

Most importantly, the project is built on approaches and innovations developed through existing NRM frameworks, which have well-established regional priorities. Australia's national NRM framework is itself built on an integrated, regional planning process that focuses not only on economic sustainability but also on ecologically sustainable management, biodiversity and water quality.

Each regional plan is founded on inclusive community-based engagement involving community working groups as well as separate stakeholder and science reference groups, ensuring that all sectors are involved in setting regional priorities, and that information is available from the earliest stages. The regional plans themselves integrate both national and state-level statutory programmes with non-statutory programmes, and identify targets and actions beyond regulation.

The Wet Tropics NRM provides the framework for the aggregation of AFOLU activities using existing international and national methodologies. This grounding in existing robust systems also allows for capacity building and development of new methodologies, and refinement of existing methodologies. The Wet Tropics project aggregates activities across the Wet Tropics landscape, reducing transaction costs.

The aim of the Australian government's proposed Carbon Farming Initiative is to provide the financing mechanism through legislating clear rules for the recognition of carbon credits for the international and national markets.

Keeping rules simple while demanding high integrity standards will be a major challenge for Australia. Frameworks and rules which are too restrictive may come at the cost of an overall environmental outcome.

# References

Australian Government (2010) 'Australian National Greenhouse Accounts: National Inventory Report 2008'. Department of Climate Change and Energy Efficiency.

Australian Government (2011) *Carbon Farming Initiative detail* [Online]. Available at: www.climatechange.gov.au/en/media/whats-new/carbon-farming-initiative-detail.aspx (Accessed: 31 January 2011).

ClimateWorks Australia (2010) 'Low Carbon Growth Plan for Australia'.

Commonwealth of Australia (2005) *Interim Biogeographic Regionalisation of Australia 6.1* [Online]. Available at: www.environment.gov.au/parks/nrs/science/bioregion-framework/ibra/index.html (Accessed: 11 August 2011). Canberra Department of Sustainability, Environment, Water, Population and Communities.

Dale, A., McDonald, G. and Weston, N. (2008) 'Integrating Effort for Regional Natural Resource Outcomes: the Wet Tropics Experience', in Stork, N.E. and Turton, S.M. (eds) *Living in a Dynamic Tropical Forest Landscape*. Blackwell Publishing, Carlton, Victoria, pp. 398–410.

Department of Environment and Resource Management (2009) 'Land cover change in Queensland 2007–08: A Statewide Landcover and Trees Study (SLATS) Report'. Brisbane, p. 97.

Department of the Environment and Heritage (2005) 'National Carbon Accounting Toolbox'. Australian Greenhouse Office, Canberra.

Eady, S., Grundy, M., Battaglia, M. and Keating, B. (eds) (2009) *An Analysis of Greenhouse Gas Mitigation and Carbon Sequestration Opportunities from Rural Land Use*. CSIRO, St Lucia QLD.

Environmental Protection Agency (2007) 'Regional Ecosystem Description Database (REDD) Version 5.2', updated November 2007. Environmental Protection Agency, Brisbane.

Garnaut, R. (2008) 'The Garnaut Climate Change Review: Final Report'. Commonwealth of Australia, Canberra.

Hartley, M.J. (2002) Rationale and methods for conserving biodiversity in plantation forests, *Forest Ecology and Management*, vol. 155, pp. 81–95.

IPCC (2006) *IPCC Guidelines for National Greenhouse Gas Inventories, Prepared by the National Greenhouse Gas Inventories Programme*, Eggleston, H.S., Buendia, L., Miwa, K., Ngara, T. and Tanabe, K. (eds). Institute for Global Environmental Strategies, Kanagawa, Japan.

Keenan, R., Lamb, D., Woldring, O., Irvine, T. and Jensen, R. (1997) Restoration of plant biodiversity beneath tropical tree plantations in Northern Australia, *Forest Ecology and Management*, vol. 99, no. 1, pp. 117–131.

Marcot, B.G. (2007) Biodiversity and the lexicon zoo, *Forest Ecology and Management*, vol. 246, pp. 4–13.

McDonald, G. and Weston, N. (2004) *Sustaining the Wet Tropics: A Regional Plan for Natural Resource Management*. Rainforest CRC and FNQ NRM Ltd, Cairns.

McKinsey & Company (2009) 'Pathways to a Low-Carbon Economy: Version 2 of the Global Greenhouse Gas Abatement Cost Curve'. McKinsey & Company, p. 192.

Richards, G.P. and Evans, D.M.W. (2009) 'Full Carbon Accounting Model (FullCAM), National Carbon Accounting System, Version 3.13.8 (Research Edition)'. Australian Greenhouse Office, Canberra.

Robins, L. and Dovers, S. (2007) 'NRM regions in Australia: the "Haves" and the "Have Nots"', *Geographical Research*, vol. 45, pp. 273–290.

Robledo, C. and Blaser, J. (2008) *Key issues on land use, land use change and forestry (LULUCF) with an emphasis on developing country perspectives,* Intercooperation, An Environment and Energy Group Publication. UNDP, Bern, Switzerland.

Sayer, J., Chokkalingam, U. and Poulsen, J. (2004) The restoration of forest biodiversity and ecological values, *Forest Ecology and Management,* vol. 201, pp. 3–11.

Steffen, W., Burbidge, A., Hughes, L., Kitching, R., Lindenmayer, D.B., Musgrave, W., Stafford Smith, M. and Werner, P.A. (2009) 'Australia's biodiversity and climate change: a strategic assessment of the vulnerability of Australia's biodiversity to climate change. A report to the Natural Resource Management Ministerial Council commissioned by the Australian Government'. CSIRO Publishing, Canberra.

Trumper, K., Bertzky, M., Dickson, B., van der Heijden, G., Jenkins, M. and Manning, P. (2009) *The Natural Fix? The role of ecosystems in climate change,* A UNEP rapid response assessment. United Nations Environment Programme, UNEP-WCMC, Cambridge.

UNFCCC (2008) *Approved simplified baseline and monitoring methodology for small-scale agroforestry – afforestation and reforestation project activities under the clean development mechanism* [Online]. Available at: http://cdm.unfccc.int/UserManagement/FileStorage/LXB75FO38Z9NW1IEGH6V0TSUKD4JYM (Accessed: 9 October 2009). Clean Development Mechanism of the United Nations Framework Convention on Climate Change.

van Oosterzee, P. (ed) (2009) *Assessment of Australia's Terrestrial Biodiversity 2008. Report prepared by the Biodiversity Assessment Working Group of the National Land and Water Resources Audit for the Australian Government, Canberra.* Department of the Environment, Water Heritage and the Arts, Canberra.

Verchot, L.V., van Noordwijk, M., Kandji, S., Tomich, T., Ong, C., Abrecht, A., Mackensen, J., Bantilan, C., Anupma, K.V. and Palm, C. (2007) Climate Change: linking adaptation and mitigation through agroforestry, *Mitig Adapt Strat Glob Change,* vol. 12, pp. 901–918.

Vorosmarty, C.J., McIntyre, P.B., Gessner, M.O., Dudgeon, D., Prusevich, A., Green, P., Glidden, S., Bunn, S.E., Sullivan, C.A., Reidy Liermann, C. and Davies, P.M. (2010) Global threats to human water security and river biodiversity, *Nature,* vol. 467, pp. 555–561.

Wentworth Group (2009) *Optimising Carbon in the Australian Landscape* [Online]. Available at: www.wentworthgroup.org/docs/1270%20Optimising_Terrestial_Carbon-9bfinal.pdf. Sydney Wentworth Group of Concerned Scientists

Williams, J.A., Beeton, R.J.S. and McDonald, G.T. (2005) 'Means to ends: success attributes of regional NRM', in Kungolos, A., Brebbia, C.A. and Beriatos, E. (eds) *Proceedings of Second International Conference on Sustainable Development And Planning: Sustainable Development and Planning II.* Bologna, Italy, September 2005, pp. 691–720.

World Bank (2010) '10 Years of Experience in Carbon Finance. Insights from working with the Kyoto mechanisms'. The World Bank.

# 18

# SUSTAINING CHINA'S AGRICULTURE IN A CHANGING CLIMATE

## A multidisciplinary action through UK–China cooperation

*Yuelai Lu, David Powlson, David Norse,*
*David Chadwick, Declan Conway, Brian Ford-Lloyd,*
*Nigel Maxted, Pete Smith, and Tim Wheeler*

## Introduction

In many respects the development of agriculture in China over the last 50 years is an outstanding success story. In China 22% of the world's current population are fed from only 7% of the world's agricultural land. From 1995 to 2005, the total number of undernourished people in the world increased by 48 million, while in China, it decreased by 16 million (FAO 2009). Per capita availability of grain in China increased from 326 kg in 1980 to 399 kg in 2008, meat from 9 kg to 42 kg, while total population increased from 987 million to 1.33 billion in the same period (NBSC 2009).

This achievement contributed significantly to global food security and poverty alleviation (IFAD 2011), however, the environmental costs of China's food security success have been high. Agriculture in China in 2007 contributed from 19% to 22% of the nation's total greenhouse gas (GHG) emission, which is now the largest in the world (see Norse *et al.*, this volume). Agriculture in China has also surpassed industry as the largest polluter of water, discharging 44%, 57% and 67% (13.2, 2.7 and 0.3 Mt), respectively, of the nation's total chemical oxygen demand (COD), nitrogen (N) and phosphorus (P) into water (Xinhua 2010). Overuse of N fertilizer has caused acidification of soils used for intensive Chinese agricultural production (Guo *et al.* 2010), thus threatening China's long-term food security and environmental safety. As "sustainable intensification" of agricultural systems worldwide is essential if the world's population

203

is to be fed (Royal Society 2009), there is an urgent need for changes in China's agricultural policies and practices to halt the environmental damage resulting from China's current version of intensification.

Food self-sufficiency (at the 95% level) is a central goal of agricultural policy in China. Chinese agriculture needs to feed more than 1.3 billion people now and 1.5 billion by 2030 (Feng 2007). Chinese agriculture also faces other resource constraints, including:

- maintaining food self-sufficiency on a shrinking arable land base due to urbanization and road building
- impacts of climate change
- agriculture pollution, including non-point source pollution
- water scarcity and poor water quality
- biodiversity loss.

This chapter describes a number of projects undertaken through the UK–China Sustainable Agriculture Innovation Network (SAIN) to tackle these challenges.[1]

## UK–China cooperation in sustainable agriculture—SAIN

SAIN is a long-term inter-governmental initiative established in 2008 as a vehicle to deliver UK–China collaboration in sustainable agriculture, funded by the United Kingdom's Department of Environment, Food and Rural Affairs (Defra), the Department for International Development (DFID) and the Chinese Ministry of Agriculture. SAIN was established based on the following rationale:

- Sustainable agriculture is a common concern of UK and China that has global implications.
- Sustainability can only be achieved through partnership, not only between the UK and Chinese governments, but also among all stakeholders within the countries.
- More needs to be done to bring policy makers, researchers, extension staff, farmers and other stakeholders together and to ensure that policy making is better informed and technical interventions are better targeted.
- The two countries have policy and research experience that should be shared, but is not because there is no appropriate mechanism in place.
- Cooperation between the two countries will benefit other developing countries through technical transfer and knowledge sharing.

SAIN's aim is to contribute to the achievement of a resource-efficient, low-carbon economy and environmental sustainability by improving the evidence base for policy development. SAIN will do this by:

- stimulating innovative thinking and research on all aspects of sustainable agriculture and its relation to the local, national and global economy;
- communicating information on sustainable agriculture issues and opportunities for change, and disseminating best practices to farmers, policy makers, and businesses;
- contributing to global sustainability through wider sharing of experience between developed and emerging economies.

At the outset, the following initial focus areas were identified as areas in which China and the UK had common interests and where it was thought that there were particular benefits from shared working:

- *Nutrient management*: to apply research and better communication tools to improve soil and crop nutrient management and reduce non-point source pollution.
- *Bioenergy*: to increase use of agricultural biomass and livestock manure for biogas, liquid biofuels and organic fertilizer production.
- *Climate change mitigation and adaptation*: to address the interface between agriculture and climate change, including the way agriculture will be affected by, and therefore need to adapt to, climate change, and the ways in which agriculture contributes to GHG emissions.
- *Circular agriculture*: to provide policy advice on how the concept of the circular economy can be applied to agriculture by exploiting the opportunities for greater recycling, waste minimization, and more efficient use of water and other critical resources.

## Research projects under SAIN

Following the establishment of SAIN, a number of projects were started during the next 1–2 years to address the main gaps in research and policy making as identified by key stakeholders in the two countries. The common objectives of SAIN projects are to address the issues of increased resource use efficiency, particularly N fertilizers and manures; reduce pollution from agricultural production to air and water, including GHG emissions reduction and to increase carbon sequestration; and communicate information on agricultural climate change mitigation, adaptation and options to policy makers as well as to farmers. In the future, and as knowledge gaps are filled and new challenges or opportunities emerge, the focus areas of SAIN will evolve and develop.

In the following, the approaches and activities of the current SAIN projects are briefly described and their potential contribution to sustainable agriculture and food security summarized.

# Improved nitrogen fertilizer management

## *The problems to be addressed*

- N fertilizer production accounts for >70% of fossil energy inputs to agriculture in China, therefore a reduction is essential for progress to a low-carbon agricultural economy.
- N fertilizer production in China releases about 235 Mt $CO_2$ plus 39 Mt $CO_2$e as nitrous oxide ($N_2O$), a potent GHG with 298 times the greenhouse warming potential of $CO_2$.
- Additional $N_2O$, at least 150 Mt $CO_2$e, is emitted from the use of fertilizers and manures in agriculture.
- These emissions represent over 20% of total $CO_2$ emissions from China and about 25% of global $N_2O$ emissions from agriculture.
- N from agriculture is a major cause of eutrophication in China's surface waters and regional seas.
- Large reductions in N use are achievable: it is estimated that cuts of *at least* 25%, and probably considerably more, are possible through application of current knowledge (Ju *et al.* 2009).
- N savings will involve greater recycling and waste minimization—key objectives in China's plans to establish a circular economy.

## *Project objectives*

The aims of this project are to foster changes in agricultural policies that will improve advice and incentives to Chinese farmers on sustainable fertilizer use to ensure progress to a low-carbon agricultural economy. They will be achieved by:

- estimating GHG emissions associated with the manufacture and use of N fertilizer in China using life cycle assessment to identify opportunities for improving energy and N use efficiency and decreasing emissions, thus lowering the carbon intensity of food production (see Baker and Murray, this volume);
- reviewing current and emerging technologies for increasing efficiency of use of N fertilizer, N from manure and manufactured organic fertilizers, building on results from the Chinese Ministry of Agriculture's project "Fertilizer Recommendations and Soil Testing";
- Assessing improved means of communicating information on rational use of N fertilizers and manures to farmers and extension staff;
- Providing information to policy makers on the GHG savings possible from improved management of N in agriculture and the ways of achieving this.

Thus the overall objective of this project is to mitigate climate change by decreasing unnecessarily large GHG emissions from agriculture caused by overuse of N fertilizer. This comprises $CO_2$ from production of fertilizers and $N_2O$ emissions, direct and indirect, when fertilizer is applied to land. It is expected that mitigation will be achieved through a combination of improved practices, application of new technologies and changes in policies to incentivize these changes and discourage the overuse of N fertilizer. This might be achieved through changes in pricing or subsidies and by changes in environmental regulations.

## Improved manure management

### *The problems to be addressed*

- Over-application of nutrients to crops via a combination of both live-stock manures and inorganic fertilizer increase risks of eutrophication of surface water bodies and emissions of harmful gases to the atmosphere (mainly ammonia and nitrous oxide).
- Little or no account is made of nutrients in livestock manures in nutrient planning in Chinese farms.
- There has been considerable research in the UK on the availability and value of nutrients (N, P, K) in livestock manures and the principles applied can be utilized in China.
- In addition to the direct use of manures and other organic co-products such as anaerobic digestate, there is much interest in China in the development of organic fertilizers through composting procedures. This can add financial value to the material but the nutrient content and availability still needs to be assessed if knowledge-based advice is to be given to farmers.

### *Project objectives*

- To assess to what extent the nutrients in livestock manure, composted manure and anaerobic digestate are taken into account when applied to soil in selected provinces in China when planning nutrient supply to a range of crops.
- Collate the information available to assist farmers in managing the nutrients in these organic resources effectively.
- Explore the major barriers to utilizing the nutrients in these organic resources in an integrated way with inorganic fertilizers.
- Recommend future investment to reduce these barriers and maximize use of nutrients in these organic resources, thus reducing reliance on inorganic fertilizers.

This project is expected to contribute to mitigation of climate change in two ways. First, changes in the handling, storage and applications of manures is likely to decrease direct emissions of $N_2O$, and decrease indirect emissions by reducing volatilization of ammonia. Ammonia gives rise to $N_2O$ emissions when it is transformed microbially after being redeposited onto soil or water. Second, by improving knowledge of the quantity and availability of N and other nutrients in manures so that fertilizer applications can be reduced accordingly. Lack of information on this, by farmers and advisers, appears to be a serious problem currently.

## Estimates of future agricultural greenhouse gas emissions and mitigation in China

### *The problem to be addressed*

* The challenge of reducing agricultural GHG emissions and increasing soil carbon sinks in China, whilst maintaining food security for its very large population.
* The need for an evidence base, policy advice and decision support tools to allow policy implementation and knowledge exchange among scientists, policy makers and farmers.

### *Project objectives*

* Provide the evidence base, policy advice and decision support tools to reduce agricultural GHG emissions and increase soil carbon sinks in China, whilst maintaining food security.
* Develop a national and regional picture of economic abatement potential from Chinese agriculture.
* Explore behavioural or incentive barriers associated with obvious high potential in mitigation (and low-cost) measures that are not being adopted.
* Assess applicability of mitigation strategies to decrease livestock and manure emissions for different farm types.
* Create a whole-China model of mitigation potential for livestock and manure emissions, also considering pollution swapping.
* Provide policy advice on cost-effective mitigation options for soil C sequestration, and for reducing GHG emissions from croplands (dry and paddy), grasslands and livestock.
* Produce a database, journal publications, decision support tools and policy briefings on GHG emissions and GHG mitigation options in China's agriculture.

Outputs from this project will help to identify the major agricultural practices where GHG emissions reductions are possible, and identify barriers to improved management. This will assist in prioritizing policies and actions.

## Harmonising adaptation and mitigation for agriculture and water

### *The problems to be addressed*

Recent research on climate change in China suggests that the interactive effects of climate change and other socio-economic drivers could lead to significant decreases in total production by the 2040s (Xiong *et al.* 2009a). Water availability plays a particularly significant role in limiting potential crop production, due to the combined effects of higher crop water requirements and increasing demand for non-agricultural use of water (Xiong *et al.* 2009b). Successful adaptation policies based on sustained improvements in agricultural technology and crop yields will be essential for China to produce enough to keep pace with population growth and the effects of other drivers such as land-use change. Such production-oriented policy goals should not ignore wider issues of sustainability, such as the intensity of fossil fuel and water use in the sector.

### *Project objectives*

The overall objective of the project is to estimate the 'carbon cost' of adaptation to future climate change in terms of water use in agriculture. Specifically, the project will address the following aim:

- assess and describe the main impacts of climate change on agriculture in China and derive adaptation policy scenarios to sustain agricultural production in China;
- developing preliminary estimates of energy consumption in agricultural water use, using case study data (see output Rothausen and Conway, 2011);
- link adaptation policies with energy use. The project will focus on a time horizon out to the 2030s, and use China's current national planning to provide the framework for the definition of socio-economic and policy scenarios.

The output of this project will highlight the carbon implications of adaptation and explore pathways for reducing emissions through harmonising mitigation and adaption policies.

## Conservation of crop wild relative (CWR) diversity

### *The problems to be addressed*

CWR are plant species closely related to crops, including their wild ancestors. CWR are recognized as critical resources for mitigating the impact of

209

climate change because they are likely to provide the genetic diversity and adaptation needed to breed crops with greater resistance to the environmental changes brought about by the changing climate. CWR diversity is threatened directly by climate change, as well as indirectly by increasing food insecurity resulting from climate change and other socio-economic factors that could push agriculture into increasingly marginal land, resulting in an erosion of diversity. The challenges facing us are to ensure effective conservation of important resources on the one hand, and to enhance food security on the other through rational and sustainable use of plant genetic resources.

### *Project objectives*

- Production of a full inventory of CWR of China using a systematic approach previously developed and applied in Europe.
- Identification of priority CWR species based on food security importance, i.e., economic importance of related crop, use potential for climate change mitigation, and IUCN Red List threat status.
- 'Gap' and climate change analysis to identify conservation needs for selected high priority crop gene pools (including rice, soybean, foxtail millet, grape, Kiwi fruit, poplar and citrus fruits).
- Publication of crop gene pool conservation strategies, including briefing papers for policy makers.
- Evaluation of CWR using novel genomic approaches to provide improved access to CWR genetic diversity for use in crop improvement, with a focus on genes likely to confer adaptation to climate change.
- Development of an online information system to provide access to the CWR inventory and associated conservation and evaluation data.

This project is mainly concerned with adaptation to climate change through the maintenance of crop breeding material for future use. However, some aspects could directly contribute to mitigation. For example, methane is a powerful GHG that is emitted from flooded soil used for growing paddy rice. Much research has been directed at ways of decreasing methane emissions from the rice system. It is known that there is variation between rice varieties in their methane emission under a given set of environmental conditions (e.g., Inubushi *et al.* 2011). So ensuring a wide availability of rice wild relatives may facilitate future breeding of lower-emission varieties.

## Agricultural adaptation to climate change in China

### *Problems to be addressed*

Mitigation and adaptation are the two closely linked aspects of agricultural response to climate change, and most categories of adaptation options for climate change have positive impacts on mitigation (Smith and Olesen

2010). Adaptation to climate change therefore forms an important component of SAIN's overall approach to a sustainable agriculture. Climate change is a key driver of change in agricultural systems in China and is closely linked with the causes and alleviation of poverty. However, because of the complexity of agricultural systems, many key knowledge gaps remain, such as: crop responses to elevated $CO_2$ conditions; farmer perception of climate change and their communication and adaptation strategies; and crop disease threats. Through a structured programme of knowledge exchange using research training and established networks developed through previous research, this project will address the skills needs of two major stakeholder groups—adaptation researchers in China and Chinese farming communities vulnerable to climate change—in order to enhance UK–China collaboration targeted to improve the adaptation of Chinese agriculture to climate change.

### *Project objectives*

The overall objective of the project is to foster joint research and knowledge exchange between Chinese and UK researchers concerned with adaptation of agricultural systems to climate change and the promotion of sustainable food systems. The specific objectives are:

- communicating issues of sustainability of farming systems to key stakeholders: farming communities, researchers and policy makers;
- exploring information flows within farming communities concerned with the perception of climate change and possible adaptation options, and linking these with research;
- building capacity of researchers and farming communities to adapt to climate change.

## Conclusion—the ways SAIN makes a difference

SAIN projects contribute to sustainable agriculture and food security through putting research into practice. This includes technology development and implementation, as well as policy making. SAIN projects adopted a number of measures to achieve this objective.

### *Support tools and evidence for policy makers and practitioners*

SAIN projects emphasize the development of support tools and provision of evidence for policy making and guided implementation of good practices. Some examples of the tools to be developed include:

- Inventory of technologies for improving the management of N (from chemical fertilizers and manures) to increase efficiency of use and decrease losses.

211

- GHG emissions inventory for various cropping systems.
- Economic assessment of the marginal abatement cost of agricultural GHG mitigation in China.

Outputs such as these are explicitly designed to assist or guide either agricultural practitioners (both farmers and their advisers) and policy makers in changing policies and practices in such ways that favour a more low-carbon development path. This is one element of an overall strategy that comprises both mitigation and adaptation.

### *Knowledge sharing*

One of the important aspects of SAIN projects is knowledge sharing and mutual learning. This includes comparison studies between the UK and China, as well as modification and application of technologies developed in the UK and Europe to the Chinese context. Examples include:

- nutrients and manure management technologies and policies in both China and the UK;
- models for inventory and systematic *in situ* and *ex situ* conservation of important CWR diversity;
- building on UK and European database experience to examine how GHG emissions vary with different cropping and livestock systems in different regions.

### *Capacity building*

Through implementing joint projects, Chinese collaborators will be trained and research capacity increased. For example, Chinese scientists will be trained to apply in botanically-rich China the crucial methodologies and principles of inventory, conservation and management of CWR resources developed and applied in the UK. A 'training the trainers' approach will be used in capacity-building activities so that key researchers and leaders within farming communities will be empowered to take the role of trainer in subsequent capacity-building activities after the projects finish.

### *Communication*

Various communications approaches will be used in SAIN projects, including:

- policy briefs
- dialogue with policy makers
- farmer field schools
- scientific publication.

Whatever innovative communication methods have been used in the different projects, an important outcome will be delivery of information to both farmers and policy makers on the applicability and options for using new techniques that are tailored to their specific economic, social and geographical situations.

## Note

1 SAIN's current projects are led by David Chadwick (Rothamsted Research North Wyke), Declan Conway (University of East Anglia), Brian Ford-Lloyd (University of Birmingham), Nigel Maxted (University of Birmingham), David Powlson (Rothamsted Research), Pete Smith (University of Aberdeen), Tim Wheeler (University of Reading), Zhang Fusuo (China Agricultural University), Lin Erda (Chinese Academy of Agricultural Sciences), Kang Dingming (China Agricultural University), Shen Qirong (Nanjing Agricultural University) and Pan Genxing (Nanjing Agricultural University). More information about SAIN can be found at: www.sainonline.org/English.html.

## References

FAO (2009) 'The state of food insecurity in the world 2009'. Food and Agriculture Organization of the United Nations, Rome, Italy.

Feng, Z. (2007) Future food security and arable land guarantee for population development in China, *Population Research*, vol. 31, no. 2, pp 15–29.

Guo, J.H, Liu, X.J., Zhang, Y., Shen, J.L., Han, W.X., Zhang, W.F., Christie, P., Goulding, K.W.T., Vitousek, P.M. and Zhang, F.S. (2010) Significant acidification in major Chinese croplands, *Science*, vol. 327, no. 5968, pp. 1008–1010.

IFAD (2011) 'Rural poverty report 2011. New realities, new challenges: New opportunities for tomorrow's generation'. International Fund for Agricultural Development, Rome, Italy.

Inubushi K., Cheng W., Mizuno T., Lou, Y., Hasegawa, T., Sakai, H. and Kobayashi, K. (2011) Microbial biomass carbon and methane oxidation influenced by rice cultivars and elevated $CO_2$ in a Japanese paddy soil, *European Journal of Soil Science*, pp. 62, 69–73 doi: 10.1111/j.1365–2389.2010.01323.x.

Ju X., Xing, G., Chen, X., Zhang, S., Zhang, L., Kiu, X., Cui, Z., Yin, B., Christie, P., Zhu, Z. and Zhang, F. (2009) Reducing environmental risk by improving N management in intensive Chinese agricultural systems' *PNAS of the USA*, vol. 106, pp. 3041–3046.

National Bureau of Statistics of China (NBSC) (2009) 'China statistic yearbook 2009'. China Statistics Press, Beijing, China.

Rothausen, S.G.S.A. and Conway, D. (2011) Greenhouse-gas emissions from energy use in the water sector. *Nature Climate Change* 1, 210–219. DOI: 10.1038/nclimate 1147.

Royal Society (2009) 'Reaping the benefits: Science and the sustainable intensification of global agriculture'. *Royal Society Policy Report*, November 2009.

Smith, P. and Olesen, J.E. (2010) Synergies between the mitigation of, and adaptation to, climate change in agriculture, *The Journal of Agricultural Science*, vol. 148, pp. 543–552 DOI: 10.1017/S0021859610000341.

Xinhua (2010) *Bulletin of First National Pollution Survey* [Online]. Available at:

http://news.xinhuanet.com/politics/2010-02/09/content_12960555.htm (Accessed: 11 August 2011).

Xiong, W., Conway, D., Lin, E. and Holman, I. (2009a) Future cereal production in China: The interaction of climate change, water availability and socio-economic scenarios, *Global Environmental Change*, vol. 19, no. 1, pp. 34–44.

Xiong, W., Conway, D., Lin, E. and Holman, I. (2009b) Potential impacts of climate change and climate variability on China's rice yield and production, *Climate Research*, vol. 40, pp. 23–35, doi: 10.3354/cr00802.

# Part III

# GHG MEASUREMENT AND ACCOUNTING

# 19

# FARM-SCALE GREENHOUSE GAS EMISSIONS USING THE COOL FARM TOOL

## Application of a generic farming emissions calculator in developing countries

*Jonathan Hillier, Pete Smith, Tobias Bandel,*
*Stephanie Daniels, Daniella Malin, Hal Hamilton, and*
*Christof Walter*

## Introduction

Emissions are a function of diverse factors such as management, climate, geography, and technology, so it may not be fruitful to search for "one-size-fits-all" mitigation policies, since they will not be optimal for the majority of farms or cropping systems (Smith *et al.* 2008). As such, there has been interest in the development of methods and software tools which estimate emissions at the farm scale (where management decisions are often made). Table 19.1 provides a summary of the types of methods available for agricultural greenhouse gas (GHG) emissions quantification.

Large uncertainties surround emissions quantification, particularly in respect to the dependence of emissions on factors that vary over small spatial scales. By nature, agricultural systems are heterogeneous, even at small scales, which means that for certain components of a product life cycle there is always significant uncertainty. As an example, the nitrification and denitrification reactions leading to nitrous oxide ($N_2O$) emissions, although known to be functions of soil structure, soil nitrate concentration, and soil water content, are difficult to characterise due to physical variation even at small temporal and spatial scales (i.e., soil heterogeneity and weather). As such, prediction of $N_2O$ emissions can, in theory, always be improved with additional information, although the quality of prediction is difficult to quantify since even measurement techniques are

Table 19.1 Summary of agriculture emissions quantification methods

| Complexity | Models | Data requirements | Aggregation level/uncertainty | Notes |
|---|---|---|---|---|
| Tier 1 | IPCC Tier 1 default factors | Limited land use and management activity data (e.g., N application rates; acres under no-till); little soil delineation; animal populations Low data inputs | Typically large spatial units; national scale; annual resolution; highest uncertainty when applied at project scale | Suitable for rough overviews and where limited data is available (e.g., indirect $N_2O$ emission factor from leaching) |
| Tier 2 | Hybrid approaches – using process or empirical models to develop region-specific empirical equations with emission factors | Intermediate spatial/temporal scale input data; land-use and activity data scaled to the spatial unit of analysis (tillage types, animal classes, N fertilizer type; crop type); Requires longer-term scientific data to develop empirical models or calibrate process models | Finer spatial and temporal resolution than above; can achieve reasonable uncertainty due to 'averaging' of modelled results | Can be suitable for project-based accounting and inventory roll-ups to national scale; application will depend on available scientific and management data |
| Tier 3 | Process-based models | Spatially explicit fine-scale data for model variables; detailed land-use and management histories; fine-scale soil maps and daily/weekly climate data; requires extensive scientific information to calibrate models at this scale; field measured data for estimating uncertainty is often limiting factor | Finest spatial scale with representation of environmental and management variables at the individual farm level | Suitable for small-scale applications where local variability can be managed; model parameterization and testing can be done; collection of land-use and verified activity data obtained; systems will be needed to make advanced modelling approaches accessible to project developers |

| Sampling and Measurement | Highest data requirements; costly to measure and variability high; long sampling intervals and crediting periods for soil carbon; can have best precision | Site scale; may be sub-daily if micrometeorological techniques are used to estimate near-continuous gas emission rates, or every few years with soil carbon stock change; uncertainty can be high if not applied correctly | Level of errors may become overwhelming in sites/ projects with high variability without tight sampling and statistical design; can be most costly to implement |

*Source:* Reproduced from Olander *et al.* 2011.

subject to significant error. The Intergovernmental Panel on Climate Change (IPCC) classifies quantification methods into three tiers. For fertilizer-induced $N_2O$ emissions, the Tier 1 method supposes that 1% of applied nitrogen (N) is emitted as $N_2O$-N, regardless of soil and climate characteristics. Detailed Tier 3 methods incorporating process-based soils emissions models such as DAYCENT (Del Grosso *et al.* 2006) and DNDC (Li *et al.* 1994) provide more refined estimates, but require a more in-depth understanding of emissions processes to operate and interpret. The fertilizer-induced emissions model of Bouwman *et al.* (2002) is a good illustration of a Tier 2 model. It acknowledges that emissions vary as a function of soil texture, soil pH, soil drainage, climate, etc., but it only requires broad characterisations of these variables—for example soil or climate classes—as inputs. So it provides some refinement over Tier 1 methods but has the potential to reach a wider user base than Tier 3 methods.

In this chapter we describe the development and application of an agricultural GHG emission calculator called the Cool Farm Tool, which can be viewed within this framework as an integrated Tier 2 model. We give a concise overview of the first version of the tool (V1.0), and then describe several new features in an upcoming release. Use of the tool will be demonstrated in a study of Kenyan smallholder coffee production.

## The Cool Farm Tool Version 1.0

Version 1.0 of the Cool Farm Tool was commissioned from the University of Aberdeen by Unilever as a user-focused decision-support tool, and released in April 2010. The goal was to package a single tool giving the best estimate of GHG emissions, whilst using only the data which is typically available to (or easily obtainable by) the average farmer. The software integrates several established "off-the-shelf" empirical models for GHG emissions to give an overall emissions estimate as a function of current (and in some cases previous) farming practice. The tool provides a tailored figure for GHG emissions based on specific management practices and enables the user to explore the most appropriate GHG mitigation options available to them with the management levers they have.

Details of the Cool Farm Tool Version 1.0 can be found in the documentation at www.unilever.com/aboutus/supplier/sustainablesourcing/tools/?WT.LHNAV=Tools, and in Hillier *et al.* (2011). It is open source and freely available under a creative commons licence.

Since autumn 2009, the software has been piloted in a project called Cool Farming Options (CFO) coordinated by the Sustainable Food Lab (http://sustainablefoodlab.org/projects/climate). The assessment involves 17 companies, non-governmental organizations (NGOs) and supply chains in 14 countries (Europe, USA, India, Africa, South and Central America).

Version 1.0 was designed to consist of four separate data entry sections: *Crop Management, Livestock, Field Energy Use*, and *Primary Processing*.

1   Geographic variables and **Crop Management**: This first section incorporates models for fertilizer (synthetic and organic) induced emissions; production (from ecoinvent Centre 2007), liming and urea hydrolysis (IPCC 2006); direct and indirect $N_2O$ emissions (FAO/IFA 2001; Bouwman 2002); and emissions from pesticide production (Audsley 1997). The effects of land-use change or crop management practice (i.e., tillage or organic amendments) on soil carbon (C) stocks are also considered by annualising the observations of Ogle *et al.* (2005), and assuming that C stock changes will occur over a 20-year period.

2   **Livestock**: In this section, emissions from enteric fermentation and manure management were estimated in relation to diet and manure management systems, largely based on an IPCC Tier 1 methodology (IPCC 2006).

3   **Field Energy Use:** for the use of energy directly related to cultivation of the crops up to harvest; for example, tillage, spraying, harvesting, irrigation, and materials transport. This includes a model for fuel use from farm machinery operations (mostly derived from ASABE 2006).

4   **Primary Processing and Storage:** is typically for operations such as washing, loading, and storage of harvested products. Most figures come from GHG Protocol (2003), or from ecoinvent for renewable electricity emissions.

The reporting section of the tool presents first a general summary of emissions from all components and then a more detailed breakdown of each specific section. Based on user inputs it identifies the largest sources of emissions—both overall and for each section.

The concise reporting constitutes a first-pass assessment of an agricultural production system. As well as identifying the largest emissions sources, it is also straightforward to conduct "what if?" scenario testing by exploring mitigation options either individually or in combination.

## The Cool Farm Tool Version 1.1

Employment of the tool in the pilot phase of the CFO project has led to significant developments, and a substantially revised version (1.1). Here we briefly discuss the major new features appearing in this version:

• Separation of sequestration from the rest of the Crop Management section and inclusion of living biomass in perennial cropping systems via standard allometric models.

• Substantial revision of the Livestock tab to enable Tier 2 estimation of

221

emissions (IPCC 2006), and emissions from feed mix (partly derived from Hillier *et al.* 2009).

The IPCC Tier 1 method is used to give an approximate estimate of emissions from enteric fermentation and manure management (as in V1.0). This estimate can, however, be improved if the farmer has a good knowledge of daily food intake of the animals and if their feed mix is well characterised. For dairy cows there is a module which allows dry matter intake to be estimated as a function of milk production and the option to correct for fat and protein content.

- Expansion of fertilizer databases to include figures from Bhat *et al.* (1994) and Fertilisers Europe (pers. comm., 2010; based on Brentrup and Palliere 2008; production values are reviewed and approved for future publication in the ELCD core database, http://lca.jrc.ec.europa.eu/lcainfohub/datasetArea.vm)
- Inclusion of a wider range of animal manures (DEFRA 2010)
- Accounting for crop residue management, i.e., treatment of the non-harvested part of the crop (IPCC 2006; Brown *et al.* 2009)
- Methane emissions from wastewater management (IPCC 2006)
- Transport of feed, produce, or other materials off the farm. Standard figures are used for transport via road, rail, air or ship from GHG protocol (2003).

## Example: smallholder coffee production in Kenya

The Sustainable Food Lab has facilitated a Cool Farm Tool assessment within the framework of the three-year Sangana Public Private Partnership (PPP) on climate change adaptation and mitigation in the Kenyan coffee sector. This partnership is between Sangana Commodities Ltd., the Kenyan subsidiary of ECOM, and the German Technical Cooperation (GIZ) (Duration: 10/2008–09/2011; Further partners: 4C Association, World Bank, Tchibo GmbH). The assessment involves a large network of smallholder coffee farmers organised into the Baragwi Farmers' Cooperative Society (BFCS). BFCS counts approximately 16,000 members with the average farm size of 0.5 to 1 acre of coffee grown with limited shade trees. Many of the farmers have a small number of livestock, the manure from which they utilise as fertilizer for coffee.

Given the large population size of the Baragwi system, Sangana's technical team first identified a farm typology in order to group farmers with a typical suite of cultivation practices such as rates of production per tree and shade management. The team grouped the farmers into three management levels and by three existing agro-ecological zones. Within each sample they collected data for approximately six farms per cluster to provide a snapshot of the net emissions for each group.

Table 19.2 illustrates a data summary typical of the requirements of the Cool Farm Tool.

*Table 19.2* Data requirements for GHG emissions assessment with the Cool Farm Tool

| | |
|---|---|
| Location | Baragwi |
| Country | Kenya |
| Production area | 0.36 ha |
| Fresh product | 1,600 kg |
| Finished product | 1,600 kg |
| Climate | Tropical |
| Soil texture | Medium |
| Soil organic matter | 1.72–5.16 |
| Soil moisture | Moist |
| Soil drainage | Good |
| Soil pH | < = 5.5 |
| Fertilizers | 266 kg/ha incorporated calcium ammonium nitrate, 5.3 tonnes incorporated compost (typical 1% N), 266 kg/ha NPK 17:17:17 |
| Pesticide applications | 4 |
| Crop residue | *unclear* |
| Trees | |
| *Coffee* | 1,389 trees per hectare. Diameter at 15 cm from ground: this year 6 cm, last year 6 cm |
| *Shade (generic)* | 81 trees per hectare. Diameter at breast height: This year 20 cm, last year: *unclear* |
| *Other* | 40 trees per hectare. Diameter at breast height: This year 30 cm, last year: *unclear* |

After completing the data input, a summary of emissions output is provided in table and graph form (Figure 19.1). In this example, a significant amount comes from fertilizer production, and a larger amount from direct and indirect $N_2O$ emissions from its application. The production of pesticides for plantation use also contributes to the total emissions. Data collection on crop residues and tree diameters from the previous year are currently unclear so we were not able to quantify the portion of the emissions that is attributable to these factors at this time.

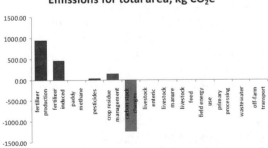

**Emissions for total area, kg CO₂e**

*Figure 19.1* Summary graph from the Cool Farm Tool.

## Discussion

The assessment within the Sangana PPP has provided insight into the value and challenges of carbon accounting in smallholder systems. The cost of visiting many distant farms to collect the required soil, tree, and residue data is significant and an experienced agronomist is needed to ensure quality. Thus the expense and effort of data collection does not in all cases warrant the associated returns. In the case of Sangana, the stakeholders were interested in piloting the Cool Farm Tool for use in concert with the development of the 4C Climate Module, which provides guidance for climate-smart coffee production. Therefore, in this case, integrating data collection for the Cool Farm Tool into existing Internal Control Systems for voluntary standard systems is found to be a viable exercise. But key data points which present unique challenges for use of the Cool Farm Tool in developing countries' smallholder coffee systems include the soil analysis, measurement of crop residues and diameter of shade trees, and the categorisation of shade tree varieties.

Due to the heavy cost of data collection in smallholder systems it is critical that potential users of the Cool Farm Tool in developing countries clearly identify the purpose of the carbon accounting exercise. The Cool Farm Tool provides an inexpensive, nimble scenario mapping tool to identify the most promising cultivation practices within a system. However, balancing cost-effectiveness of data collection with the associated value of the results is still the salient challenge.

The intention of the Cool Farm Tool is to aid farmers in identifying promising mitigation options on their farms. For example, although it is understood in general that certain uses of synthetic chemical inputs contribute to the farm gate emission of produce, the relative contributions can vary significantly with product and geography. In other assessments on arable crops in developed countries we have observed many cases where there is a higher dependence on synthetic fertilizer or pesticide use than in this study. Cooling and storage can also be significant sources of emissions—in these cases the use of renewable energy sources can substantially reduce emissions.

The exercise for the Kenyan coffee system illustrates the utility of conducting the farmer clustering exercise and data collection, but also illuminates the challenges of drawing solid management recommendations and assuring data quality. For example, from the data collected in this system, it is clear that the relationship between productivity and fertilisation rates is an area that warrants more analyses. We know that reduced fertilizer usage will reduce the emissions per unit area of land, however since there is usually a yield penalty associated with this, it is key to locate the "sweet spot" between productivity and emissions.

In the assessments we have conducted so far using the Cool Farm Tool, a common theme is that management of crop residues is one of the most important levers that producers have to reduce their emissions. The balance with residue recycling is that although the treatment of residues typically represents a source of emissions, the recycling—e.g., through residue or compost incorporation—tends to have the effect of increasing soil carbon and nitrogen stocks. There are differences in emissions between residue management options, for example fully aerobic composting emits much less methane than anaerobic composting systems. Details of residue management practice may make the difference between residue usage either beneficial or detrimental to the GHG emissions of the farming system.

There are several advantages to providing a free open-source software tool. First, it allows the scientific community to easily review and contribute to the development of the tool. Coupled with the user-friendly interface it encourages uptake by consumers, farmers, growers, and companies. It promotes engagement of motivated decision-makers in understanding the sources of emissions and exploration of mitigation options. Often, the land users will have a better understanding of the agronomy and constraints of their systems than experts on emissions quantification. For networks of smallholder farmers, the cost of GHG quantification is most often in the data collection and farmer education. Access to a free, open-source tool allows them to allocate scarce resources to these priority areas. This should help to target mitigation options for specific systems and survey data from the Cool Farming Options assessment will allow empirical explorations of the balance between emissions and management practices for a range of crops, climates, and geographies.

# References

ASABE (2006) 'Agricultural machinery management data', *American Society of Agricultural and Biological Engineers Standard ASAE EP496.3*, pp. 385–390. ASABE, St Joseph, MI.

Audsley, E. (Coordinator) (1997) 'Harmonisation of environmental life cycle assessment for agriculture', *Final Report, Concerted Action AIR3-CT94–2028*. European Commission, DG VI Agriculture, 139 pp.

Bhat, G.M., English, B.C., Turhollow, A.F. and Nyangito, H.Z. (1994) 'Energy in synthetic fertilizers and pesticides: Revisited', Pub. No. ORNL/Sub/90–99732/2. Environmental Science Division, Oak Ridge National Laboratory, Oak Ridge, TN.

Bouwman, A.F., Boumans, L.J.M. and Batjes, N.H. (2002) Modeling global annual $N_2O$ and NO emissions from fertilized fields, *Global Biogeochem Cycles*, vol. 16, no. 4, 1080, doi:10.1029/2001GB001812.

Brentrup, F. and Palliere, C. (2008). GHG Emissions and Energy Efficiency in European Nitrogen Fertiliser Production and Use. Proceedings No: 639. The International Fertiliser Society, York, UK.

Brown, S., Cotton, M., Messner, S., Berry, F. and Norem, D. (2009) 'Methane avoidance from composting', *Issue paper for the Climate Action Reserve*.

Del Grosso, S.J., Parton, W.J., Mosier, A.R., Walsh, M.K., Ojima, D.S. and Thornton, P.E. (2006) DAYCENT national-scale simulations of nitrous oxide emissions from cropped soils in the United States, *J Environ Qual*, vol. 35, pp. 1451–1460.

DEFRA (2010) *Fertiliser Manual (RB209)* [Online]. Available at: www.defra.gov.uk/ publications/files/rb209-fertiliser-manual-110412.pdf (Accessed: 10 August 2011).

ecoinvent Centre (2007) *ecoinvent data v2.0. ecoinvent reports No. 1–25* [Online]. Available at: www.ecoinvent.org. Swiss Centre for Life Cycle Inventories, Düben-dorf.

FAO/IFA (2001) 'Global estimates of gaseous emissions of $NH_3$, NO and $N_2O$ from agricultural land', First Version. Published by FAO and IFA, Rome, Italy.

GHG Protocol (2003*) Emissions Factors from Cross-Sector Tools* [Online]. Available at: www.ghgprotocol.org/calculation-tools/all-tools (Accessed: 7 December 2010).

Hillier, J.G., Hawes, C., Squire, G., Hilton, A., Wale, S. and Smith, P. (2009) The carbon footprints of food crop production, *International Journal of Agricultural Sustainability*, vol. **7**, pp. 107–118.

Hiller, J.G., Walter, C., Malin, D., Garcia-Suarez, T., Mila-i-Canals, L. and Smith, P. (2011) A farm-focused calculator for emissions from crop and livestock produc-tion, *Journal of Environmental Modelling and Software* (in press).

IPCC (2006) *IPCC 2006 revised good practice guidelines for greenhouse gas inventories* [Online]. Available at: www.ipcc-nggip.iges.or.jp/public/2006gl/index.html. *Intergovernmental Panel on Climate Change (IPCC), Institute for Global Environmental Strategies*, Tokyo, Japan.

Li, C., Frolking, S. and Harriss, R. (1994) Modeling carbon biogeochemistry in agricultural soils, *Global Biogeochemical Cycles*, vol. 8, no. 3, pp. 237–254.

Ogle, S.M., Breidt, F.J. and Paustian, K. (2005) Agricultural management impacts on soil organic carbon storage under moist and dry climatic conditions of tem-perature and tropical regions, *Biogeochemistry*, vol. 72, pp. 87–121.

Olander, L.P., Eagle, A.J., Baker, J.S., Haugen-Kozyra, K., Henry, L.R., Murray, B.C. and Jackson, R.B. (2011) 'Assessing greenhouse gas mitigation opportunities and implementation options for agricultural land management in the United States', *Technical Working Group on Agricultural Greenhouse Gases (T-AGG) report*.

Paustian, K., Babcock, B.A., Hatfield, J., Lal, R., McCarl, B.A., McLaughlin, S., Mosier, A., Rice, C., Robertson, G.P., Rosenberg, N.J., Rosenzweig, C., Schlesin-ger, W.H. and Zilberman, D. (2004) 'Agricultural mitigation of greenhouse gases: Science and policy options', *CAST (Council on Agricultural Science and Tech-nology) Report*, R141 2004, pp. 120.

Martino, D., Cai, Z., Gwary, D., Janzen, H., Kumar, P., McCarl, B., Ogle, S., O'Mara, F., Rice, C., Scholes, B. and Sirotenko, O. (2007) 'Agriculture', in *Climate Change 2007: Mitigation, Contribution of Working Group III to the Fourth Assessment Report of the Intergovernmental Panel on Climate Change*, edited by Metz, B., Davidson, O.R., Bosch, P.R., Dave, R. and Meyer, R.L. Cambridge University Press, Cambridge, UK.

Smith, P., Martino, D., Cai, Z., Gwary, D., Janzen, H.H., Kumar, P., McCarl, B., Ogle, S., O'Mara, F., Rice, C., Scholes, R.J., Sirotenko, O., Howden, M., McAl-lister, T., Pan, G., Romanenkov, V., Schneider, U., Towprayoon, S., Wattenbach, M. and Smith, J.U. (2008) Greenhouse gas mitigation in agriculture, *Philosophi-cal Transactions of the Royal Society, B.*, vol. 363, pp. 789–813.

# 20

# USING BIOGEOCHEMICAL PROCESS MODELS TO QUANTIFY GREENHOUSE GAS MITIGATION FROM AGRICULTURAL MANAGEMENT

*Lydia P. Olander with contributions from*
*Stephen Del Grosso, Karen Haugen-Kozyra,*
*R. César Izaurralde, Daniella Malin, Keith Paustian, and*
*William Salas*

## Introduction

To conduct inventories and monitor climate change mitigation in agriculture, we need to be able to quantify the greenhouse gas (GHG) impacts resulting from different agricultural management activities. While we have a rough, but relatively clear picture about the most important opportunities for mitigating agricultural greenhouse gases (Smith *et al.* 2008, see Seeberg-Elverfeld and Tapio-Biström, this volume), we have considerably less clarity on how to quantify the changes due to mitigation. The purpose of this chapter is to review process-based biogeochemical modeling as one approach to cost-effective and reliable quantification of agricultural greenhouse gases (GHGs) in agriculture.

The Intergovernmental Panel on Climate Change (IPCC) has developed default factors, average rates of GHG flux for various processes, which are used for estimating GHG emissions or sequestration associated with agricultural practices, other land uses and land-use change at the national level,[1] but these methods become less accurate as spatial scale decreases from national to regional to farm and they do not account for many of the management practices that are expected to reduce emissions (e.g., changing fertilizer type). Thus, the default factors are not sensitive to management changes that farmers would implement on the ground.

227

The other option, measuring changes in carbon pools or GHG fluxes in the field is expensive. Soil carbon (C) is extremely variable, thus detecting small changes against a large pool of stored carbon can require numerous samples and can be expensive. Field measurements of nitrous oxide ($N_2O$) and methane ($CH_4$) with current chamber and tower methods (Olander *et al.* 2011a) are expensive and difficult to use, and thus not ready for widescale implementation.[2]

Given these difficulties, modeling can be a cost-effective way to quantify emissions and sequestration of GHGs. Two approaches exist:

1   regression analysis, which uses statistical analysis of prior emissions data to produce an emission factor that can be applied to a larger geographic region (examples include Millar *et al.* 2010 and Rochette *et al.* 2008); and
2   process-based biogeochemical modeling at a site or regional level, which uses prior data and knowledge about GHG dynamics together with information about local environmental and management conditions to simulate net emissions.

Regression approaches do not capture the farm-level effects of spatial and temporal variability on GHG dynamics. Thus these regional averages are more appropriate and accurate for a program with widespread participation in the region. Regressions are developed for individual practices (e.g., tillage) and thus do not capture the potential interactions of practices, the marginal benefits or trade-offs, which may not have additive outcomes.

Process-based biogeochemical models can simulate GHG dynamics under a range of changing environmental (soil physical properties, climate, topography, previous land management) and management (cropping, livestock, manure, grazing practices) variables, while capturing temporal and spatial variability. These models were developed for systems analysis of management on agricultural productivity and sustainability and adapted to assess farm level and watershed effects (Parton *et al.* 1987; Gassman *et al.* 2010). These models can work at the site-level scale, and are calibrated and tested using data from long-term controlled experiments and field observations. Most models were initially developed with data from agricultural systems in temperate regions in developed countries, but have since been calibrated and applied to other countries and climates. They can produce estimates of GHG change in response to changes in land use or management reasonably well, but require significant environmental and agricultural data inputs and detailed site knowledge. They can also model interactions of multiple changes in crops and practices over a complex landscape, if the necessary data are available. Remote sensing or limited field data are used to calibrate models for specific sites or regions. This combination of site-specific data and modeling is likely the best for developing large-scale programs.

While biogeochemical process models can be used for individual sites or farms, substantial data is needed to run at a high accuracy and gain the benefits of a full, site-level model run. Management data can come from farmers or project developers, but data on environmental variables (soils, climate) which are available in industrialized countries may be more difficult to gather in developing countries. Running models at the site level allows the flexibility to tailor the model to site and project, but can make verification and validation more difficult. The alternative is to run the models at a coarser resolution (e.g., 10,000 ha), scaled to the ecozones, or areas of similar climate and soil type, and to develop very localized default factors for each combination of crops and practices relevant to that region, an approach similar to regression analysis. This can result in simple application and easy verification, but reduces flexibility in addressing the particulars of the project, especially if they include activities outside those included in the modeled default factors.

Even where there are extensive high-quality national and regional databases and numerous long-term agricultural research sites, data is a limiting factor in validating the model and quantifying the uncertainty of the outcomes.[3] Modelers are developing ways to improve confidence in the results to advance agricultural mitigation programs, including an approach for repeat sampling of regional references sites that may be transferable to countries with poorer data (CAST, 2004; Paustian et al. 2006; van Wesemael et al. 2011). Some of the most significant data limitations in developing countries are very coarse daily weather data and the absence of widespread collection of land management data, which can limit regional assessments. Soils maps are generally available, however, and refinements of this data are under way in most regions (see www.africa-soils.net/).

While there will be challenges in using biogeochemical process models in developing countries that lack good data, especially concerning smallholder agriculture and subsistence practices, a number of efforts are already under way to extend the models' use around the world. The CENTURY and DayCent models have been used around the world for research studies and global simulations were conducted to assess GHG mitigation scenarios at very course resolution (2 degrees) (Abdalla et al. 2010; Del Grosso et al. 2009; Stehfest et al. 2007; Leite et al. 2004). The DeNitrification-DeComposition (DNDC) model has been used more internationally, including in developing countries, than in the US. It is being used at scales ranging from site level with detailed inputs to the national level where uncertainties in inputs are assessed (Beach et al. 2008; Tang et al. 2006; Pathak et al. 2005; Cai et al. 2003).

Three of the most commonly used process-based models are summarized below, based on a review conducted by the Technical Working Group on Agricultural Greenhouse Gases (T-AGG) program for the United States (Olander *et al.* 2011b).

## Summary of three biogeochemical process models

This chapter provides a detailed look at three biogeochemical process models that are widely used in the United States and Canada to quantify GHGs from agriculture and other land uses. These models are parameterized for use in North America. The three models covered here are DAYCENT, DNDC, and the Erosion Productivity Impact Calculator (EPIC[4])/Agricultural Policy Extender (APEX) (Table 20.1). The information below was gathered directly from members of the modeling teams. We catalogue the GHGs, management practices, and crops included in these models, as well as the input data required to run them. We also share some insights into the accuracy and precision of these models.

The models all quantify changes in plant and soil C storage and fluxes and on-site $N_2O$ emissions (Table 20.2). For changes in off-site $N_2O$ emissions, the models estimate nitrogen leaching and volatilization loss rates, which can then be combined with the IPCC Tier 1 emissions factor to determine indirect $N_2O$ emissions from these sources. Only DNDC has fully modeled methane fluxes at this point, while EPIC is the only model that includes GHGs from energy and fuel use impacted by changes in management (e.g., fertilizer production, fuel use). Most models cover important management practices (applicable globally, to both developed and developing countries), but a subset of these, particularly those related to nitrous oxide, methane management, and biochar, need further testing and calibration (Table 20.3). The models include a wide variety of crop types, which vary somewhat by model[5] (Olander *et al.* 2011b). Because the models have historically been used for commodity crops, other crops are often missing, especially many commonly grown by smallholders in the tropics. However, there has recently been a significant effort to expand the models to include these other crops. The models require a wide variety of data inputs[6] (Olander *et al.* 2011b). In the US, many of the data inputs can be found or extrapolated from national databases, while others are commonly collected by or available to farmers. A few data inputs may require new measurements. While less of this information will be readily available in many developing countries, the new soils map, global climate maps, and information from local land managers will help. It may also be possible to develop a relatively simple network of sampling sites to increase data accuracy.

A somewhat different model is the National Aeronautics and Space Administration—Carnegie-Ames-Stanford Approach (NASA-CASA) model, which simulates net primary production (NPP) and $CO_2$ emission from respiration of plants and organisms at regional to global scales. This model calculates monthly terrestrial NPP based on the concept of light-use efficiency, modified by temperature and moisture stress. Model outputs

*Table 20.1* Description of the major biogeochemical process models capable of quantifying GHG fluxes for the agricultural sector in the US

| Model | Description |
|---|---|
| DAYCENT* | DAYCENT simulates exchanges of carbon, nutrients, and trace gases among the atmosphere, soil, and plants in daily time steps. Flows of C and nutrients are controlled by the amount of C in the various pools, the N concentrations of the pools, abiotic temperature/soil water factors, and soil physical properties related to texture. It is built using the CENTURY model which runs on monthly time steps. www.nrel.colostate.edu/projects/century5/ Contact: Stephen Del Grosso |
| DNDC** (DeNitrification-DeComposition) | DNDC is a family of models for predicting plant growth, soil C sequestration, trace gas emissions and nitrate leaching for cropland, pasture, forest, wetland, and livestock operation systems. The core of DNDC is a soil biogeochemistry model simulating thermodynamic and reaction kinetic processes of C, N and water driven by the plant and microbial activities in the ecosystems. www.dndc.sr.unh.edu/ Contact: William Salas or Changsheng Li |
| EPIC*** (Erosion Productivity Impact Calculator) | EPIC (Environmental Policy Integrated Climate) is a comprehensive terrestrial ecosystem model capable of simulating many biophysical processes as influenced by climate, landscape, soil, and management conditions. Salient processes modeled include growth and yield of numerous crops as well as herbaceous and woody vegetation; water and wind erosion; and the cycling of water, heat, carbon, and nitrogen. The carbon algorithms in EPIC are based on concepts used in the Century model. EPIC also uses a process-based algorithm to estimate $N_2O$ flux during denitrification and $N_2O$ and NO fluxes during nitrification. http://epicapex.brc.tamus.edu/ Contact: César Izaurralde or Jimmy Williams |
| APEX*** (Agricultural Policy Extender) | APEX is the watershed version of EPIC. It contains all of the algorithms in EPIC plus algorithms to quantify the hydrological balance at different spatial resolutions (farms to large watersheds) under different land cover and land use. The fate of eroded carbon and nitrate can be traced through the entire watershed. www.brc.tamus.edu/apex.aspx http://epicapex.brc.tamus.edu/ |

*Notes*
  *Description contributed by Steven Del Grosso.
 **Description contributed by Bill Salas.
***Description contributed by César Izaurralde.

*Table 20.2* Greenhouse gases measured in three of the biogeochemical process models that can quantify GHG emissions from land use

| GHG measured | DAYCENT | DNDC | EPIC/APEX |
|---|---|---|---|
| Electricity, fuel, and input energy | No | No | Yes |
| Soil carbon Sequestration | yes (Tier 3) | yes (Tier 3) | yes (Tier 3) |
| $N_2O$* | yes (Tier 3): leaky pipe approach to $N_2O$ emissions–calculated on basis of % of N mineralization subject to soil environment conditions | yes (Tier 3): soil Eh and microbial population dynamics | yes (Tier 3): based on electron flow, oxygen availability, and competitive inhibition. |
| $CH_4$ | uptake Only | yes (Tier 3) | in progress |

*Notes*
Fields indicated in *italics* are included in the model, but special considerations are noted.
*On-site $N_2O$ emissions are included directly in the models. For off-site $N_2O$ the models can estimate nitrogen lost through leaching and volatilization which can then be combined with the IPCC emissions factor to calculate off-site $N_2O$ emissions.

include the response of net $CO_2$ exchange and other major trace gases in terrestrial ecosystems to interannual climate variability in a transient simulation mode.[7] The model has been primarily used for global and large-scale regional simulations (Potter *et al.* 1999; Potter *et al.* 1998), but some consultants to farmers in the US are beginning to use it at a farm-level scale.[8] Another common model, RothC, has been used more in Europe but only included carbon dynamics, thus it cannot be used to quantify net GHG changes which include nitrous oxide and methane.

The three models (DAYCENT, DNDC, and EPIC/APEX) have many similarities, and a few critical differences. These include:

- All three models can be used at a Tier-2 or -3 level for quantifying Carbon Dioxide ($CO_2$) and $N_2O$ emissions and C removal (sequestration).
- DNDC simulates whether soils are aerobic or anaerobic and methane ($CH_4$) emissions from moisture-saturated soils and $CH_4$ uptake by soils; DAYCENT only models $CH_4$ uptake in non-saturated soils, but is working on an emissions model; and EPIC/APEX currently does neither, but is working on incorporating both.
- APEX places EPIC into a spatial context, where it can model hydrological flows using algorithms similar to those used in the Soil and Water Assessment Tool (SWAT) model[9] and thus estimate transport and deposition of soil sediment, nutrients, and pesticides. They are working on adding estimates of indirect (off-site) $N_2O$ emissions. The other two models can estimate nitrogen losses from leaching and volatilization, which can be used to calculate $N_2O$ losses using an IPCC emissions factor, but the models do not do this directly. While all three models cover common agricultural practices, there are a few practices that each of them is not yet set up to run.
- All models can manipulate quantity of irrigation, but DAYCENT does not currently allow different types of irrigation (flood, sprinkler, drip). DNDC and EPIC/APEX do include irrigation type.
- All the models can control the amount of fertilizer. DAYCENT only includes nitrate versus ammonia fertilizers, while the other models have multiple types (~7). DAYCENT also does not have the ability to change method or placement of fertilization, but the other models do (e.g., surface vs. injection).
- Nitrification inhibitors are in DAYCENT and DNDC, but not yet in EPIC/APEX.
- Nitrous oxide emissions from manure, irrigation, and other management can be predicted using DAYCENT, DNDC, and EPIC, but not yet from APEX, which uses empirical equations to predict denitrification.
- DNDC is the only model that currently includes $CH_4$ emissions, thus it is the only one of the models that can look at the effects of water management in rice cultivation.

Table 20.3 Management activities included in three of the biogeochemical process models that can quantify GHG emissions from land use

| Management practice* | DAYCENT | DNDC | EPIC/APEX |
|---|---|---|---|
| Conventional to conservation till | yes | yes | yes |
| Conventional to no-till | yes | yes | yes |
| Conservation till to no-till | yes | yes | yes |
| Switch from irrigated to dry land | yes | yes | yes |
| Use winter cover crops | yes | yes | yes |
| Eliminate summer fallow | yes | yes | yes |
| Intensify cropping (more crops/year) | yes | yes | yes |
| Switch annual crops (change rotations) | yes | yes | yes |
| Include perennial crops in annual crop rotations | yes | yes *(new crops will need to be calibrated)* | yes *(new crops will need to be calibrated)* |
| Short-rotation woody crops | yes | yes | yes |
| Irrigation improvements (drip, supplemental…) | yes—model simulates irrigation, but can't distinguish types | yes (DNDC distinguishes sprinkler, flood, and drip; manual or automatic based on water stress) | yes (different types; manual or automatic based on water stress) |
| Agroforestry (windbreaks, buffers, etc.) | yes | yes (by compartmentalizing the fields) | yes—both in EPIC and APEX |
| Herbaceous buffers | *possible but has not been tested* | *possible but has not been tested* | yes—explicit in APEX |
| Application of organic materials (esp. manure) | yes | yes | yes (beef, dairy, swine, poultry) |
| Application of biochar | *possible but has not been tested* | *possible but has not been tested* | *under consideration* |
| Reduce N application rate | yes | yes | yes |
| Change fertilizer N source | yes *(only distinguished NO$_3$ from NH$_4$)* | yes (7 distinct chemical fertilizer types) | yes—single and compound fertilisers |
| Change fertilizer N timing | yes | yes | yes—flexible application based on N test or plant N stress |
| Change fertilizer N placement | No | yes (user-prescribed depth, only limited testing) | yes—(broadcast, banding) |

| Practice | | | |
|---|---|---|---|
| Use nitrification inhibitors | *yes (limited testing)* | *yes (limited testing)* | *not tested yet* |
| Improved manure application to soils management ($N_2O$) | *yes (amount and type)* | *yes (amount and type)* | *not tested yet* |
| Irrigation management for $N_2O$ | *yes (only amount)* | *yes (only amount)* | *not tested yet* |
| Manage histosols to reduce GHG emissions | no | *yes (new application in CA, needs calibration)* | *not tested yet* |
| Drainage on croplands, $N_2O$ and $CH_4$ | *yes ($CH_4$ not included)* | yes | *EPIC/APEX can simulate drainage; no test of drainage and $N_2O$ and $CH_4$* |
| Rice water management for $CH_4$ | no | yes | no |
| Improved grazing management, range | yes | yes | yes |
| Improved grazing management, pasture | yes | yes | yes |
| Fertilizing grazing lands | yes | yes | yes |
| Irrigation management for grazing lands | yes | yes | yes |
| Species composition on grazing lands | *model represents vegetation mix, not species* | *no–users would have to define special mix* | yes (up to 10 species) |
| Grazing land fire management | yes | yes | yes |
| Rotational grazing | yes | *yes (grass model requires calibration and testing of physiological response to grazing intensity)* | *yes (new grazing model in APEX)* |
| Manure management (lagoon, compost, etc.) | no | *yes (new manure model with enteric fermentation requires more testing, continued development)* | *yes (in APEX)* |
| Transition to natural land (forests, native grasslands, wetlands) | yes | *yes (wetland/forest DNDC)* | *yes (in APEX) –in development (have not done wetlands)* |

*Notes*

Fields indicated in *italics* are included in the model, but special considerations are noted.

* Inclusion of a management practice and variations on those activities (e.g., seven chemical fertilizer types), means that the models include a process to estimate impacts of the practice, but does not guarantee that the science is fully developed. For example, biochar and fertilizer types are active areas of research with little scientific consensus on the basic process and outcomes of implementing the practice.

- DNDC has a model for manure management, which includes intensive management systems and enteric fermentation. APEX includes a model for extensive grazing and confined area feeding. According to Gassman *et al.* (2010), up to 10 herds of groups of animals can be simulated with APEX, but only one herd can occupy a subarea at any given time. Livestock can rotate among subareas. Animals may be confined to a feeding area. Grazing can occur throughout the year or periodically according to limits. When no more grazing material is available, the owner can provide supplemental feeding.
- The models differ slightly in the inputs required to run them, but all of them can use estimates or national databases to fill in most variables where site-specific information is not available. The quality of these national data and estimates vary. (More details in T-AGG assessment report; Olander *et al.* 2011a).

## Accuracy of the models

Accuracy is related to model error and uncertainty. Sources of model error can be partitioned into two categories: errors due to uncertainty in model inputs and errors due to model structure. Errors from model input uncertainty occur because model inputs are not precisely known and can be estimated by using a Monte Carlo approach, which involves randomly drawing model inputs from a probability distribution function for key model inputs and performing a series of simulations. Once the models are calibrated they can be used to adjust expectations and crediting ratios in programs. Errors related to model structure result from the fact that equations in the models are imperfect representations of the real world processes that result in GHG emissions. These errors can be estimated by statistically quantifying the agreement between model outputs and field measurements.

Various statistics should be used when evaluating model performance because they each have strengths and weaknesses. For example, the correlation coefficient quantifies how well model outputs are correlated with measurements, but it is not influenced by model bias. For example, model outputs could be perfectly correlated with measurements ($r^2 = 1.0$) but highly biased if each model outputs is twice as high as the measured value. Model evaluation is also dependent on the variable of interest, the reliability of measured data, and the scale of model application. For example, grain yields are more accurately and precisely measured than GHG fluxes, so model errors also tend to be smaller for grain yields than GHG emissions. Scale dependency is complicated. When results from many model simulations are aggregated spatially and temporally, errors tend to shrink as scale increases. However, model errors for small plots of land can be small if all important inputs are well known. For example, individual landowners can provide detailed information on land-use history and land management and use on-site sampling where other data bases are lacking. This can provide high accuracy for a small scale. However, gathering all of this detail can be time consuming and expensive and raises questions of data consistency across sites. As a result many programs look to regional modeling where detailed data is not known and uncertainty analysis of data inputs will be critical.

If we want to better understand and compare the uncertainties associated with models we need to have a parallel assessment of the models. Comparisons of model outcomes to field observations show that model accuracy increases as the number of observations increase, suggesting that models are good at estimating the average outcome across a landscape (Olander *et al.* 2011b). A side-by-side assessment of the models with comparable methods that explore both the structural uncertainty and the input uncertainties does not yet exist. The modelers are interested in developing such a comparison over the next few years.

## Current use of these models

DayCent/Century:
- foundation for decision support tools CometVR[10] and CometFarm (under development);
- used to develop input data to run Forest and Agriculture Sector Optimization Model with Greenhouse Gases (FASOMGHG), an economic model used by USDA and EPA to assess policy options;
- used by researchers around the world to investigate the impacts of land use and climate change on GHG fluxes and nutrient cycling;
- US National GHG Inventory.

DNDC:
- effort to expand coverage of non-commodity crops;
- being tested for targeted crops (e.g., tomato) for efforts to account for GHG emission along an entire product supply chain (e.g., tomato-based products);
- adopted by dairy and swine industries in the US to assess air emissions;
- used for international and multinational inventories and mitigation studies (e.g., Nitro-Europe);
- basis for web decision support tools (e.g., http://riceghg.info);
- used by researchers for quantifying GHG emissions from natural and restored wetlands.

EPIC/APEX:
- EPIC, core model of the DOE Carbon Sequestration in Terrestrial Ecosystems Consortium;
- EPIC, core model of the DOE Great Lakes Bioenergy Research Center;

- EPIC, core model of GLOBIOM to conduct global runs of 18 major crops under 4 different managements;
- EPIC, core model for building Nutrient Trading Tool, a decision support tool, in collaboration with USDA-ARS;
- APEX, one of the core models for the USDA-NRCS CEAP project.

## User-friendly versions of these models

Efforts are also under way to develop user-friendly decision support tools based on process models. The DayCent (CENTURY) model has been used to develop the COMET-VR tool (Paustian *et al.* 2010; Paustian *et al.* 2009) (www.cometvr.colostate.edu/) to quantify soil carbon sequestration potential from various management practices. This tool is currently being updated into COMET-FARM, which is a whole farm or ranch GHG emission estimation tool that uses DayCent for estimating soil emissions and uptake of $CO_2$ and $N_2O$ (and other models for livestock and other on-farm emissions). APEX has been used by USDA in the development of the Nutrient Trading Tool (http://ntt.tarleton.edu/nttwebars/) to track nitrogen impacts of agricultural practices on water quality, but could potentially also be used to quantify GHG impacts. An additional decision-support system for EPIC is under development by US government researchers with support from NASA. There is also an ARCGIS version of APEX. DNDC has a simplified version that requires only eight data inputs (Willey and Chameides, 2007). It also has a user-friendly interface to the full model that has been prototyped as online decision support tool (see http://riceghg.info/ and http://nugget.sr.unh.edu).

## Conclusion

Process-based biogeochemical models have been under development and undergoing improvement for 30 years and can provide a robust tool for quantifying GHG impacts of agricultural management if sufficient data and research are available. In the US data availability allows farm-level use of these models, but it is critical to develop user interfaces (decision support tools) that imbed acquisition of critical management data and allow consistent and appropriate use of these complex models. These decision support tools are under development and will be available soon. Where this level of resolution (farm-scale quantification for 40 different management practices) is not possible, due to insufficient environmental input data or missing research on relevant management practices or cropping systems, which is likely the case in many developing countries, it may be possible to develop regional emissions factors using these models or regression approaches that are climate and soil (ecozone) specific with sufficient confidence to move forward. The use of models can allow extrapolation of research results to regions of similar climate and soils even if no research has been done in that location. The proposal noted above for a network of soil sampling sites is to help improve such extrapolations across regions. Work is already under way to apply biogeochemical models for agriculture in developing countries across diverse land management applications. The Carbon Benefits Project, which is working with the same team that develops and runs the CENTURY and DayCent models, is developing an approach to quantify carbon for sustainable land management activities at national to project scales using a web-accessible system (see www.unep.org/climatechange/carbon-benefit/Home/tabid/3502/Default.aspx).

## Acknowledgements

We would like to thank the David and Lucile Packard Foundation for support of the Technical Working Group on Agricultural Greenhouse Gases which has made this work possible.

## Notes

1 In 2000 and 2003 the International Panel on Climate Change (IPCC) published IPCC Good Practice Guidance for Land Use, Land Use Change, and Forestry (www.ipcc.ch/publications_and_data/publications_and_data_reports.htm), and in 2006, IPCC Guidelines for National Greenhouse Gas Inventories (www.ipcc-nggip.iges.or.jp/public/2006gl/index.html). All default emissions and removal factors can be found in the IPCC Good Practice Guidance and National Greenhouse Gas Inventory Guidelines (IPCC 2000, 2003, 2006). These and additional emission factors submitted by countries are also posted on the IPCC Emission Factor Database (EFDB) (www.ipcc-nggip.iges.or.jp/EFDB/main.php).
2 Greater discussion of measurement issues with additional references can be found in Technical Working Group on Agricultural Greenhouse Gases (T-AGG) assessment report (Olander et al. 2011a).
3 More details on existing data and data gaps available in T-AGG assessment report.
4 EPIC is also known as Environmental Policy Integrated Climate.
5 See Table 4 in Olander et al. (2011b), available from http://nicholasinstitute.duke.edu/ecosystem/land/pubsland/ecosystem/t-agg.
6 See Table 5 in Olander et al. (2011b), available from http://nicholasinstitute.duke.edu/ecosystem/land/pubsland/ecosystem/t-agg.
7 The team responsible for this model was not available to comment or provide more details.
8 CASA EXPRESS CQUEST http://geo.arc.nasa.gov/sge/casa/index.html; CQUEST online tool (slightly more limited in scope and customisability): http://sgeaims.arc.nasa.gov/website/cquest/viewer.htm.
9 http://swatmodel.tamu.edu/.
10 www.cometvr.colostate.edu/.

# References

Abdalla, M., Jones, M., Yeluripati, J., Smith, P., Burke, J. and Williams, M. (2010) Testing DayCent and DNDC model simulations of N₂O fluxes and assessing the impacts of climate change on the gas flux and biomass production from a humid pasture, *Atmospheric Environment*, vol. 44, no. 25, pp. 2961–2970.

Beach, R.H., DeAngelo, B.J., Rose, S., Li, C., Salas, W. and Del Grosso, S.J. (2008) Mitigation potential and costs for global agricultural greenhouse gas emissions, *Agricultural Economics*, vol. 38, pp. 109–115.

Cai, Z., Sawamoto, T., Li, C., Kang, G., Boonjawat, J., Mosier, A., Wassmann, R. and Tsuruta, H. (2003) Field validation of the DNDC model for greenhouse gas emissions in East Asian cropping systems, *Global Biogeochemical Cycles*, vol. 17, no. 4, 18–1–18–10.

CAST (2004) 'Climate change and greenhouse gas mitigation: challenges and opportunities for agriculture'. Council for Agricultural Science and Technology, Ames, IA.

Del Grosso, S.J., Ojima, D.S., Parton, W.J., Stehfest, E., Heistemann, M., DeAngelo, B. and Rose, S. (2009) Global scale DAYCENT model analysis of greenhouse gas emissions and mitigation strategies for cropped soils, *Global and Planetary Change*, vol. 67, pp. 44–50.

Gassman, P., Williams, J.R., Wang, X., Saleh, A., Osei, E., Hauck, L.M., Izaurralde, C. and Flowers, J.D. (2010) Invited review article: The agricultural policy environmental EXtender (APEX) model: An emerging tool for landscape and watershed environmental analyses, *American Society of Agricultural and Biological Engineers*, vol. 53, no. 3, pp. 711–740.

IPCC (2000) *Good Practice Guidance and Uncertainty Management in National Greenhouse Gas Inventories* [Online]. Available at: www.ipcc-nggip.iges.or.jp/public/gp/english/. Intergovernmental Panel on Climate Change, Geneva, Switzerland.

IPCC (2003) *Good Practice Guidance for Land Use, Land-Use Change and Forestry* [Online]. Available at: www.ipcc-nggip.iges.or.jp/public/gpglulucf/gpglulucf_files/0_Task1_Cover/Cover_TOC.pdf. Intergovernmental Panel on Climate Change, Geneva, Switzerland.

IPCC (2006) *Guidelines for National Greenhouse Gas Inventories* [Online]. Available at: www.ipcc-nggip.iges.or.jp/public/2006gl/index.html. Intergovernmental Panel on Climate Change, Geneva, Switzerland.

Leite, L.F.C., Sa Mendonca, E., Machado, P.L.O.A., Filho, E.I.F. and Neves, J.C.L. (2004) Simulating trends in soil organic carbon of an Acrisol under no-tillage and disc-plow systems using the Century model, *Geoderma*, vol. 120, no. 3–4, pp. 283–295.

Millar, N., Robertson, G.P., Grace, P.R., Gehl, R.J. and Hoben, J.P. (2010) Nitrogen fertilizer management for nitrous oxide (N₂O) mitigation in intensive corn (maize) production: An emissions reduction protocol for US Midwest agriculture, *Mitigation and Adaptation Strategies for Global Change*, vol. 15, pp. 185–204.

Olander, L.P., Eagle, A.J., Baker, J.S., Haugen-Kozyra, K., Murray, B.C., Kravchenko, A., Henry, L.R. and Jackson, R. (2011a) 'Assessing greenhouse gas mitigation opportunities and implementation strategies for agricultural land management in the United States', *T-AGG report [Online]*. Available at: http://nicholasinstitute.duke.edu/ecosystem/t-agg. Nicholas Institute for Environmental Policy Solutions, Durham, North Carolina, USA.

Olander, L.P, Haugen-Kozyra, K., Malin, D., with contributions from Del Grosso, S., Izaurralde, C., Malin, D., Paustian, K. and Salas, W. (2011b) 'Using Biogeochemical Process Models to Quantify Greenhouse Gas Mitigation from Agricultural

Management Projects', *T-AGG report [Online]*. Available at: http://nicholasinstitute.duke.edu/ecosystem/land/pubsland/ecosystem/t-agg. Nicholas Institute for Environmental Policy Solutions, Durham, North Carolina, USA.

Parton, W.J., Schimel, D.S., Cole, C.V. and Ojima, D.S. (1987) Analysis of factors controlling soil organic matter levels in Great Plains grasslands, *Soil Science Society of America Journal*, vol. 51, pp. 1173–1179.

Pathak, H., Li, C., and Wassman, R. (2005) Greenhouse gas emissions from Indian rice fields: Calibration and upscaling using the DNDC model, *Biogeosciences*, vol. 2, no. 2, pp. 113–123.

Paustian, K.H., Antle, J.M., Sheehan, J. and Paul, E.A. (2006) 'Agriculture's role in greenhouse gas mitigation'. *Pew Center on Global Climate Change*, Arlington, VA.

Paustian, K., Ogle, S. and Conant, R.T. (2010) 'Quantification and decision support tools for agricultural soil carbon sequestration', in World Scientific: Hillel, D. and Rosenzweig, C. (eds), *Handbook of Climate Change and Agroecosystems: Impacts, Adaptation and Mitigation*, pp. 307–41.

Paustian, K., Brenner, J., Easter, M., Killian, K., Ogle, S., Olson, C., Schuler, J., Vining, R. and Williams, S. (2009) Counting carbon on the farm: Reaping the benefits of carbon offset programs, *Journal of Soil and Water Conservation*, vol. 64, no. 1, pp. 36A–49A.

Potter, C.S., Davidson, E.A., Klooster, S.A., Nepstad, D.A., De Negereiros, G.H. and Brooks, V. (1998) Regional application of an ecosystem production model for studies of biogeochemistry in Brazilian Amazonia, *Global Change Biology*, vol. 4, no. 3, pp. 315–333.

Potter, C.S., Klooster, S. and Brooks, V. (1999) Interannual variability in terrestrial net primary production: Exploration of trends and controls on regional to global scales, *Ecosystems*, vol. 2, no. 1, pp. 36–48.

Rochette, P., Worth, D.E., Huffman, E.C., Brierley, J.A., McConkey, B.G., Yang, J., Hutchinson, J.J., Desjardins, R.L., Lemke, R. and Gameda, S. (2008) Estimation of $N_2O$ emissions from agricultural soils in Canada. II. 1990–2005 Inventory, *Canadian Journal of Soil Science* vol. 88, pp. 655–669.

Smith, P., Martino, D., Cai, Z., Gwary, D., Janzen, H., Kumar, P., McCarl, B., Ogle, S., O'Mara, F., Rice, C., Scholes, B., Sirotenko, O., Howden, M., McAllister, T., Pan, G., Romanenkov, V., Schneider, U., Towprayoon, S., Wattenbach, M. and Smith, J. (2008) Greenhouse gas mitigation in agriculture, *Philosophical Transactions of the Royal Society B: Biological Sciences* vol. 363, pp. 789–813.

Stehfest, E., Heistermann, M., Priess, J.A., Ojima, D.S. and Alcamo, J. (2007) Simulation of global crop production with the ecosystem model DayCent, *Ecological Modelling*, vol. 209, no. 2–4, pp. 203–219.

Tang, H., Qui, J., Van Ranst, E., Li, C. (2006) Estimations of soil organic carbon storage in cropland of China based on DNDC model, *Geoderma*, vol. 134, no. 1–2, pp. 200–206.

Van Wesemael, B., Paustian, K., Andren, O., Cerri, C., Dodd, M., Etchevers, J., Goidts, E., Grace, P., Katterer, T., McConkey, B., Ogle, S., Pan, G. and Siebner, C. (2011) How can soil monitoring networks be used to improve predictions of organic carbon pool dynamics and $CO_2$ fluxes in agricultural soils?, *Plant and Soil*, vol. 288, no. 1–2, pp. 247–259.

Willey, Z. and Chameides, B. (2007) 'Harnessing farms and forests in the low-carbon economy: How to create, measure, and verify greenhouse gas offsets', pp. 1–229, Duke University Press, Durham, NC, and London.

# 21

# AN EMISSIONS-INTENSITY APPROACH FOR CREDITING GREENHOUSE GAS MITIGATION IN AGRICULTURE

## Reconciling climate and food security objectives in the developing world

*Justin S. Baker and Brian C. Murray*

### Introduction

In this chapter we discuss the use of an output or intensity-based crediting methodology for greenhouse gas (GHG) offsets in agriculture that directly accounts for productivity and all sources of emissions and sequestration from the production system.[1] Such an accounting scheme could provide greater monetary incentives to producers than methodologies that do not credit the potential indirect emissions reductions achieved if productivity improvements in one system lower agricultural expansion rates and management intensity elsewhere (which we refer to throughout this chapter as positive leakage). As recent studies suggest, productivity enhancements per unit area in agriculture and forestry through technological improvements or increased on-farm production intensity can provide a viable source of GHG mitigation by compensating for extensive expansion that brings new land into production (Burney *et al.* 2010; Hertel *et al.* 2009; Jackson and Baker 2010).

Although agricultural intensification is generally thought to increase emissions from the agricultural sector through higher levels of input use and more intense soil cultivation, we assert that productivity improvements can directly and indirectly reduce emissions from production and land-use changes. However, opportunities for emissions reduction through de-intensification, land set-asides and enhanced carbon sequestration should not be abandoned, especially if such activities improve productivity on low-

242

intensity systems, or only marginally decrease productivity on high-intensity systems. In the context of agricultural GHG mitigation in the developing world, an output-based offset (OBO) accounting methodology is extremely flexible, as it credits both reductions in emissions per unit area and productivity gains, thus capturing two important policy goals in the international agricultural community.

## Advantages of an OBO methodology

An OBO accounting scheme can potentially contribute to agricultural productivity improvement goals, sustainable development in rural communities, and reduced-carbon agriculture in the developing world. If implemented at a global scale, it is possible that OBO protocols could provide an additional source of carbon finance for projects that have proven successful at improving productivity through new practices or agricultural infrastructure.

Allowing offset credits to be generated for emissions intensity improvements addresses concerns that current agricultural offsets based on area accounting may favor practices that either take land out of production or lower the productivity of agriculture to create offsets, thereby reducing food and fiber production or shifting it to other locations. To the extent that declining or shifting production undermines food security, many parties will deem this unacceptable. Moreover such a pattern would be self-defeating if production and emissions were simply shifted elsewhere (i.e., leakage).

Baker *et al.* (2010) show that afforestation of cropland or pasture can provide substantial GHG mitigation potential in the U.S. However, taking land out of agricultural production raises a number of concerns, including higher food prices and overall food security. Food prices and stagnating agricultural productivity are significant determinants in global malnourishment trends, so persistently higher prices and lower production levels could exacerbate commodity price spikes and global hunger concerns. If an offset market recognized output-based offsets, then the commodity price and secondary economic effects of moving a portion of land out of agricultural production would be relaxed by productivity gains elsewhere in the system.

Additionally, moving to an OBO accounting framework can expand the range of activities or practices that generate offsets by including yield-enhancing management activities that might not be considered GHG reducing in isolation (including technological improvement that boosts production efficiency, or more intense forms of production that increase yield per unit area). While this might sound counter-intuitive as intensification can lead to higher emissions on a cultivated parcel of land, research shows that agricultural intensification can contribute to mitigation goals at the landscape level if by growing more per unit of land, agricultural conversion of forests decreases (Burney *et al.* 2010). This is especially true

243

in tropical regions, where yields following deforestation are lower, and emissions are higher relative to temperate zones, with carbon lost per unit of food gained from deforestation being nearly three times as great for tropical agricultural expansion (West *et al.* 2010).

Additionally, some offset practices that are considered too costly due to low expected GHG gains relative to the cost of adoption could be encouraged under an OBO scheme if the practice is both GHG-reducing and yield improving. Here, credits generated through an OBO approach would outweigh those from the typical area-based offsets (ABO) approach if emissions reductions were accompanied by yield improvement, meaning such practices could be more economically competitive under OBO. However, more information on the costs and yield effects of these practices is needed before concluding that they would be spurred by an OBO system. Further insight on such practices is provided in subsequent sections.

## Basics of an output-based accounting approach

All offset mechanisms calculate emission reductions as the difference between actual emissions and a baseline estimate. The baseline may capture the emissions generated under historic practices for an appropriate cohort group, emissions expected under business-as-usual (BAU) practices or some other proxy, such as best practices. The basic method is: (1) establish baseline farming practices, (2) calculate baseline emissions resulting from baseline practices, and (3) establish emission reduction relative to baseline emissions. Existing or potential protocols would extend this to include discounts for permanence and leakage and possibly uncertainty (see Murray *et al.* 2007).

Many offset programs currently use an area-based metric where tonnes of $CO_2$ equivalent (t $CO_2$e) are reported per unit area (hectare).[2] Output-based accounting takes this process a step further by establishing a baseline for on-farm yields (total production per hectare). This metric is represented in t $CO_2$e per unit output (tons of corn and soybeans) accounting for GHG emissions and returns to production. Output-based offsets therefore capture the GHG intensity resulting from adoption of the new practice or practices.[3]

Agricultural yields have grown in many regions over time (though not everywhere), which also needs to be factored into the mechanism. An OBO metric would likely have a baseline emissions rate that changes over time, meaning that subsequent yield enhancements will have to outweigh anticipated productivity growth rates. A way to handle this would be to establish baseline yield growth for specific commodities consistent with current growth projections published by national agricultural research agencies or other independent sources. By incorporating total output directly into the crediting metric, increasing output increases credit, all else equal, thus encouraging practices that maintain or increase yield and total output, while providing weaker incentives for merely pulling land out of production. Since leakage is caused by reduced production that accompanies an offset activity, a yield-based metric incorporates this effect, potentially eliminating the need for a leakage discount.

## Explanation of the OBO approach

The OBO system could be implemented subject to the following steps:

1   Under a mitigation program, each year farmers commit to use specific practices on a given land area expected to decrease emissions for a given crop or livestock product.
2   Baseline emissions intensity (GHG tons per unit of output) for each commodity produced will be established by the crediting authority for the current year and some number of years into the future. Baselines could be set in two distinct ways, including a project-by-project basis, where each project is charged with developing its own specific baseline (often called the "project-specific" approach). A project-specific baseline offers more flexibility in the types of potentially creditable projects, but presents high transaction costs in verification, monitoring, or modeling. Alternatively, a baseline could be derived from broader estimates that capture emissions rates from normal—or perhaps best—practices and applies this as a baseline for all production activities (often called the "performance standard" approach). A regional baseline might exclude those producers with emissions or production profiles that fall outside the regional norm (for example, a marginally productive farm with high baseline emissions intensity).
3   At the end of each year, yields from each product will be tallied and actual emissions for each commodity will either be directly measured or modeled from farm- or field-level information.
4   Credits for the farm's mitigation activity will be determined by product, determined by the difference between the actual emissions intensity (average emissions/average yield) and regional baseline emissions intensity multiplied by the observed production for each product.
5   The total credits generated by the farm equal the sum of the credits generated for each participating product (via Step 4).

## Formation of metrics

The output-based approach builds off the baseline emissions intensity per unit output for each crop in the system as follows:

$$eB_{it} = \frac{EB_{it}}{YB_{it}}$$

Where:

$EB_{it}$ = baseline *projected* net emissions (including sequestration) under BAU practices for commodity $i$ in year $t$
$YB_{it}$ = baseline *projected* yield

Projected net emissions, $EB_{it}$, could be based on modeling emissions from standard agricultural practices expected in the future. Baseline yields, $YB_{it}$, could be modeled with agronomic models or by taking observed crop yields in the production region or at the individual farm, and extrapolating projected yield growth rates over time, producing a different projected base yield for all $t$.

A farmer enters into an offset agreement committed to improve on that intensity target $eB_{it}$. They will undertake actions to reduce their actual emissions intensity:

$$eA_{it} = \frac{EA_{it}}{YA_{it}}$$

Here, the terms are the same as above, substituting $A$ for actual (vs. $B$ for baseline). Thus, emissions intensity reduction can be accomplished either by reducing net emissions (numerator), increasing yields (denominator), or both relative to baseline levels.

Under OBO, total credits are generated as a product of actual output of the crop or livestock yield $i$ and realized difference in intensity. Total production is yield multiplied by area in production, denoted by: $Q_{it} = YA_{it} * AA_{it}$. The creditable offsets under OBO, by crop, are as follows:

$$CreditsO_{it} = Q_{it}[eB_{it} - eA_{it}]$$

The total credits for the farm is the sum earned across all $N$ commodities for which the offset actions are taken:

$$Total\_CreditsO_t = \sum_{i=1}^{N} CreditsO_{it}$$

An OBO scheme also allows one to account for emissions displacement and to directly impose a leakage adjustment (credit) for abatement activities that reduce (enhance) yield. Crediting based on improvements in emissions intensity is built on two foundations: (1) onsite changes in emissions from the farm undertaking the mitigation practice, and (2) the influence of onsite changes in yield *on production and emissions elsewhere*. The first element directly reflects the emission consequences of the mitigation action (e.g., increase in soil carbon sequestered, reduced $N_2O$ emissions from changes in fertilizer applications). This is analogous to ABO crediting—emissions effects attributed to the area in question.

The second effect is indirect in that it captures the secondary effects of changes in site productivity. Operating under the assumption that the overall consumption demand for each agricultural commodity $i$ is fixed, the onsite yield improvement for the farm in question implies that less area or less intense forms of production are needed elsewhere to produce this aggregate level of output. The reduced area and management intensity elsewhere means that emissions decline elsewhere. Murray and Baker (2011) provide a more detailed discussion on the emissions displacement effect and how one might incorporate a leakage adjustment directly into an OBO accounting methodology.

## Opportunities for OBO crediting and examples of current projects

As mentioned, the OBO crediting approach is extremely flexible due its applicability to a wide range of production activities (including activities or technology adoption options applicable to smallholder operations), including:

1   Emissions reduction or productivity improvements from shifts along the intensive margin, or changes in input use, which could include increased input use and higher on-farm emissions as long as productivity gains outweigh emissions increases. Alternatively, a shift in intensity might imply reduced input use where emissions reductions outweigh marginal productivity losses.
2   Activities that alter the production frontier, or technological improvements that improve productivity regardless of baseline production intensity (input use). Examples might include different seeds or soil amendments that enhance productivity (such as biochar).
3   Activities that entirely change the emissions profile of a production system per unit area by increasing soil carbon sequestration, regardless of intensity in other input use not related to soil management (such as alternative tillage practices).
4   Activities that reduce movement to the extensive margin (such as agricultural deforestation) or take marginally productive and highly emitting land out of production, thus boosting the carbon balance of the whole land-use system; or
5   Some combination of the above.

In general, as the OBO approach could provide credits for a number of projects that standard area-based metrics would not and there are currently pilot projects on the ground in the developing world where use of an OBO crediting scheme would be beneficial.

In a recent article, Palm et al. (2010) examine the relationship between productivity enhancement and carbon sequestration for increased fertilizer use, agroforestry, and green manure cover cropping interventions in maize cropping systems in sub-Saharan Africa. Results show that use of fertilizers can slow extensification substantially and induce reforestation; all three mitigation activities can enhance yields. If OBO were applied to these projects, landowners would be subsidized at a higher rate than under a standard area or activity-based crediting scheme.

There are numerous examples of other activities that could benefit from an OBO crediting scheme by rewarding emissions reduction and productivity gains simultaneously. Perhaps the most prominent example is genetic modification of crops to enhance yield, especially under suboptimal growing conditions; technological improvement continues to advance on this front, and additional carbon finance could push research and development even further (Tester and Landridge 2010). Improved pest management techniques or other forms of crop protection can also increase productivity.

Use of iron silicate fertilizer in rice production systems has been shown to enhance yields while significantly reducing methane emissions (Ali *et al.* 2008). Precision agriculture techniques can enhance yields and reduce the need for excessive inputs by optimally distributing input use to a farm system with varying topography and soil quality (Gebbers and Adamchuk 2010). The Danish consulting group Nordeco has facilitated a number of sustainable agriculture activities that would be eligible for OBO crediting, including rice intensification and improved livestock management. The World Bank funds several similar projects, including the Kenya Agricultural Carbon Project (KACP), which is designed to increase the carbon uptake and soil quality of the agricultural landscape and increase maize yields by promoting manure management, cover cropping, residue management, and tree planting (Seeberg-Elverfeldt and Tapio-Biström, 2010, see also Lager and Nyberg, and Woelcke, this volume). Conservation International and Starbucks have paired up to promote sustainable coffee production to reduce deforestation and improve productivity in North Sumatra (see chapter by Semroc *et al.*). These are a few examples of many, and increased emphasis on food security and climate mitigation in the international community will likely result in additional government or philanthropic funding of such efforts. For more information on current mitigation projects in the developing world, refer to Seeberg-Elverfeldt and Tapio-Biström (2010).

## Limitations of the OBO approach

While an OBO methodology presents many distinct advantages that have been highlighted above, there are potential limitations of such an approach that also warrant attention. In particular, an output-based system is focused primarily on decreasing the emissions intensity of agriculture, rather than reducing total emissions. Emissions intensity could improve significantly, but if emissions increase proportionately more than productivity, total emissions from the sector go up. The secondary effects of expanding the area of production may be minor relative to the direct emission-reducing effect, but this is an empirical matter that should be carefully considered. Also, allowing agricultural sector emissions to increase somewhat to achieve food security may be acceptable if reductions in the other sectors of the economy (or other parts of the world) are cut dramatically, but the main purpose of an offset-based system should be to help cut aggregate global emissions.

In addition to the concern over total emissions, there is also the concern that emission intensity reductions in agriculture create more emissions upstream in the production or provision of inputs. Consider the emissions reduction equations reported earlier. As indicated, to reduce emissions intensity, several approaches can be taken. One could reduce on-farm emissions (decreasing numerator), improve yield under status quo practices (increasing the denominator), or could adopt practices that reduce emissions and simultaneously boost yield. However, another option is to increase farm management intensity through higher levels of input use, which would almost certainly increase per-hectare emissions relative to the baseline if the emissions from all inputs were included. If these upstream emissions are captured in a life cycle analysis and broader cap-and-trade program, this may not be a problem since these emissions will count against the cap and thus require emission reductions elsewhere in the system. But if these upstream sources are not capped, then OBO protocols could have a counterbalancing increase in emissions upstream and should be designed to limit such perverse incentives whenever possible.

Further exposure to yield volatility is another disadvantage. Under an OBO approach credits are directly a function of the *realized* yield, rather than expected yield, yet agricultural yields fluctuate greatly given the general variability in weather inputs and differences in land quality. The farmer may undertake actions to enhance the expected yield and thereby reduce GHG intensity, all else equal. But if weather or other factors cause lower realized yields, this could raise actual yields and therefore emissions intensity above the baseline level, leading to a loss in credits. The reverse scenario also holds—if unanticipated factors lead to a large increase in yields, this mechanism could assign an excessively large number of credits to the farm. Measuring offset success based on yield inherently adds an additional source of volatility to a process that already has a potentially high degree of uncertainty. The rules of the offset program will need to determine whether farmers are liable for underperforming relative to the GHG threshold and thus incur debits (payments) entailing potential losses.

Finally, there are limitations of regional, performance standard baselines. As mentioned, the lack of data at the farm scale requires the use of a regional baseline in an OBO framework, though this could discourage participation by those farms that are the highest emitters. If a marginal producer has higher-than-average emissions matched by lower-than-average yield, this could make it difficult for the producer to participate in an OBO scheme if their emissions intensity still falls above the regional baseline after the mitigation action is taken.

One potential problem with the use of a regional baseline in an OBO approach is that a producer with relatively low baseline emissions could participate in the offset market and be subsidized for maintaining status-quo practices (meaning the offset credits generated would be non-additional), though this problem persists regardless of the offset accounting methodology as long as a regional baseline is used. Nevertheless, the distinction between marginal and average emissions intensity is very relevant and deserves further attention.

## Conclusions

In summary, the OBO approach has certain limitations that deserve further attention, but in general it provides a more flexible approach for crediting GHG reductions from agriculture, while incentivizing productivity improvements. Offset projects need to be viewed as a way to reduce emissions at the farm level, but should also take into consideration the impact these farm-level actions have on overall productivity and food security as well as the proper accounting for offsite leakage. Use of OBO accounting for crediting agricultural GHG abatement internalizes on-farm productivity changes, thus reconciling both climate and food security goals.

## Notes

1 The term "offset" is often used to describe mitigation from emission sources not limited by a cap or some other source of regulation that can be credited against emissions from regulated sources. Most GHG mitigation policies do not directly regulate agricultural emissions, so the term "offset" is often a term of art used synonymously with mitigation in agriculture. The methods here could just as easily be applied to agriculture as a directly capped sector.
2 While some offset protocols for agriculture do include some form of output-based accounting (e.g., Alberta's protocols for $N_2O$ and livestock management), some existing methodologies are area-based, including Voluntary Carbon Standard, American Carbon Registry, and Chicago Climate Exchange (ACR 2010; CCX 2009; Government of Alberta 2010).
3 While expressed as tons of $CO_2$ equivalent, these measures generally include reductions of nitrous oxide ($N_2O$) and methane ($CH_4$), which are then transformed to $CO_2$ based on relative global warming factors of each gas.

## References

Ali, M.A., Lee, C.H. and Kim, P.J. (2008) Effect of silicate fertilizer on reducing methane emission during rice cultivation, *Biology and Fertility of Soils*, vol. 44, no. 4, pp. 597–604.

American Carbon Registry (ACR). (2010) *The American carbon registry methodology for emission reductions through changes in fertilizer management: version 1.0* [Online]. Available at: www.americancarbonregistry.org/carbon-accounting/emissions-reductions-through-changes-in-fertilizer-management (Accessed: June 2010).

Baker, J.S., McCarl, B.A., Murray, B.C., Rose, S.K., Alig, R.J., Adams, D., Latta, G., Beach, R. and Daigneault, A. (2010) Net farm income and land use under a U.S. greenhouse gas cap-and-trade, *Policy Issues*, vol. 7, pp. 17.

Burney, J., Davis, S.J. and Lobell, D.B. (2010) Greenhouse gas mitigation by agricultural intensification, *PNAS*, vol. 107, no. 26, pp. 12052–12057.

Chicago Climate Exchange (CCX). (2009) *CCX carbon offset protocol: Agricultural best management practices – continuous conservation tillage and conversion to grassland soil carbon sequestration* [Online]. Available at: www.chicagoclimatex.com/docs/offsets/CCX_Conservation_Tillage_and_Grassland_Conversion_Protocol_Final.pdf.

Gebbers, R. and Adamchuk, V.I. (2010) Precision agriculture and food security, *Science*, vol. 327, pp. 828–831.

Government of Alberta. (2010) *Nitrous oxide emissions reductions from farming operations: Draft for public comment* [Online]. Available at: http://carbonoffsetsolutions.climatechangecentral.com/files/microsites/OffsetProtocols/ProtocolReviewProcess/3rdRoundPublicReview/NERP_v1_Draft_Public_Review__July14_10_.pdf.

Hertel, T.W., Golub, A.A., Jones, A.D., O'Hare, M., Plevin, R.J. and Kammen, D.M. (2010) Effects of US maize ethanol on global land use and greenhouse gas emissions: Estimating market-mediated responses, *BioScience*, vol. 60, no. 3, pp. 223–31.

Jackson, R.B. and Baker, J.S. (2010) Opportunities and constraints for forest climate mitigation, *BioScience*, vol. 60, no. 9, pp. 698–708.

Murray, B.C., Sohngen, B.L. and Ross, M.T. (2007) Economic consequences of consideration of permanence, leakage and additionality for soil carbon sequestration projects, *Climatic Change*, pp. 80: 127–143.

Murray, B.C. and Baker, J.S. (2011) 'An output-based intensity approach to crediting greenhouse gas mitigation in agriculture: Explanation and policy implications', *Greenhouse Gas Measurement and Management*, Forthcoming.

Palm, C.A., Smukler, S.M., Sullivan, C.C., Mutuo, P.K., Nyadzi, G.I. and Walsh, M.G. (2010) Identifying potential synergies and trade-offs for meeting food security and climate change objectives in sub-Saharan Africa, *PNAS*, vol. 107, no. 46, pp. 19661–19666.

Seeberg-Elverfeldt, C. and Tapio-Biström, M.-L. (2010) 'Global survey of agricultural mitigation projects', *Mitigation of Climate Change in Agriculture* (MICCA), Food and Agriculture Organization.

Tester, M. and Landridge, P. (2010) Breeding technologies to increase crop production in a changing world, *Science*, vol. 327, pp. 818–822.

West, P.C., Gibbs, H.K., Monfreda, C., Wagner, J., Barford, C.C., Carpenter, S.R. and Foley, J.A. (2010) Trading carbon for food: Global comparison of carbon stocks vs. crop yields on agricultural land, *PNAS*, vol. 107, no. 46, pp. 19645–19648.

# 22

# EMERGING TECHNIQUES FOR SOIL CARBON MEASUREMENTS

*Debora M.B.P. Milori, Aline Segnini,*
*Wilson T.L. Da Silva, Adolfo Posadas, Victor Mares,*
*Roberto Quiroz, and Ladislau Martin Neto*

## Introduction

In the face of climate change and increasing carbon dioxide ($CO_2$) levels in the atmosphere, the global carbon cycle, soil organic carbon (SOC) sequestration, and the role of different world biomes as potential sources and sinks of carbon are receiving increasing attention (Feller and Bernoux 2008). Carbon (C) sequestration in plant and soil systems offers an opportunity for mitigating the greenhouse effect (Lal 2004) but the relationship between soil carbon stocks (CS) and carbon fixation in natural and anthropic vegetation remains one of the least studied issues. Emphasis has been placed on measurements of carbon fixation by forests or measurements of carbon emissions following land-use changes. As agriculture represents one of the major land-use systems, there is a close relationship between soil carbon (SC) capture and emissions and the allocation of land for agriculture. Many management practices can increase soil CS as well as above-ground carbon in biomass, including soil conservation practices (e.g., no-tillage, reduced tillage, terracing), incorporation of crop residues, increases in cropping intensity and fertilization, and conversion of cropland to permanent grasslands or forests. Although the Clean Development Mechanism (CDM) protocol has prioritized only above-ground carbon sequestration via afforestation and reforestation (A/R), the soil might represent an even larger carbon sink. In this respect, one of the challenges for the incorporation of agriculture into post-Kyoto agreements (after 2012) is to develop simple and effective methodologies for measuring, monitoring and verifying SC in cropping, farming and land-use systems.

The Instrumentation Center of the Brazilian Agricultural Research Corporation (EMBRAPA) has worked to address this challenge. EMBRAPA's team has used spectroscopic techniques, with high potential for field application, to analyze whole soil samples without sample preparation. Particularly, positive results have been obtained using Laser-Induced Breakdown Spectroscopy (LIBS), Laser-Induced Fluorescence Spectroscopy (LIFS) and Near-Infrared Spectroscopy (NIRS). Methodologies using LIBS and NIRS are being developed for soil C quantification and LIFS equipment was developed to evaluate C stability in soils. A partnership between EMBRAPA and the International Potato Center (CIP) has been conducive to the assessment of soil CS and C stability in different agro-ecosystems in the Southern Peruvian Andes, Kenya and Brazil, using state-of-the-art non-destructive analytical methods. This chapter adds to our own work on emergent techniques through a review of the recent literature on the subject.

## SC measurements

Determination of total carbon by measuring $CO_2$ emitted from the oxidation of organic C and thermal decomposition of carbonate materials is a common laboratory technique. However, these traditional methods have limitations, e.g., they produce toxic residues, are time consuming and labor intensive. When weight change rather than $CO_2$ emitted is used, discrepancies in results among laboratories are common (Kimble et al. 2001). The weight loss-on-ignition method tends to overestimate soil organic matter (SOM) due to loss of inorganic constituents, primarily hydrated clays, during the heating process, affected by ignition temperature and sample size (Cambardella et al. 2001). Consequently, the accuracy of the quantification is dependent on the reference method used (Segnini et al. 2008). For SOC, total organic carbon (TOC) and elemental analyses (C, H, N, S) methods enhance both accuracy and reproducibility.

The advantages of LIBS and NIRS to quantify SC (Cremers et al. 2001; Madari et al. 2006; Da Silva et al. 2008; Christy 2008; Ferraresi 2010; Martin et al. 2010) include: limited sample preparation, process quickness, low cost, detection of metals and portability.

Climate uncertainty and expected changes demand new methods delivering quick carbon quantification and qualitative organic matter (OM). With timely and reliable OM information better assessments of agricultural sustainability and the carbon cycle are feasible. The advent of *in situ* methods for SC determination are important due to the comparatively rapid and potentially cost-effective benefits of these methods, the reduction in sampling and laboratory errors and their potential to minimize soil disturbance while increasing the ability to analyze large areas. Advanced field carbon analysis methods should be capable of providing repetitive, sequential measurements for evaluation of spatial and temporal variation at a scale that was previously unfeasible. Thus, spectroscopy becomes a clean and rapid alternative approach for this purpose (Ferraresi 2010).

Qualitative information can be provided by laboratory spectroscopic methods, e.g., Nuclear Magnetic Resonance (NMR), Electron Paramagnetic Resonance (EPR), Fourier Transform Infrared (FTIR), NIRS, Raman, Ultraviolet-visible (UV-vis) Absorption and Fluorescence. However, only NIRS and fluorescence spectroscopy has the potential to be transformed into field techniques to produce quality SOM indicators.

In the following sections, we review the literature and summarize the bases of new SC determination techniques and present some of our findings in different ecosystems.

## Carbon quantification and soil bulk density by NIRS

Infrared (IR) radiation refers broadly to that part of the electromagnetic spectrum between visible and microwave regions. IR spectroscopy is one of the most powerful tools available to the chemist for identifying organic and inorganic compounds because, with the exception of a few homomolecular molecules such as $O_2$, $N_2$ and $Cl_2$, all molecular species absorb infrared radiation.

NIRS is a reproducible and low-cost method that characterizes materials according to their reflectance in the NIR spectral region. However, NIRS has only relatively recently been tested for quantitative analysis of soils (Reeves and McCarty 2001) and it has a high potential for evaluating and characterizing large areas, quickly, reliably and economically. NIRS has reduced the need for chemically and other instrumentally based methods of sample analysis (Madari *et al.* 2006; Shepherd and Walsh 2006). Calibrations are based on spectral information of the target analyte in representative samples. Chemometrics tools are already necessary for quantitative analysis, since the equipment response is limited and in several spectral regions.

Madari *et al.* (2006) showed that both Diffuse Reflectance Infrared Fourier Transform (DRIFT) and NIRS, when combined with chemometric methods could be useful for quantifying total carbon, nitrogen, sand and clay in soil samples. These techniques could also be calibrated to estimate parameters such as soil aggregation indices, which are not directly measurable by other means. Ferraresi (2010) evaluated the NIRS and DRIFTS on SOM, microbial biomass and texture determination in Brazilian soils. Clay, silt, sand, total organic carbon and microbial carbon contents as well as microbial quotient were accurately quantified by IR spectroscopy. Janik *et al.* (2007) showed a good correlation between mid-IR and total and particulate organic carbon.

Soil bulk density is needed to convert organic carbon content to mass of organic carbon per unit area. However, measurements of soil bulk density are labour-intensive, costly and tedious. The use of NIRS for characterizing the particle size distribution of SOM through determinations based on bulk soil spectra has been suggested with promising results (Barthes *et al.* 2008).

254

Christy (2008) presented an on-the-go spectrophotometer for *in situ* measurement of reflectance spectra and evaluated the potential of the system for making real-time predictions of various soil attributes using NIRS. A review by Cécillon *et al.* (2009) showed that NIRS can provide information related to many chemical and some physical and biological soil properties. The spectra can be used as an integrated measure of soil quality, e.g., to classify sites according to their degradation status or for monitoring the effect of a management factor. However, several technological limitations, i.e., the need for robust quantification models based on several soil characteristics data (site, classification, texture, land use, etc.), and more user-friendly and cheaper devices for *in situ* analysis, must be addressed for broad application of this method.

### *LIBS and its potential to perform analysis* in situ

LIBS is an emerging analytical technique based on atomic and ionic emission of elemental sample constituents that provides qualitative information about the sample composition. LIBS performs multi-elemental direct analysis, dispensing the pre-treatment of samples, while showing potential for *in situ* analyses (Cremers and Radziemski 2006).

The large number of elements in soil precludes the use of calibration standards for LIBS, thus researchers adapt univariate models constructed with classical least-squares regression (Trevizan *et al.* 2008). Nowadays, Artificial Neural Network (ANN) has been used for calibrating LIBS spectroscopy with good performance to classify and predict carbon and metals in soil (Ferreira *et al.* 2008; 2009).

Quantitative carbon measurements in temperate soils have been reported by several research groups (Cremers *et al.* 2001; Martin *et al.* 2010). However, there is only one study for tropical soil using a portable LIBS system (Da Silva *et al.* 2008).

Portable LIBS equipment makes it possible to carry out quantitative field soil analysis. The main advantage is the fast and low-cost assessment with scanty sample preparation.

Ferreira *et al.* (2008) determined Cu concentration in tropical soil samples using a portable LIBS system calibrated with ANN. Cross-validation was applied, following ANN training, for prediction accuracy verification. The ANN showed good efficiency for Cu predictions (R = 0.96), despite features of the portable instrumentation employed. The proposed method presented a limit of detection of $2.3 \, mg/dm^3$ of Cu and a MSE of 0.5 for the predictions.

Da Silva *et al.* (2008) calibrated a portable LIBS system for quantitative measurements of carbon in whole soil samples from the Brazilian Savanna region. Simple and multivariate linear regressions plus cross-validation showed correlation coefficients higher than 0.91, indicating the potential of using portable LIBS systems for quantitative carbon measurements in tropical soils.

*Chemical structure of OM assessed by portable LIFS equipment*

NMR and EPR Spectroscopic techniques are mostly used to characterize more stable fractions of SOM, i.e., humic substances (HS). These analyses require extraction and chemical fractioning of HS from the soil (e.g., humic acids—HA, fulvic acids and humin). This chemical process results in the production of dangerous residues and causes modifications in OM structure (Feller and Beare 1997). In order to render SOM analysis viable at the closest to natural state possible and to carry out quick and low-cost *in situ* analyses, methodologies using LIFS have been developed. LIFS is one of the most sensitive devices available for analytical purposes. It is relatively easy to implement, phenomenologically straightforward and well investigated, and largely non-invasive and consequently can be useful for environmental applications. Milori *et al.* (2006) developed a methodology using LIFS to assess the effect of soil management on the humification index of SOM at different depths. The excitation at near ultraviolet-blue radiation is more resonant with rigid and complex OM structures (Milori *et al.* 2002; Martin-Neto *et al.* 2009). Thus, the area under the fluorescence emission normalized by carbon content of soil was defined as the SOM humification index ($H_{LIF}$). The technique was used to assess the SOM humification index in different soil layers of the Brazilian Savanna (Cerrado) region and how it was affected by land management. In the conventional tillage treatment, $H_{LIF}$ was rather uniform along the profile, which was consistent with the homogeneity imparted by tillage disturbances on the 0–20 cm layer of these soils. The $H_{LIF}$ for the native Cerrado and no-tillage soils showed a tendency to increase from top to deeper layers as the proportion of the particulate OM fraction, and thus of labile compounds like carbohydrates and peptides, decreased in depth (Milori *et al.* 2006). Similar results were observed by González-Perez *et al.* (2007) for subtropical Brazilian soils. The results concurred with other spectroscopic techniques such as Fluorescence spectroscopy of dissolved HS, EPR and NMR (González-Pérez *et al.* 2007; Dieckow *et al.* 2009).

Segnini *et al.* (2011) studied CS and stability in five Peruvian agro-ecologies, along a 1,000 km transect covering the arid Pacific coast, the Andean high plateau, and the Amazonian rainforest. Coffee plantations in the Amazon and alfalfa under irrigation in the dry valleys presented larger CS (83 mg/ha) than primary rainforests. The dry lowlands showed the lowest CS (40 mg/ha). SOC increased with elevation in the arid environments. In the high plateau potato systems, low CS (47 mg/ha) were found. The soils in both the Amazonian site and the dry valleys presented lower $H_{LIF}$, when compared to other agro-ecologies. $H_{LIF}$ increased with soil depth due to the presence of recalcitrant carbon, while at the surface the presence of labile carbon dominates as a result of a constant input of plant residues.

Undoubtedly, the humification index is a highly relevant factor in agro-ecosystems studies. LIFS allows affordable analyses of SOM humification since the technique measures stable carbon, which can be extremely useful in studies on carbon sequestration by soils and mitigation of $CO_2$ emissions.

Segnini *et al.* (2010a) used EPR and a portable LIFS system developed by EMBRAPA Instrumentation for assessing SOM stability in whole soil samples from permanent and seasonally flooded wetlands in the Peruvian Andes (Figure 22.1). The results obtained by both techniques were comparable ($R^2 = 0.88$) and were reliable indicators of increased or reduced SOM stability through its humification index. Results from the portable LIFS system also showed significant correlations when compared to both the bench LIFS system ($R^2 = 0.88$) and EPR ($R^2 = 0.62$), showing the feasibility of using a portable LIFS system and representing an improvement in speed and convenience for SOM stability evaluation across the landscape (Figure 22.2). The soil characterization can be done in the field with whole soil samples, thus avoiding transporting them through complex terrains like the Andean mountains. In another study, Segnini *et al.* (2010b) successfully used the portable LIFS equipment and powder soil instead of

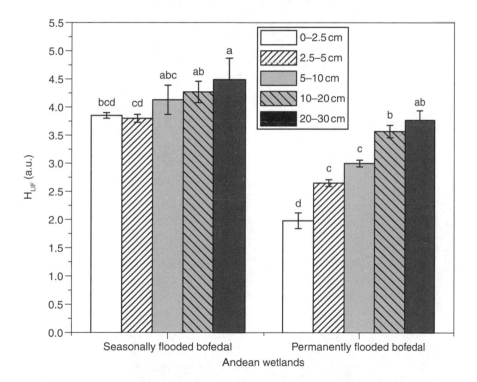

*Figure 22.1* Humification index ($H_{LIF}$) of soils determined through LIFS of soils sampled at different depths in two areas of the Peruvian Andes: seasonally and permanently flooded bofedales. Different letters denote statical differences.

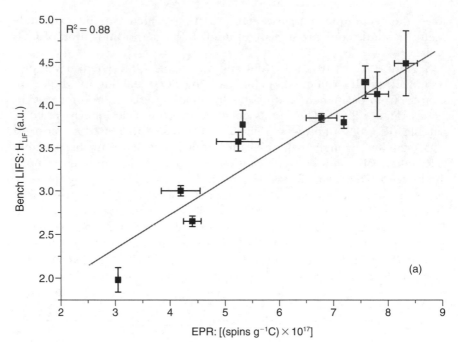

*Figure 22.2* (a)

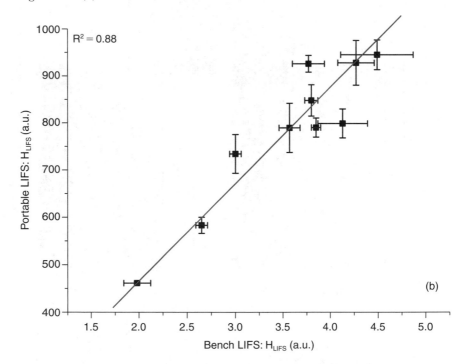

*Figure 22.2* (b)

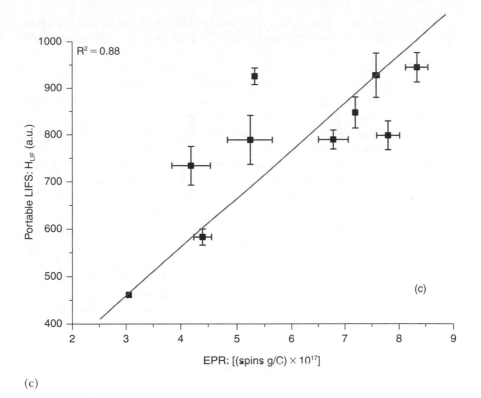

(c)

*Figure 22.2* Data showing the close relationship between (a) the concentrations of semiquinone-type free radicals, as measured by EPR and the results of the $H_{LIF}$, as measured by LIFS; (b) bench and portable LIFS systems and (c) portable LIFS and EPR in the assessment to humification index of bofedales from the Peruvian Andes.

soil pellets for conducting a comparative SC analysis of cropping systems and native vegetation areas in different agro-ecosystems in Kenya. The results showed wide variations in the levels and stability of carbon stored in the soil, depending on factors such as land use, crops grown, water content, elevation, native vegetation and agricultural practices. In the future, it will be possible to link the portable systems with GPS to reliably map soil C quantity and quality (EMBRAPA 2001).

## Conclusions

Relevant results from some emerging techniques based on NIRS, LIBS and LIFS, which measure SC quantity and stability were presented. The development and application of portable equipment systems, based on spectroscopic methods, provide opportunities for *in situ* measurements of

SOM contents, in different ecosystems. Databases on soil C contents and quality can be enriched through the utilization of the aforementioned apparatuses due to the reduced cost per sample, portability, speed and accuracy. With larger global coverage and quality of data the capacity to model carbon balance as a function of management, climate variation and land use and the feedback to the atmosphere will be improved. Hence, the uncertainties created by interpolation in data-scarce environments and the negative consequences of the policies made based on interpolated results are likely to be reduced.

The evidences presented also surmount the main difficulty alluded to for leaving out soil CS from the IPCC accords: the lack of methods to measure SC levels and stability on site in undisturbed soil samples. Until recently, such measurements were only possible within the laboratory, using sophisticated and expensive equipment. With the techniques described, it is possible to measure, monitor and verify carbon contents in whole soil samples in situ with equipment that is portable, affordable and reliable. Furthermore, SOM stability should be considered since it is crucial to assess its permanence in the soil. This assessment is done using soil C dynamics models which can be better parameterized with the stability measures now feasible.

# References

Barthes, B.G., Brunet, D., Hien, E., Enjalric, F., Conche, S., Freschet, G.T., d'Annunzio, R. and Toucet-Louri, J. (2008) Determining the distributions of soil carbon and nitrogen in particle size fractions using near-infrared reflectance spectrum of bulk soil samples, *Soil Biology and Biochemistry*, vol. 40, no. 6, pp. 1533–1537.

Cambardella, C.A., Gajda, A.M., Doran, J.W., Wienhold, B.J. and Kettler, T.A. (2001) 'Estimation of particulate and total organic matter by weight loss-on-ignition', in Lal, R., Kimble, R.M.J., Follett, R.J. and Stewart B.A. (eds) *Assessment Methods for Soil Carbon*. Lewis Publishers, CRC Press, Boca Raton, Fl.

Cécillon, L., Barthes, B.G., Gomez, C., Ertlen, D., Genot, V., Hedde, M., Stevens, A. and Brun, J.J. (2009) Assessment and monitoring of soil quality using near-infrared reflectance spectroscopy (NIRS), *European Journal of Soil Science*, vol. 60, no. 5, pp. 770–784.

Christy, C.D. (2008) Real-time measurement of soil attributes using on-the-go near infrared reflectance spectroscopy, *Computers and Electronics in Agriculture*, vol. 61, no. 1, pp. 10–19.

Cremers, D.A., Ebinger, M.H., Breshears, D.D., Unkefer, P.J., Kammerdiener, S.A., Ferris, M.J., Catlett, K.M. and Brown, J.R. (2001) Measuring total soil carbon with laser-induced breakdown spectroscopy (LIBS), *Journal of Environmental Quality*, vol. 30, no. 6, pp. 2202–2206.

Cremers, D.A. and Radziemski, L.J. (2006) 'History and fundamentals of LIBS', in Miziolek, A.W., Palleschi, V. and Schechter, I. (eds), *Laser-Induced Breakdown Spectroscopy (LIBS): Fundamentals and Applications*, Cambridge University Press, New York, NY.

Da Silva, R.M., Milori, D.M.B.P., Ferreira, E.C., Ferreira, E.J., Krug, F.J. and Martin-Neto, L. (2008) Total carbon measurement in whole tropical soil sample, *Spectrochimica Acta Part B*, vol. 63, no. 10, pp. 1221–1224.

Dieckow, J., Bayer, C., Conceição, P.C., Zanatta, J.A., Martin-Neto, L., Milori, D.M.B.P., Salton, J.C., Macedo, M.M., Mielniczuk, J. and Hernani, L.C. (2009) Land use, tillage, texture and organic matter stock and composition in tropical and subtropical Brazilian soils, *European Journal of Soil Science*, vol. 60, no. 2, pp. 240–249.

EMBRAPA – Empresa Brasileira de Pesquisa Agropecuária, Milori, D.M.B.P., Martin-Neto, L., Vaz, C.M.P. and Bagnatto, V.S. (2001) 'Sensor de qualidade de matéria orgânica de solos', Brazilian patent no PI0106477–0.

Feller, C. and Beare, M.H. (1997) Physical control of soil organic matter dynamics in the tropics, *Geoderma*, vol. 79, no. 1–4, pp. 69–116.

Feller, C. and Bernoux, M. (2008) Historical advances in the study of global terrestrial soil organic carbon sequestration, *Waste Management*, vol. 28, no. 4, pp. 734–740.

Ferraresi, T.M. (2010) 'Espectroscopias de infravermelho próximo e médio na quantificação de atributos do solo, com ênfase na matéria orgânica e na biomassa microbiana', p. 118. Dissertação (Mestrado em Ciências, ênfase em Quimica Analitica). Instituto de Quimica de São Carlos (IQSC), Universidade de São Paulo (USP), São Carlos, SP.

Ferreira, E.C., Milori, D.M.B.P., Ferreira, E.J., Da Silva, R.M. and Martin-Neto, L. (2008) Artificial neural network for Cu quantitative determination in soil using a portable laser induced breakdown spectroscopy system, *Spectrochimica Acta Part B*, vol. 63, no. 10, pp. 1216–1220.

Ferreira, E.C., Anzano, J.M., Milori, D.M.B.P., Ferreira, E.J., Lasheras, R.J., Bonilla, B., Montull-Ibor, B., Casas, J. and Martin-Neto, L. (2009) Multiple response optimization of laser-induced breakdown spectroscopy parameters for multi-element analysis of soil samples, *Applied Spectroscopy*, vol. 63, no. 9, pp. 1081–1088.

González-Pérez, M., Milori, D.M.B.P., Colnago, L.A., Martin-Neto, L. and Melo, W.J. (2007) A laser-induced fluorescence spectroscopic study of organic matter in a Brazilian Oxisol under different tillage systems, *Geoderma*, vol. 138, no. 1–2, pp. 20–24.

Janik, L.J., Skjemstad, J.O., Shepherd, K.D. and Spouncer, L.R. (2007) The prediction of soil carbon fractions using mid-infrared-partial least square analysis, *Journal of Australian Soil Research*, vol. 45, pp. 73–81.

Kimble, J.M, Lal, R. and Follett, R.F. (2001) 'Methods for assessing soil C pools', in Lal, R., Kimble, J.M., Follett, R.F. and Stewart, B.A. (eds) *Assessment Methods for Soil Carbon*. Lewis Publ, Boca Raton, FL.

Lal, R. (2004) Soil carbon sequestration to mitigate climate change, *Geoderma*, vol. 123, no. 1–2, pp. 1–22.

Madari, B.E., Reeves III, J.B., Machado, P.L.O.A., Guimarães, C.M., Torres, E. and McCarty, G.W. (2006) Mid- and near-infrared spectroscopic assessment of soil compositional parameters and structural indices in two Ferralsols, *Geoderma*, vol. 136, no 1–2, pp. 245–259.

Martin, M.Z., Labbé, N., André, N., Wullschleger, S.D., Harris, R.D. and Ebinger, M.H. (2010) Novel multivariate analysis for soil carbon measurements using laser-induced breakdown spectroscopy, *Soil Science Society of America Journal*, vol. 74, no. 1, pp. 87–93.

Martin-Neto, L., Milori, D.M.B.P., Da Silva, W.T.L, Simões, M.L. (2009) 'EPR, FTIR, Raman, UV-visible absorption and fluorescence spectroscopies in studies of NOM', in Senesi, N., Xing, B. and Huang, P.M. (eds), *Biophysico-Chemical Processes Involving Natural Nonliving Organic Matter in Environmental Systems*, Wiley.

Milori, D.M.B.P., Martin-Neto, L., Bayer, C., Mielniczuk, J. and Bagnato, V.S. (2002) Humification degree of soil humic acids determined by fluorescence spectroscopy, *Soil Science*, vol. 167, no. 11, pp. 739–749.

Milori, D.M.P.B., Galeti, H.V.A., Martin-Neto, L., Diekow, J., González-Peréz, M., Bayer, C. and Salton, J. (2006) Organic matter study of whole soil samples using laser-induced fluorescence spectroscopy, *Soil Science Society of America Journal*, vol. 70, no. 1, pp. 57–63.

Reeves III, J.B. and McCarty, G.W. (2001) Quantitative analysis of agricultural soils using near infrared reflectance spectroscopy and fibre-optic probe, *Journal of Near Infrared Spectroscopy*, vol. 9, no. 1, pp. 25–34.

Segnini, A., Santos, L.M., Da Silva, W.T.L., Borato, C.E., Melo, W.J., Bolonhezi, D. and Martin-Neto, L. (2008) Estudo comparativo de métodos para a determinação da concentração de carbono em solos com altos teores de Fe (Latossolos), *Química Nova*, vol. 31, no. 1, pp. 94–97.

Segnini, A., Posadas, A., Quiroz, R., Milori, D.M.B.P., Saab, S.C., Vaz, C.M.P. and Martin-Neto, L. (2010a) Spectroscopic assessment of soil organic matter in wetlands from the high Andes, *Soil Science Society of America Journal*, vol. 76, no. 6, pp. 2246–2253.

Segnini, A., Posadas, A., Quiroz, R., Claessens, L., Gavilan, C., Milori, D.M.B.P. and Martin-Neto, L. (2010b) 'Carbon stocks and spectroscopic assessment of carbon stability in Kenyan soils'. Poster presented at ASA–CSSA–SSSA International Annual Meetings 31 October–3 November 2010, Long Beach, CA.

Segnini, A., Posadas, A., Quiroz, R., Milori, D.M.B.P., Vaz, C.M.P. and Martin-Neto, L. (2011) 'Comparative assessment of soil carbon stocks and stability in different agroecologies in southern Peru', *Journal of Soil and Water Conservation* vol. 66, no. 4, pp. 213–220.

Shepherd, K.D. and Walsh, M.G. (2006) 'Diffuse reflectance spectroscopy for rapid soil analysis', in Lal, R. (ed) *Encyclopedia of Soil Science: Second Edition*.

Trevizan, L.C., Santos Jr., D., Samad, R.E., Vieira Jr., N.D., Nomura, C.S., Nunes, L.C., Rufini, I.A. and Krug, F.J. (2008) Evaluation of laser induced breakdown spectroscopy for the determination of macronutrients in plant materials, *Spectrochimica Acta Part B*, vol. 63, no. 10, pp. 1151–1158.

# 23

# CARBON ACCOUNTING FOR SMALLHOLDER AGRICULTURAL SOIL CARBON PROJECTS

*Matthias Seebauer, Timm Tennigkeit, Neil Bird, and Giuliana Zanchi*

## Introduction

For most countries in sub-Saharan Africa, the impact of climate change on food security is a major livelihood and security concern. Considering the low level of emissions, countries are much more concerned with responding to rapid climate change and increasing climate resilience than with reducing or removing emissions caused by industrialized countries. However, because 89% (Smith *et al.* 2007) of agricultural mitigation potential is related to soil carbon sequestration, which has strong synergies with food security (FAO 2009a), and because of the growth in climate finance for providing mitigation services to the industrialized world, agricultural mitigation has generated interest in sub-Saharan Africa (see respective NAMA submissions under the Copenhagen Agreements (FAO 2009b)). Pursuing these mitigation opportunities will require quantifying climate benefits according to verified carbon standards (such as those in the Verified Carbon Standard (VCS) and Climate, Community and Biodiversity Alliance (CCBA)) or monitoring, reporting and verification (MRV) standards related to public finance mechanisms in the framework of Nationally Appropriate Mitigation Actions (NAMAs) to capture related benefits.

This chapter describes approaches to account for changes in soil carbon over time, and gives an example of the soil carbon accounting approach used in the Kenya Agricultural Carbon Project (KACP) (see also Lager and Nyberg). The authors are involved in developing this soil carbon sequestration project including methodology development, MRV system development and soil carbon modelling.

## Carbon accounting approaches

Developers of soil carbon (C) projects must be able to document and accurately quantify soil organic carbon (SOC) stock changes to meet carbon offset and measurement standards that are currently evolving at the national and international level. Soil carbon has specific attributes that make it challenging to account for fluctuations. The first challenge is the high spatial variability of soil carbon. Carbon stocks in the soils vary as a function of soil texture, parental material, landscape position, drainage, plant productivity and bulk density, which makes it difficult to quantify changes in SOC stocks over time (e.g., Cambardella *et al.* 1994; Robertson *et al.* 1997; VandenBygaart 2006). Second, the amount of carbon present in the soil (i.e., background) relative to the change rate is typically high, thus there is a low 'signal-to-noise' ratio (see Milori *et al.* for discussion of soil carbon measurement). Carbon accounting needs to be designed to handle these attributes to quantify fluctuations with sufficient accuracy while also minimizing costs to meet the needs of different carbon buyers. We discuss three carbon accounting approaches (see Conant *et al.* 2011 for a more detailed review of carbon accounting approaches):

*   direct measurement of soil carbon stock changes over time based on micro-plot sampling design;
*   activity- or practice-based estimation of SOC sequestration using regional or country-specific default values;
*   combined activity monitoring with soil carbon modelling to derive local applicable default values.

The approaches differ in terms of their cost and accuracy, which each vary with agro-ecological conditions, farming systems and data availability. In the Kenya Agricultural Carbon Project we used the third approach, as the aim of the project is to develop a methodology that is scientifically robust, but also practical for heterogeneous smallholder farming systems. Coupled with process-based models that describe the behaviour of carbon in the soil (see Olander *et al.*, this volume) and published research on how activities influence SOC to verify model results, this approach is capable of estimating SOC sequestration rates and associated uncertainty (Ogle *et al.* 2009).

In contrast, direct measurement of soil carbon is highly accurate, but due to variability and the small annual changes in SOC is expensive to use in any context other than research projects. As smallholders often manage diverse crops in patchy patterns, the number of samples required to capture variability in smallholder agricultural systems is large, hence soil sampling protocols are costly, since the costs of collecting soil samples, which often requires very sophisticated technology and extremely well trained staff, cannot be reduced.

Activity-based estimation also has drawbacks for project-level quantification. According to the IPCC Good Practice Guidelines, activity-based estimation can be used where published studies indicate how changes in management practices affect SOC. The change in soil C over time is estimated by multiplying the initial (or "reference") stock of C by the change in C observed due to a given activity ("stock change factor," found from published studies) and then adding any other C inputs resulting from land use or other organic matter amendments. The calculation is performed for each major soil type to account for differences in initial C stocks. A limitation of this approach is that it is applicable only for broader regions, for instance on a country level, since the studies of management impacts on SOC stocks are sparse enough that relying upon them to estimate sequestration rates for a specific project location would lead to substantial uncertainty (Conant *et al.* 2011).

## Soil carbon accounting through activity-based monitoring and modelled default values in the Kenya Agricultural Carbon Project

In this section we describe the activity baseline and monitoring survey (ABMS), which is part of the sustainable agricultural land management (SALM) methodology. The methodology is currently undergoing validation under the VCS (the methodology is available on the VCS website; see also chapter by Swickard and Nihart). The methodology enables agricultural activities in the baseline to be assessed and adoption of SALM practices to be monitored as a proxy for carbon stock changes estimated using the activity-based model (Figure 23.1).

See Lager, Nyberg and Woelke in this volume for introduction to the KACP.

We designed the ABMS to be based on a representative sample of farm households, interviewed periodically throughout the project's lifetime based on a structured questionnaire. Questions target current as well as future management practices to anticipate the adoption of improved SALM practices and monitor the adoption (ex-post) of practices in the project.

Sampling is driven by the nature of the prevailing site and farming systems and the accuracy requirements of the carbon accounting standard. In the Kenya project, we used stratified random sampling as the most efficient approach. For the first survey in 2010 we selected 200 farms based on a pre-survey of 30 farms to evaluate the variance and to determine the minimum sampling size. The ABMS will be repeated periodically over 20 years with annual intervals in the first 10 years and 3–5-year intervals onwards. This sampling frame reflects the proportionally higher adoption rates during the first half of the project. To account for an increasing number of adopters of SALM practices over time, the sample size increases

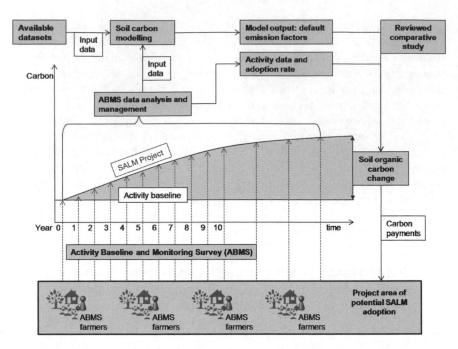

*Figure 23.1* Activity-based modelling in the Kenya Agricultural Carbon Project (KACP).

by an additional 5% per year within the first 10 years. To detect bias in the permanent sample farms, an additional 20 farms are selected and monitored every year.

Figure 23.2 presents the results of the first ABMS conducted in 2010 for each of the two project locations, Kisumu and Kitale. It illustrates that the ABMS is a strong tool for monitoring agricultural practices and input and output cycles, for instance of biomass, as well as for revealing livelihood issues directly linked with agricultural subsistence farming. A typical farm household in Kisumu held 0.7 ha of land, of which 0.5 ha is classified as agricultural land. This land area was not sufficient to feed the average of six household members, and more than 45% of farmers indicated that they have to deal with food insecurity for more than six months per annum. The main cultivated crop was maize (97% of the farmers in Kisumu), beans and various other crops, with high variability among farm households. The average annual yield of maize was estimated to be around 1.3 tonnes per hectare with significantly higher yields during the first growing season from April to September than the second season. Only 6% of farmers left residues on the field, reflecting the limited quantities of biomass available and opportunity cost of using biomass as a soil amendment. Maize stover, in particular provides the principle livestock feed in Kisumu (Kapkiyai *et al.* 1999). Using residues in stall-feeding systems

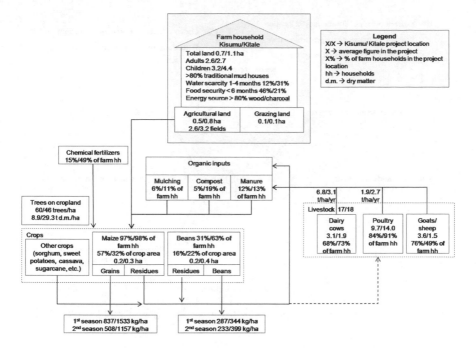

*Figure 23.2* Analysis of the activity baseline and monitoring survey (ABMS) results for an average farm household for Kisumu and Kitale sites in the Kenya Agricultural Carbon Project (KACP) (source: adapted from Kapkiyai *et al.* 1999).

linked to simple biogas digesters and subsequent placement of manure on fields would be one alternative to attain both residue-related food security and soil carbon sequestration benefits. Only 12% of the manure from the average of 17 livestock heads per ha was distributed back to the cropland and only 5% of farmers used composting. To facilitate production of more biomass, the project is providing seeds to establish fodder crops on fallow land and boundaries. (e.g., Calliandra and Sesbania).

Figure 23.3 illustrates the current and planned adoption of different SALM practices in the project. The figures are based on the 2009 ABMS survey and reflect the perceptions of the farmers in this year regarding current practices and the potential adoption of practices in the future. In the figure the potential is expressed as the difference between current and future practices. The highest potential is to apply composting and return the composted manure to the fields, but direct mulching and terracing of fields also show adoption potential. Interestingly, the adoption rates estimated from the farmers' interviews turned out to be realistic and conservative based on follow-up surveys in the subsequent cropping season. However, considering a project lifespan of 20 years, these first results have to be considered preliminary and only follow-up ABMS will draw a clear picture on the long-term adoption of SALM practices.

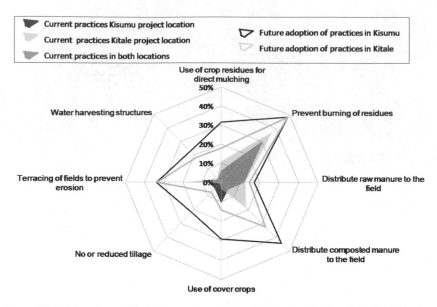

*Figure 23.3* Adoption of SALM practices in the Kenya Agricultural Carbon Project (KACP).

Referring to the flow of carbon accounting in Figure 23.1, ABMS results defined the activity data and adoption rate as well as provided local data to run the carbon model, which results in improved default emission factors for the respective sites. This will be further elaborated in the next section.

### The modelling of SOC changes

The Rothamsted C soil decomposition model (RothC) and the CENTURY ecosystem model are the most widely used models to predict soil carbon stock changes in Africa (Kaonga and Coleman 2008) (see discussion of process-based models in Olander *et al.*). Both the RothC and CENTURY models have been shown to be useful tools for predicting changes in soil C stocks under Kenyan conditions, provided a time series of measured soil C is available for evaluation of the model (Kamoni *et al.* 2007a). The RothC model proved to be more suitable for complex smallholder farming systems with limited data availability in the project site. The input data required to run the model are monthly rainfall, pan evaporation, air temperature, soil cover, as well as monthly input of plant residues and manure in tonnes of carbon per hectare. Further, the soil clay content is required which is very influential regarding the model outputs. The model generally partitions incoming organic material into decomposable plant material (DPM) and resistant plant material (RPM). This decomposability is set with standard DPM/RPM ratios, which resulted in a reasonable estimation of annual plant C inputs to the soil in sole maize and tree fallow stands (Kaonga and Coleman 2008).

Possible applications of the model with regard to SALM practices include modifying amounts of organic inputs (plant residues, manure); soil cover changes by, for instance, introducing cover crops; and slowing the decomposition rate of SOC through minimum tillage. In the Kenya project, the soil carbon change resulting from organic inputs, including plant residues and manure within maize fields, was modelled and quantified. Maize is the most evenly distributed crop spatially among the potentially adopting farmers; other crops are too heterogeneously scattered, leading to high statistical variance. In addition, the mitigation benefits from improved management on non-maize plots would be very low due to the limited area covered.

The change in soil C stock due to improved management practices is produced by the additional organic inputs and their decomposition over time, which is influenced by environmental factors. Therefore, only SOC changes and not total SOC stocks were modelled to produce default emission factors. The default emissions factors for six different maize management systems are summarized in Table1. Next, using the activity data from the ABMS, the actual adoption of these systems in the project will be assessed and the emission factors applied accordingly. This means it is necessary to identify, for instance, how many farmers leave a certain percentage of residues in the field and apply manure directly to the field (Maize system 5).

*Table 23.1* Range of soil carbon sequestration values in t $CO_2$ per ha and year for different maize systems in the project locations based on low-, medium- and high-yield responses

| Maize system | Residues used for mulching | Manure applied | Composted manure applied | Kisumu, both seasons t $CO_2$ ha/yr | Kitale, both seasons t $CO_2$ ha/yr |
|---|---|---|---|---|---|
| 1 | no | no | no | 0 | 0 |
| 2 | no | yes | no | 0.73–4.47 (1.91) | 0.37–1.91 (0.84) |
| 3 | no | no | yes | 1.8–10.82 (4.66) | 0.84–4.62 (2.05) |
| 4 | yes | no | no | 0.11–0.37 (0.22) | 0.22–0.7 (0.44) |
| 5 | yes | yes | no | 0.84–4.84 (2.13) | 0.55–2.6 (1.28) |
| 6 | yes | no | yes | 1.91–11.18 (4.84) | 1.06–5.32 (2.49) |

*Note*
The median value is given in brackets; mean environmental parameters are assumed (mean precipitation, mean clay content); (N = 200 farms).

The organic residue input data for the model were provided by the ABMS indirectly since the average fresh yield of maize can be converted to amount of residues produced in t C/ha on the basis of IPCC conversion equations (Table 11.2, IPCC 2006). Thus, the main driver for the amount of residues is the yield response of maize in the project. Likewise, the raw manure and composted manure inputs in tC/ha is calculated by applying IPCC factors to the average amount of farm animals per hectare (Table 10.4, IPCC 2006).

The remaining model inputs are drawn from available databases. Data on soil types and its clay content are derived from regional soil databases, e.g., Harmonized World Soil Database (FAO/IIASA/ISRIC/ISSCAS/JRC 2009) and climate parameters are provided from local weather stations.

To account for different conditions within the project, such as significantly differing clay contents, and more importantly, different levels of organic input (high inputs versus low inputs), default soil carbon sequestration values for low, medium and high sequestration conditions were modelled.

## Managing uncertainty

There are three broad sources of uncertainty associated with the modelling, namely related to land use and management activities, environmental data and SOC emission factors. The uncertainty in the activity-based crop monitoring contributes to uncertainty in the soil carbon model-based estimate in a linear fashion. Starting at the field level, the sampling procedure of the ABMS needs to be analyzed with regard to random errors associated with the sampling design and systematic errors related to the interview situation. The sample size was determined on a required precision level of 15% of the true value of the mean at the 95% confidence interval. The minimum number of ABMS farmers was guided by the need to ensure that the data within each stratum is normally distributed considering the Central Limit Theorem that for samples of a sufficiently large size, the distribution of sample means is approximately normal.

Questionnaire-based surveys include potential errors due to bias from respondents, the questionnaire design, the interviewer, and the environment or specific circumstances under which the interview is taking place. To reduce uncertainty or error, the ABMS needs to be conducted by trained extension or community workers able to conduct plausibility checks, by triangulating the answers received with their own observations. In the Kenya project, a standard ABMS procedure was developed with best practice guidance based on an extensive pre-test among 30 farmers and the identification of potential error sources to minimize such risks. As mentioned earlier, 10% of the sample size will be tested annually to minimize bias among the ABMS farmers.

After in-depth plausibility checks of the recorded ABMS data, the mean values, standard deviation and standard errors of residue and manure production were calculated. Further, the lower and upper bounds of the confidence interval were calculated for each model input parameter.

During the next step, soil model response was calculated with the minimum and maximum values of the input parameters. The range of model responses demonstrates the uncertainty of the soil modelling.

In addition, the project has defined the following rules, which are reflected in the corresponding SALM Methodology: Below 15% uncertainty of the soil model, no deduction of the estimated total project offsets is required, whereas between 15% and 30% a deduction needs to be calculated and above 30% uncertainty the sampling size of the input parameters needs to be increased accordingly.

Finally, to validate the performance of the RothC model in the specific project settings in Western Kenya, the project and the methodology required a comparative validation study to demonstrate that use of the RothC model is appropriate for the IPCC climatic regions in which the project is situated. A study conducted by Kamoni *et al.* (2007b) evaluated the ability of RothC and Century models to estimate changes in SOC resulting from varying land use or management practices for the climate and soil conditions in Kenya. At a long-term experimental site similar to the climatic conditions of the Kenya project locations (semi-humid) they found that the model showed a fair-to-good fit compared to measured data (Kamoni *et al.* 2007b).

## Conclusions

Designing a monitoring system to measure and quantify soil carbon changes that meets the level of accuracy required in a carbon offset standard, which is often developed for and by experts in data-rich countries, is a challenge for less developed countries. Kenya, for instance, can be regarded as one of the few relatively data rich countries in sub-Saharan Africa. With regard to soil carbon research, however, there are only two long-term research monitoring trials which are not representative for all agro-ecological conditions in the country. Consequently, this would mean that substantial inputs—both financially and technically—are needed to develop a soil carbon project under these circumstances. The approach described in this chapter proved to be efficient in terms of data collection since the ABMS can be easily integrated into existing project surveys (e.g., social baseline survey) with relatively little efforts. Nevertheless, the crop diversity and variability in smallholder farming systems remains a challenge since sufficient modelling data is only available for mixed maize farming systems. However, carbon accounting and monitoring must adhere to the principles of relevance, completeness, consistency, transparency, accuracy and conservativeness (IPCC 2006) to ensure true and fair accounting. Conservativeness is important for projects where accuracy may not be fully attained and may serve as a moderator to accuracy to maintain the credibility of project greenhouse gas (GHG) quantification (VCS 2010). Monitoring systems in smallholder land-use carbon projects should be designed to achieve multiple benefits apart from carbon accounting. Above all, they need to be transparent for the farmers who actively reduce GHG emissions in the project area to ensure ownership of sequestered carbon and to create fair distribution of revenues. Further, carbon monitoring should support project implementation, extension and impact monitoring. Monitoring can be used to identify specific training needs and priority interventions for extension, particularly during early stages of the project. General livelihood and socio-economic impact monitoring is also important considering that for the farmer, increasing climate resilience and crop production are the main project benefits apart from mitigation benefits. Activity monitoring entails engaging the farmer, provides crucial information to improve extension and self-learning structures and creates an environment of commitment to the project among farmers.

More research is required to improve our current understanding and to evaluate existing carbon monitoring approaches. Well-designed and controlled long-term research experiments, designed and executed by well-trained researchers, are necessary to achieve this.

# References

Cambardella, C.A., Moorman, T.B., Novak, J.M., Parkin, T.B., Turco, R.F. and Konopka, A.E. (1994) Field-scale variability of soil properties in central Iowa soils, *Soil Science Society of America Journal*, vol. 58, pp. 1501–1511.

Conant, R.T., Ogle, S.M., Paul, E.A. and Paustian, K. (2011) Measuring and monitoring soil organic carbon stocks in agricultural lands for climate mitigation, *Frontiers in Ecology and the Environment*, vol. 9, no. 3, pp. 169–173.

Food and Agriculture Organization of the United Nations (2009a) 'Food security and agricultural mitigation in developing countries: Options for capturing synergies'. FAO, Rome, Italy.

Food and Agriculture Organization of the United Nations (2009b) 'Agriculture, food security and climate change in post-Copenhagen process'. FAO Information Note.

FAO/IIASA/ISRIC/ISSCAS/JRC (2009) *Harmonized World Soil Database (version 1.1)* [Online]. Available at: www.iiasa.ac.at/Research/LUC/External-World-soil-database/HTML/index.html (Accessed: June 2009). FAO, Rome, Italy and IIASA, Laxenburg, Austria.

Intergovernmental Panel on Climate Change (2006) 'Volume 4: Agriculture, forestry and other land uses', in Eggleston, H.S., Buendia, L., Miwa, K., Ngara, T. and Tanabe, K. (eds) *2006 IPCC Guidelines for National Greenhouse Gas Inventories*. Institute for Global Environmental Strategies, Japan.

Kamoni, P.T., Gicheru, P.T., Wokabi, S.M., Easter, M., Milne, E., Coleman, K., Falloon, P. and Paustian, K. (2007a) Predicted soil organic carbon stocks and changes in Kenya between 1990 and 2030, *Agriculture, Ecosystems & Environment*, vol. 122, pp. 105–113.

Kamoni, P.T., Gicheru, P.T., Wokabi, S.M., Easter, M., Milne, E., Coleman, K.,

Falloon, P., Paustian, K., Killian, K., Kihanda, F.M. (2007b) Evaluation of two soil carbon models using two Kenyan long term experimental datasets, *Agriculture, Ecosystems & Environment*, vol. 122, pp. 95–104.

Kaonga, M.L. and Coleman, K. (2008) Modelling soil organic carbon turnover in improved fallows in eastern Zambia using the RothC-26.3 model, *Forest Ecology and Management*, vol. 256, no. 5, pp. 1160–1166.

Kapkiyai J.J., Karanja, N.K., Qureshi, J.N., Smithson, P.C. and Woomer, P.L. (1999) Soil organic matter and nutrient dynamics in a Kenyan nitisol under long-term fertilizer and organic input management, *Soil Biology and Biochemistry*, vol. 31, no. 13, pp. 1773–1782.

Ogle, S.M., Breidt, F.J., Easter, M., Williams, S., Killian, K. and Paustian, K. (2009) Scale and uncertainty in modelled SOC stock changes for U.S. croplands using a process-based model, *Global Change Biology*, vol. 16, no. 2, pp. 810–822.

Robertson, G.P., Klingensmith, K.M., Klug, M.J., Paul, E.A., Crum, J.C. and Ellis, B.G. (1997) Soil resources, microbial activity, and primary production across an agricultural ecosystem, *Ecological Applications*, vol. 7, pp. 158–170.

Smith, P., Martino, D., Cai, Z., Gwary, D., Janzen, H.H., Kumar, P., McCarl, B., Ogle, S., O'Mara, F., Rice, C., Scholes, B. and Sirotenko, O. (2007) 'Agriculture', in Metz, B., Davidson, O.R., Bosch, P.R., Dave, R. and Meyer, L.A. (eds) *Climate Change 2007: Mitigation. Contribution of Working Group III to the Fourth Assessment Report of the Intergovernmental Panel on Climate Change*, Cambridge University Press, Cambridge, United Kingdom and New York, NY.

VandenBygaart A.J. (2006) Monitoring soil organic carbon stock changes in agricultural landscapes: Issues and a proposed approach, *Canadian Journal of Soil Science*, vol. 86, pp. 451–463.

Voluntary Carbon Standard (2010) *Voluntary carbon standard VCS 2011* [Online]. Available at: www.v-c-s.org (Accessed: 20 November 2010). Public Consultation Document, VCS Association.

# 24

# ACCOUNTING FOR QUALITY IN AGRICULTURAL CARBON CREDITS

## A Verified Carbon Standard for agricultural land management projects

*Naomi Swickard and Alison Nihart*

As interest in voluntary carbon markets has grown, stakeholders have identified the need to standardize and certify carbon credits to ensure quality across projects. The Verified Carbon Standard (VCS, formerly the Voluntary Carbon Standard), was developed to satisfy this need. VCS offers a quality assurance system for accounting for greenhouse gas (GHG) emission reductions in the voluntary carbon market. In 2009, VCS credits accounted for more than one-third of all transacted GHG credits verified to a third-party standard globally (Hamilton *et al.* 2010).

The VCS Program was established in 2005 through collaboration between the Climate Group, the International Emissions Trading Association and the World Economic Forum. The World Business Council for Sustainable Development joined the initiative as a founding partner in 2007. Today the VCS Program is supported by a Board of Directors, a small staff, and a series of advisory groups, through which the program continues to advance its work.

This chapter describes the process used by the VCS to verify the quality of GHG emissions reductions projects, and provides details on Agricultural Land Management (ALM) requirements as part of the VCS Agriculture, Forestry and Other Land Use (AFOLU) Program.

## The VCS Program

The VCS Program has four main components through which it ensures environmental integrity: the VCS Standard, a methodology approval process, independent auditing, and a registration process for issuing and listing GHG credits on the VCS registry system.

### The VCS Standard

The VCS Standard is at the core of the VCS Program. First released in 2006, the Standard is a living document that has been developed in consultation with over 1,000 stakeholders. It was formally launched in 2007, with a third major update, VCS Version 3, released in March 2011. The Standard is framed by a set of eight criteria that define the attributes of a quality assured Verified Carbon Unit (VCU): All VCUs produced under the Standard must represent GHG emission reductions that are real, measurable, additional, and permanent, as well as independently verified, conservatively estimated, uniquely numbered, and transparently listed.

To be *real*, GHG emission reductions must be accounted for after they occur, ex-post, and never before, ex-ante. To be *measurable*, reductions must be quantifiable against a credible emissions baseline using robust and established scientific practices. To be *additional*, reductions must go beyond 'business as usual' and reduce emissions that would have occurred in the absence of the project. And reductions must be *permanent*—this means land-use projects, which carry the risk that stored carbon may eventually be released to the atmosphere, must ensure GHG removals last over time.

The VCS Standard and the VCS Program Guide, available on the VCS website, lay out clear rules for meeting these criteria.

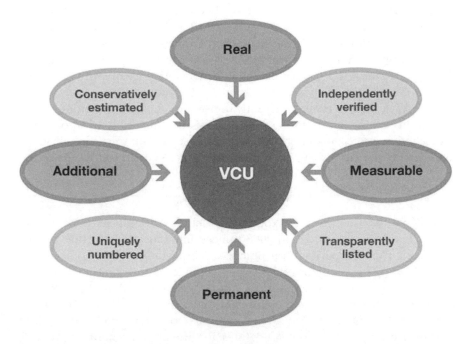

*Figure 24.1* Verified Carbon Unit (VCU) criteria.

## Methodologies

In order to support project development, the VCS Program provides a methodological approval process to ensure scientifically sound methodologies are available for project development. The process is flexible in that it allows the developer to choose a methodology already approved under the VSC Program or another approved program, such as the Climate Action Reserve or the UN Clean Development Mechanism (CDM). Alternatively, project proponents can use the VCS methodology approval process to develop new methodologies in cases where specific opportunities are afforded by new approaches or technologies to curb emissions.

## Auditing

Under the VCS Program, all project plans and GHG emission reductions and removals must be audited by a qualified, independent validation/verification body (VVB). Qualified VVBs must be accredited to audit to VCS criteria by a recognized organization such as the American National Standards Institute or by an approved program such as the CDM.

## Registry system

Once GHG emission reductions have been verified, project developers may request the issuance of Voluntary Carbon Units (VCUs). To ensure every VCU is unique and can be tracked from birth to retirement, each VCU is assigned a unique serial number so it can be tracked across its full life cycle in the VCU project database, which serves as a central clearinghouse of information on all projects and VCU issuance.

The VCS registry system currently consists of three independent registries. This system offers project proponents the flexibility of choosing a preferred registry, encouraging competition while still ensuring all VCUs and project documents are centrally listed in the VCS project database.

## The VCS AFOLU Program

The VCS expanded the VCS Program to include Agriculture, Forestry and Other Land Use (AFOLU) in 2008 to accommodate projects that reduce emissions through agricultural practices, afforestation, reforestation and revegetation, forest management and avoided deforestation. The VCS AFOLU Program was established to provide a credible and robust approach to account for and credit voluntary offset projects in the agriculture and forestry sectors, and was the first internationally recognized carbon standard to set out guidelines for the development of AFOLU projects.

Due to the potential for future carbon loss from sequestration projects, the VCS AFOLU Program makes special efforts to ensure the quality of emissions reductions from these sectors. As with all VCS projects, credits must meet the eight criteria of the Standard, and the AFOLU Program takes additional measures to address permanence and leakage.

A larger number of VCS AFOLU projects are under development, and a major 2009 EcoSecurities survey[1] of the leading carbon market participants, the VCS was found to be the most highly desired standard for forest projects. To date, the majority of registered VCS projects are energy efficiency and renewable energy projects, but the VCS project database includes a small number (4 out of 541) of AFOLU projects, with many more under development and in the process of validation and verification. Although none of the registered projects are agricultural, several methodologies are currently undergoing the approval process so that they can be used on projects in the pipeline.

As of April 2011 there are 8 approved Reducing Emissions from Deforestation and forest Degradation (REDD) and Improved Forest Management (IFM) methodologies, with an additional 10 AFOLU methodologies under development, including 4 agricultural methodologies.

### General requirements

#### Permanence

Carbon sequestered through land-based projects is subject to potential loss over time, whether intentional (e.g., timber harvesting or agricultural expansion) or unintentional (e.g., natural fires or pests). To address the risk of such reversals, the VCS has established a robust self-insurance mechanism, based on a buffer reserve, which ensures that all GHG credits issued are backed by a diverse pool of non-tradable buffer credits.

Using the AFOLU Non-Permanence Risk Tool, the project proponent assesses the specific project risks, and has this risk assessment audited as a part of the validation and verification of the project. Based on this assessment a certain percentage of the project's GHG credits are deposited into a pooled buffer account shared by all VCS AFOLU projects. These buffer credits are non-tradable and are maintained in the pooled account to cover the risk of unforeseen losses in carbon stocks no matter where they occur within the AFOLU project portfolio, and some may be returned to the project over time as projects effectively mitigate risk.

This allows the VCS Program to issue permanent VCUs to AFOLU projects, making them fungible with VCUs from all other project types, which enhances liquidity and enables the market to function effectively.

## Leakage

Projects that effectively reduce GHG emissions within their boundaries can result in increased emissions elsewhere. For example, when a piece of threatened forestland is protected, the drivers of deforestation (e.g., shifting agriculturalists or timber harvesters) can oftentimes simply move to another area and continue deforesting. Such offsite impacts are called 'leakage' and can dramatically reduce the net climate benefits generated by individual carbon projects.

The VCS requires that all projects define, mitigate, monitor and account for potential leakage. VCUs are issued based on each project's net GHG emissions reductions or removals, conservatively taking into account any leakage that may occur as a result of the project.

## Measuring and monitoring

AFOLU projects must identify significant carbon pools (e.g., above- and below-ground biomass) and their rates of change; collect and analyze data; determine the most likely and conservative baseline scenario; estimate the change in carbon stocks and GHG emissions due to the project activities, taking into account harvesting, planting or other activities; and measure and deduct any leakage attributable to the project. All of this must be done using conservative assumptions, while maintaining accuracy and reducing uncertainty to the extent possible. Over the life of the project, the actual changes in carbon stocks and emissions from project activities must be monitored using reliable and robust methods.

## Agricultural Land Management (ALM) requirements

Eligible ALM activities are those that reduce net GHG emissions on croplands and grasslands by increasing carbon stocks in soils and woody biomass and/or decreasing carbon dioxide ($CO_2$), nitrous oxide ($N_2O$) and/or methane ($CH_4$) emissions from soils. Eligible ALM activities include:[2]

### Improved Cropland Management (ICM)

- Soil carbon stocks can be increased by practices that increase residue inputs to soils and/or reduce soil carbon mineralization rates. Such practices include, but are not limited to, the adoption of no-till, elimination of bare fallows, use of cover crops, creation of field buffers (e.g., windbreaks or riparian buffers), use of improved vegetated fallows, conversion from annual to perennial crops and introduction of agroforestry practices on cropland.
- Soil $N_2O$ emissions can be reduced by improving nitrogen fertilizer management practices to reduce the amount of nitrogen added as

fertilizer or manure to targeted crops. Examples of practices that improve efficiency while reducing total nitrogen additions include improved application timing (e.g., split application), improved formulations (e.g., slow-release fertilizers or nitrification inhibitors) and improved placement of nitrogen.

- Soil $CH_4$ emissions can be reduced through practices such as improved water management in flooded croplands (in particular flooded rice cultivation), through improved management of crop residues and organic amendments and through the use of rice cultivars with lower potential for $CH_4$ production and transport.

### Improved Grassland Management (IGM)

- Soil carbon stocks can be increased by practices that increase below-ground inputs or decrease the rate of decomposition. Such practices include increasing forage productivity (e.g., through improved fertility and water management), introducing species with deeper roots and/or more root growth and reducing degradation from overgrazing.
- Soil $N_2O$ emissions can be reduced by improving nitrogen fertilizer management practices on grasslands.
- $N_2O$ and $CH_4$ emissions associated with burning can be reduced by reducing the frequency and/or intensity of fire.
- $N_2O$ and $CH_4$ emissions associated with grazing animals can be reduced through practices such as improving livestock genetics, improving the feed quality (e.g., by introducing new forage species or by feed supplementation) and/or by reducing stocking rates.

### Cropland and Grassland Land-use Conversions (CGLC)

- The conversion of cropland to perennial grasses can increase soil carbon by increasing below-ground carbon inputs and eliminating and/or reducing soil disturbance. Decreases in nitrogen fertilizer and manure applications resulting from a conversion to grassland may also reduce $N_2O$ emissions.
- Grassland conversions to cropland production (e.g., introducing orchard crops or agroforestry practices on degraded pastures) may increase soil and biomass carbon stocks.

### ALM methodology development

ALM projects have been eligible under the VCS since the release of the AFOLU requirements in 2008. However, agricultural GHG mitigation as a sector has followed behind the advancements in the forestry sector somewhat. In order for projects to move forward under the VCS, they must use an approved methodology. Generally, methodologies are developed with a project in mind. For example, the World Bank's Sustainable Agricultural Land Management (SALM) methodology is developed for their project in Kenya (see Baroudy and Hooda, this volume). However, once developed, methodologies are applicable to any relevant project. Currently the following methodologies are undergoing the methodology approval process and when approved, will be available for project development:[3]

- *VCS Methodology for Agricultural Land Management: Improved Grassland Management*, developed by Greencollar Climate Solutions—for estimating and monitoring the GHG emission reductions resulting from improved grassland management projects. The methodology identifies and accounts carbon stock changes and GHG emissions in the soil carbon, above-ground woody biomass and below-ground biomass pools, fossil fuel combustion, enteric emissions, biomass burning, and nitrogen fertilizer application.
- *Quantifying $N_2O$ Emissions Reductions in US Agricultural Crops through N Fertilizer Rate Reduction*, developed by Michigan State University—quantifies emissions reductions of $N_2O$ from US agriculture, brought about by reductions in the rate of nitrogen fertilization to annual cropping systems.
- *ALM Adoption of Sustainable Grassland Management through Adjustment of Fire and Grazing*, developed by Soils for the Future and Jadora International—applicable to projects that introduce sustainable adjustment of the density of grazing animals and the frequency of prescribed fires into an uncultivated grassland landscape.
- *Adoption of Sustainable Agricultural Land Management (SALM)*, developed by the World Bank—aims to estimate and monitor GHG emissions of project activities that reduce emissions in agriculture by applying sustainable land management practices. Carbon stock enhancement in agricultural areas in the above-ground, below-ground and soil carbon pool are achieved. The methodology is based on the project activity "Western Kenya Smallholder Agriculture Carbon Finance Project" in Kenya (see Baroudy and Hooda, this volume).

## The future of the VCS

### Lessons learned

Since the release of the AFOLU requirements in November 2008, there have been a number of key lessons learned. Many of these lessons have been factored into the updates that are integrated into the new version of the VCS—Version 3, as outlined below. The development of methodologies and projects and the application of the VCS AFOLU requirements to these on-the-ground developments have contributed a number of lessons relevant to the agriculture sector, in particular.

*Where accuracy is difficult to obtain due to the high cost of monitoring or physical limitations, a conservative approach may serve as a moderator to some uncertainty.* Agricultural projects face a number of barriers and challenges as other authors in this volume have demonstrated. A key factor for any project is the need for methods for establishing baseline stocks and emissions, and measuring and monitoring changes in stocks and emissions due to project activities over time. In accordance with recognized principles for GHG credit quality, the VCS requires that all GHG emission reductions and removals be measurable and accurate. While this implies that GHG emissions reductions and removals be directly quantifiable using recognized measurement tools, in some circumstances, the direct measurement of GHG emission reductions and removals is not feasible either due to the nature of the project activity or due to the complexity and cost involved in field-based measurements. Particularly where projects include the soil carbon pool, establishing sufficient soil sampling plots to reach an acceptable level of certainty can be cost-prohibitive for many projects. New methods of sampling offer promise for changing the economics of sampling, but in the meantime many projects will rely on a variety of analytic models to estimate changes in soil carbon.

In order to use such models for carbon accounting there is a need to ensure models are applicable to both the project activities being implemented and the agro-ecological zones in which they are located, requiring field-based research that particularly for developing countries is often lacking. For the use of such models to be seen as credible for the estimation of emissions reductions and issuance of credits, it is necessary to be able to estimate the level of uncertainty in carbon stock change estimates.

The VCS is based on a set of principles such as accuracy (reduce bias and uncertainties as far as is practical) and conservativeness (use conservative assumptions, values and procedures to ensure that net GHG emission reductions or removals are not overestimated). The VCS Association has determined that where direct measurements are not feasible, accuracy should be pursued as far as possible, but the hypothetical nature of baselines, the high cost of monitoring of some types of GHG emissions and removals, and other limitations make accuracy difficult to attain in many cases. In these cases, conservativeness may serve as a moderator to accuracy in order to maintain the credibility of project GHG quantification.

Therefore where projects are unable to attain the required level of accuracy, they must take a discount that results in a conservative estimation of GHG emissions reductions or removals. This has implications for the development of projects and methodologies, and means that in cases where the required level of accuracy may be impossible to maintain, this doesn't necessarily prohibit project development, where uncertainty can be estimated and a conservative approach taken.

*There is a need for further research, pilot projects, and data collection to facilitate the development of agricultural carbon projects, worldwide.* As noted above, where model estimates are uncertain, a discount must be taken to ensure conservative estimation of emissions reductions. Over time, as models improve, uncertainty should be reduced and credited emissions reductions may increase. In order to achieve this, there is a need for methodologies to include a means to integrate ongoing learning from pilot projects, model improvements, and increased data.

While there are a number of efforts under way in the US and Canada to improve available data, such as the USDA National Soil Carbon Assessment, efforts are needed to ensure data collection will cover the data needed to support project baselines and monitoring. Data availability in developing countries, where there is very large potential for agricultural GHG emissions reductions and removals is particularly lacking. Such projects could bring significant co-benefits as well, and should be supported through pilot projects and further data collection efforts.

*Further effort is needed to bring agricultural GHG mitigation to scale.* This is needed on a number of fronts;

1   Agriculture still lags behind forestry in international negotiations. While REDD has gained traction and Cancún saw major progress for including REDD, there is still an uphill battle to bring in the agriculture sector. Successful pilot projects may go a long way to creating the momentum needed.

2   Performance approaches (based on intensity per unit of output) have high potential, especially for certain types of agricultural projects, but such approaches also have very high data needs (see also chapter by Baker and Murray on emissions-intensity approaches, and Seebauer *et al.* for example of activity-based monitoring). While performance approaches have significant potential for reducing transaction costs, to date very few performance-based methodologies have been approved for GHG projects, and existing performance approaches in forestry (under Climate Action Reserve (CAR) for example) use a hybrid of performance and project-specific baseline and additionality methods. There is not yet a set of agreed-upon best practice guidelines for developing and setting appropriate baseline and crediting benchmarks in methodologies. The VCS has convened a Steering Committee (see below) to help address this gap, but even when complete, challenges remain to ensure sufficient data availability to set benchmarks and develop methodologies.

### VCS Version 3

VCS Version 3 was released in March 2011. The new version includes a number of updates:[4]

- Requirements for grouped projects: The 2007.1 version of the VCS allowed for grouped projects (often called aggregation), but did not set out specific requirements for such projects. Grouped projects may be of particular interest to small land holders who may not be able to reach a feasible scale to overcome the costs associated with GHG project development, in particular validation and verification costs.
- A new version of the AFOLU Non-Permanence Risk Tool. The tool now includes more detail on how risk factors are to be assessed, types of evidence required, and determines an exact withholding percentage for projects.

### *Directions for the future*

The VCS continues to expand the scope of the Program and is currently working on a number of developments, including:

- The VCS AFOLU program is being expanded to include a new project type covering the Avoided Conversion of Grasslands and Shrublands (ACoGS). The ACoGS requirements will be similar to those of REDD, but aimed at non-forest areas, in particular grasslands and shrublands. These requirements will cover several project activities, including avoiding the conversion of rangelands to croplands. The new requirements are expected in late 2011.
- The VCS convened the Steering Committee on Standardized Approaches to Baselines and Additionality with the aim to develop requirements to guide methodology developers and auditors working on standardized approaches for determining baselines and additionality. The steering committee will work to develop requirements over two years.
- A new Advisory Committee was convened in February 2011 to provide strategic input and guidance into the conceptual development of the VCS Jurisdictional and Nested REDD Initiative.[5] The initiative will produce criteria and procedures for the accounting and crediting of GHG emission reductions from REDD projects, policies and programs within entire jurisdictions. The initiative will also produce criteria for setting and using regional baselines for REDD projects, helping to ensure consistency and environmental integrity while also streamlining processes and reducing transaction costs for individual projects. The new criteria and procedures are expected by early 2012.

### Notes

1  See www.ecosecurities.com/Registered/ECOForestrySurvey2009.pdf.
2  Full rules and requirements are available at: www.v-c-s.org/program-documents/find-program-document.

3 For a full list of methodologies undergoing the approval process, see: www.v-c-s.org/methodologies/find.
4 For a full synopsis of updates, see: www.v-c-s.org/sites/v-c-s.org/files/VCS%20Synopsis%20of%20VCS%20Version%203%20Release%2C%208%20MAR%202011.pdf.
5 See www.v-c-s.org/node/296.

# Reference

Hamilton, K., Peters-Stanley, M. and Marcello, T. *Building Bridges: State of the Voluntary Carbon Markets 2010* [Online]. Available at: http://moderncms.ecosystem-marketplace.com/repository/moderncms_documents/vcarbon_2010.2.pdf. Ecosystem Marketplace

# Part IV

# CASE STUDIES
Sectors

# 25

# CONSERVATION AGRICULTURE AS A MEANS TO MITIGATE AND ADAPT TO CLIMATE CHANGE

## A case study from Mexico

*Nele Verhulst, Bram Govaerts, Ken D. Sayre, Kai Sonder,*
*Ricardo Romero-Perezgrovas, Monica Mezzalama, and*
*Luc Dendooven*

### Introduction

Conservation agriculture (CA) has been proposed as an adapted set of management principles that assures a more sustainable agricultural production. It combines the following basic principles: (1) reduction in tillage, (2) retention of adequate levels of crop residues and soil surface cover, (3) use of crop rotations. These CA principles are applicable to a wide range of crop production systems. However, the application of CA will be different in different situations. Specific and compatible management components (pest and weed control tactics, nutrient management strategies, rotation crops, appropriately-scaled implements, etc.) will need to be identified through adaptive research with active farmer involvement.

In this chapter, climate change predictions for Mexico will be discussed. Then the potential of CA as a means to mitigate and adapt to climate change will be examined for two contrasting agro-ecological environments, using research results of long-term trials. Finally, the economic potential of CA for climate change mitigation and adaption will be examined and an extension strategy will be outlined.

### Materials and methods

Long-term trials that compare tillage and residue management strategies were set up in contrasting conditions, i.e., rainfed conditions in the semi-arid highlands of Mexico and irrigated conditions in arid north-western

Mexico. The rainfed experiment was initiated in 1991 in El Batán in the semi-arid, subtropical highlands of Central Mexico (2,240 meters above sea level (masl); 19.318°N, 98.508°W). The mean maximum and minimum temperatures are 24 and 6 °C, respectively (1991–2009) and the average annual rainfall is 625 mm/yr, with approximately 545 mm falling between May and October. Short, intense rain showers followed by dry spells typify the summer rainy season and the total yearly potential evapotranspiration of 1,550 mm exceeds annual rainfall. The soil is a Haplic Phaeozem (Clayic). The irrigated experiment (initiated in 1993) is conducted near Ciudad Obregón, Sonora, Mexico (38 masl; 27.33°N, 109.09°W). The site has an arid climate with a mean annual temperature of 24.7 °C and an average rainfall of 384 mm (1971–2000). Rainfall is summer dominant and only 23% of total rainfall occurs during the wheat-growing season from November to May (Servicio Meteorológico Nacional). The soil is a Hyposodic Vertisol (Calcaric, Chromic).

## Results and discussion

### Climate change predictions for Mexico

Analyses of climate data for Central America of the past 30 to 50 years show patterns of changes consistent with a general warming (Magrin *et al.* 2007). The occurrence of warmer maximum and minimum temperatures has increased, while incidences of extremely cold temperature events have decreased (Aguilar *et al.* 2005). There is also a positive tendency for more intense rainfall events and consecutive dry days (Aguilar *et al.* 2005; Groisman *et al.* 2005). Groisman *et al.* (2005) found a substantial decrease in precipitation over the Central Plateau of Mexico over the last 30 years. In the same period, the frequency of rain events above 75 mm increased substantially by 110%.

Climate models predict that Mexico will continue to warm during this century. Averaging the predictions of 19 models with emission scenario A2 (commonly known as "business as usual" (BAU)), the predicted increase in annual average temperature by the 2050s (the period 2040 to 2069) ranges from 1.6–2.0 °C on the Yucatán peninsula to 3.1–3.5 °C in northern Mexico (Figure 25.1a). Temperatures will increase less in the coastal areas than in inland regions. Annual minimum and maximum temperatures show the same trend with higher increases in maximum temperature (between 1.6 and 4.0 °C) than in minimum temperature (between 1.6 and 2.5 °C) (data not shown). Annual precipitation is predicted to decrease in most of Mexico, with the largest decrease (10 to 25%) on the Yucatán peninsula and in north-eastern Mexico (Figure 25.1b). Only the Baja California peninsula (currently the driest region) is predicted to have substantial increases in annual rainfall (Figure 25.1b). The predicted changes in climate could severely affect agricultural production in Mexico. According

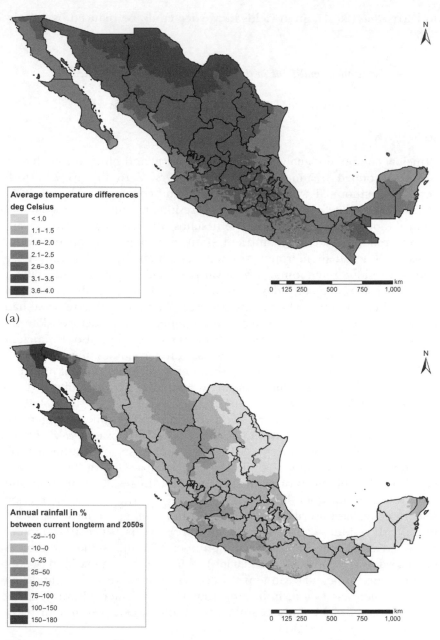

(a)

(b)

*Figure 25.1* Predicted annual (a) average temperature difference (in °C) and (b) rainfall difference (in %) between current long-term average (1951–2002) and the 2050s (average of 19 models for emission scenario A2).

to Parry *et al.* (2004), grain yields in Mexico could be reduced by 30% by 2080.

### Rainfed conditions in the semi-arid highlands of Mexico

#### Conservation agriculture as an adaptive measure

SOIL QUALITY

Principal component analysis (PCA) of chemical and physical soil characteristics grouped treatments in three clusters: Zero tillage (ZT) with residue retention, ZT with residue removal and conventional tillage (CT) (Govaerts *et al.* 2006a). Plots with ZT and residue retention were separated from the other plots by high nutrient status, whereas plots with ZT with residue removal had a low nutrient status and high Mn concentration. Plots with CT had an intermediate chemical soil quality and the difference between residue retention and removal was small (Govaerts *et al.* 2006a). The physical soil quality data showed that ZT with residue retention resulted in a good topsoil structure, whereas ZT with residue removal had a less favorable topsoil structure and high penetration resistance. Conventional tillage had the lowest penetration resistance, but also a deficient topsoil structure (Govaerts *et al.* 2006a). Additional studies showed that ZT with residue retention also had higher water infiltration and soil moisture content than ZT with residue removal and practices involving tillage (Govaerts *et al.* 2007a, 2009; Verhulst *et al.* 2011a).

The effect of tillage practice, crop rotation and residue management on biological soil characteristics was also evaluated (Govaerts *et al.* 2006b, 2007b, 2008; Chocobar-Guerra, 2010). Earthworms strongly influence soil structure and aggregation (Lavelle 1997). Chocobar-Guerra (2010) found equal number of earthworms in CT and ZT with residue retention and crop rotation, but when looking at the more determinant factor for soil structure, i.e., earthworm biomass, significantly higher biomass was found in ZT with residue retention. In 12 monitored years (since 1997), maize roots had a higher incidence of rotting under ZT with residue retention as compared to CT (Govaerts *et al.* 2006b). Wheat showed a low level of root rot incidence under all treatments. No correlation between yields and root rot incidence was found, indicating that CA had other advantages that determined the yield of wheat and maize in the target environment (Govaerts *et al.* 2005; Table 25.1).

Changes in tillage, residue management, and rotation practices also induced shifts in the number and composition of soil fauna, including pests and beneficial organisms. After more than a decade, the effects of the different management practices on groups of beneficial soil microflora (total bacteria, fluorescent *Pseudomonas*, actinomycetes, fungi and *Fusarium* spp.) were studied with different techniques (Govaerts *et al.*

2007b, 2008). Residue retention under ZT and CT induced greater microbial diversity, especially higher total bacteria, fluorescent *Pseudomonas* and actinomycetes both for maize and wheat cropping systems. The continuous uniform supply of carbon from crop residues serves as an energy source for micro-organisms (Govaerts *et al.* 2008). Retaining crop residues also increases microbial abundance. Zero tillage, as such, is not responsible for increased microflora, but rather the combination of ZT with residue retention. The favorable effects of these two components are due to increased soil aeration, cooler and wetter conditions, less temperature and moisture fluctuations and increased carbon content in the topsoil (Govaerts *et al.* 2008;). Functional diversity and redundancy are signs of increased soil health. It allows an ecosystem to remain stable when facing changes in environmental conditions (Wang and McSorley 2005).

CROP YIELD AND SYSTEM RESILIENCE

Small-scale maize and wheat farmers may expect yield improvements through ZT, appropriate rotations and retention of sufficient residues (average maize and wheat yield of CA practices was 5.94 t/ha and 5.50 t/ha), compared to the common practices of heavy tillage before seeding, monocropping and crop residue removal (average maize yield of 3.76 t/ha and wheat yield of 4.83 t/ha; Table 25.1). Again, leaving residue on the field is critical for ZT practices, since ZT with residue removal drastically reduced yields, except in the case of continuous wheat. Conventional tillage with or without residue incorporation resulted in intermediate yields. Zero tillage treatments with partial residue removal gave yields equivalent to treatments with full residue retention. This indicates that part of the residue can be removed for fodder while sufficient amounts can be retained to provide the necessary ground cover. This could make the adoption of ZT more acceptable for small-scale, subsistence farmers in systems where crop residue is used for fodder or fuel. Overall, CA practices had an average yield advantage of approximately 2.0 t/ha for maize and 0.6 t/ha for wheat compared to practices involving tillage or ZT with complete residue removal (Table 25.1).

Rainfall in the Mexican Highlands is erratic and water shortage can occur at any time during the growing season, resulting in high yield variability and changes in farmer income over the years. In ZT with residue retention, the increased aggregation and protection of the surface by the residue maximizes the recharge of the soil profile during rainfall events, which are often afternoon storms with high rainfall intensity. Additionally, the residue cover decreases evaporation. As a result, ZT with residue retention has higher soil water contents than practices with CT and ZT with residue removal, providing a buffer for dry spells during the growing season. For example, in 2009, the total amount of rainfall during the growing season was close to the long-term average (420 mm), but there

*Table 25.1* Effect of tillage practice, crop rotation, residue management on grain yield (t/ha at 12% $H_2O$) averaged since 1997 in CIMMYT's long-term sustainability trial in El Batán, Mexico

| Management practice | Maize | | | Wheat | | |
|---|---|---|---|---|---|---|
| **Core treatments** | | | | | | |
| Rotation, ZT, Keep | 5.65 | (0.03) | A | 5.96 | (0.09) | A |
| Rotation, ZT, Remove | 4.43 | (0.38) | B | 3.92 | (0.32) | F |
| Monoculture, ZT, Keep | 4.80 | (0.36) | B | 5.89 | (0.10) | AB |
| Monoculture, ZT, Remove | 2.52 | (0.06) | E | 4.95 | (0.11) | DE |
| Rotation, CT, Keep | 4.59 | (0.08) | B | 5.31 | (0.20) | CD |
| Rotation, CT, Remove | 4.31 | (0.32) | BC | 5.04 | (0.17) | DE |
| Monoculture, CT, Keep | 3.91 | (0.19) | CD | 5.53 | (0.02) | BC |
| Monoculture, CT, Remove | 3.76 | (0.04) | D | 4.83 | (0.33) | E |
| **CA treatments** | | | | | | |
| Rotation, ZT, Keep | 5.65 | (0.03) | A | 5.96 | (0.09) | A |
| Rotation, ZT, Keep M½W | 5.78 | (0.09) | A | 4.93 | (0.07) | B |
| Rotation, ZT, Keep ½ | 6.37 | (0.57) | A | 5.59 | (0.23) | A |

*Note*
Keep: all residue kept in the field; Keep M½W: all maize residue kept and 25 cm wheat stubble; Keep ½: maize stubble left until below ear and 25 cm wheat stubble.

was an extended dry period in July–August. In ZT with residue retention, the soil water content stayed above or near wilting point during the whole dry period, whereas in the other practices, values were below wilting point from 62 to 83 days after planting with severe implications for maize growth (Verhulst *et al.* 2011a). This resulted in the large yield differences, i.e., 4.71 t/ha, between the treatments with the highest and the lowest yield. These results show that the increased soil quality in CA practices results in a more resilient system than ZT with full residue removal or CT.

### Conservation agriculture to mitigate climate change

Dendooven *et al.* (2011) evaluated the effect of tillage practice and crop residue management on the net greenhouse gas flux taking into account soil C sequestration, emissions of greenhouse gases from soil, i.e., carbon dioxide ($CO_2$), methane ($CH_4$) and nitrous oxide ($N_2O$), and fuel used for farm operations (tillage, planting and fertilizer application, harvest) and the production of fertilizer and seeds. Tillage and residue management had little effect on greenhouse gases emitted from the soil. Maximum difference between the agricultural systems was 242 kg $CO_2$-C/ha/yr. However, the soil organic C content in the 0–60 cm layer was affected strongly by tillage and crop residue management. The soil organic C content was $118 \times 10^3$ kg C/ha in ZT with residue retention, approximately 40,000 kg C/ha higher than in practices involving tillage or ZT with residue removal. Taking into account the almost 20-year duration of the experiment, approximately 2,000 kg C/ha/yr was sequestered in the soil in ZT with residue retention compared to tillage or ZT with residue removal (Dendooven *et al.* 2011). Zero tillage reduced the C emission of farm operations with 74 kg C/ha/yr compared to CT. This may seem a small difference, but while the amount of C that can be sequestered in soil is finite, the reduction in net $CO_2$ flux to the atmosphere by reduced fossil-fuel use can continue indefinitely (West and Marland 2002). The net greenhouse gas flux was near neutral for ZT with crop residue retention (40 kg $CO_2$-C/ha/yr), whereas in the other management practices it was approximately 2,000 kg $CO_2$-C/ha/yr.

### Irrigated conditions in arid north-western Mexico

*Conservation agriculture as an adaptive measure*

SOIL QUALITY

Similar to the results in rainfed conditions of Central Mexico, a PCA of soil quality data from the arid north-west of Mexico distinguished three groups of tillage-straw systems: Permanent beds (PB)-straw burned, Conventionally tilled beds (CTB)-straw incorporated and PB-straw not burned (Verhulst *et al.* 2011b). The PB-straw burned system had the poorest soil quality with high electrical conductivity, sodium (Na) concentration and penetration resistance and poor self-mulching ability and aggregation. As in Central Mexico, the absence of tillage (in this case PB) without residue retention (in this case burning of residue) degraded soil quality and was not a sustainable management option. The CTB-straw incorporated system differentiated from the PB practices by its soil physical characteristics, especially the low direct infiltration and aggregate stability, indicating degradation of physical soil quality. This may have an effect on the efficiency of the use of resources, such as irrigation water. The PB with all or part of the residue retained had the highest soil microbial biomass C, total N, direct infiltration and aggregate stability and the lowest electrical conductivity and Na concentration. The practice of PB, where all or part of the residue is retained in the field, seems to be the most sustainable option for this irrigated wheat-based cropping system.

CROP YIELD AND SYSTEM RESILIENCE AND FLEXIBILITY

There have been marked annual changes in wheat yields in the long-term trial with 300 kg N/ha applied at first node (Figure 25.2; Sayre *et al.* 2005). Low wheat yields in 1995 and 2004 were the result of extended warm, cloudy periods during the first half of the crop cycles. There were no significant wheat yield differences between any of the tillage-residue management practices for the first 5 years (10 crop cycles). However, yield differences between management treatments diverged after the first five years with an overall reduction in the yield for PB-straw burned. This reflected the soil degradation found in this management practice. It seems that for irrigated cropping systems (at least for tropical, semi-tropical and the warmer, temperate areas), the application of irrigation water "hides or postpones" soil degradation associated with continuous residue burning until they reach a level that can no longer sustain high yields. Again, both full retention and partial retention of residues resulted in similar yields, indicating that for irrigated systems associated with high amounts of crop residue, large parts of the residue can be removed for other uses without inducing a decline in yield.

Yield differences between CTB-straw incorporated and PB-straw retained were small in this experiment: 7.05 t/ha in CTB vs. 7.21 t/ha in PB averaged over 1998 to 2009. In this trial, planting dates have been kept similar for all treatments (Sayre *et al.* 2005). However, in the real field situation like in the training/extension module managed in cooperation with the local farmer association, it was possible to plant the permanent beds 10–15 days earlier than the tilled beds. This resulted in a yield difference

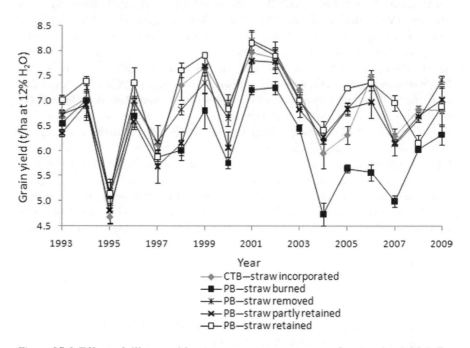

*Figure 25.2* Effect of tillage-residue management system on wheat grain yield (t/ha at 12% $H_2O$) when 300 kg N/ha are applied at the first node stage in the long-term trial, Yaqui Valley, Mexico (adapted from Sayre *et al.* (2005)). Bars indicate standard errors of the mean.

of 0.58 t/ha between PB and CTB, averaged over 2001 to 2004 (in the long-term trial, the difference between both practices was 0.12 t/ha averaged over the same period) (Sayre *et al.* 2005). Timely planting is important because in March and especially April (during grainfilling), average temperature and the frequency of heat events (>30 °C) increase substantially, affecting yield. With increasing temperatures, timely planting will become even more important in the future.

### Conservation agriculture to mitigate climate change

Studies of the net GWP of agronomic systems in the long-term trials in the Yaqui Valley are pending. In high-input systems like the one in the Yaqui Valley, N fertilizers are a significant direct source of emissions of $N_2O$ and nitric oxide and nitrogen dioxide ($NO_x$) in the field and an indirect source through fossil fuel energy consumption associated with manufacturing and transport of fertilizers (Ortiz-Monasterio *et al.* 2010). Adequate fertilizer management can play an important role in reducing $N_2O$ and $NO_x$ emissions in the field and increasing the fertilizer efficiency, thereby reducing the necessary fertilizer and associated manufacturing emissions. Sensor-based N management in wheat and maize is a new technology that uses an optical sensor, which measures the normalized difference vegetative index of the canopy. The use of this vegetative index in conjunction with an N-rich strip (a well fertilized part of the field) and a crop algorithm can be used to establish the optimum N fertilization rate (Raun *et al.* 2009). This technology optimizes N rates and minimizes the risk of excess fertilizer application. In addition, the N fertilizer is applied at the time of high demand by the crop, which in turn reduces the probabilities of generating favorable conditions for $N_2O$ emissions. Conservation agriculture may adjust C and N cycling compared to conventional systems (Govaerts *et al.* 2006c) and thus research is needed to determine whether dose, method and timing of application should be adjusted depending on management system.

The use of agricultural inputs such as irrigation carries a 'hidden' carbon cost (West and Marland 2002). As mentioned above, the degradation of physical soil quality in CTB may have an effect on the efficiency of the use of resources, such as irrigation water. The decreased infiltration in CTB might result in lower irrigation efficiency. Additionally, a better water conservation due to reduced evaporation in PB systems with residue retained may decrease the irrigation requirements compared to CTB. However, more research is needed to determine irrigation water requirements in CA systems compared to conventional systems.

### The economic potential of conservation agriculture for climate change mitigation and adaption, beyond direct incentives

In the Mexican Highlands, production parameters (production cost and returns) were determined for both the newly introduced CA practices and the conventional tillage-based practice in 2009. As mentioned above, the 2009 growing season was characterized by an extended dry period in July–August. Conservation agriculture had higher yields than the conventional practice in farmers' fields under these adverse conditions. Averaged over 6 locations (30 fields for both treatments), maize yield was 31% higher under CA (6.4 t/ha under CA vs. 4.8 t/ha under the conventional practice) (Figure 25.3a). These higher yields combined with the lower costs of CA resulted in a net return that was almost twice as high for CA as for the conventional practice averaged over the six locations: average net return was 9,940MXN/ha (US$813/ha) under CA compared to 5,019MXN/ha (US$420/ha) under the conventional practice (Figure 25.3b).

Numerous authors have demonstrated that apart from the on-farm benefits, CA also created off-site public benefits. Many soil and water conservation technologies have been implemented at very low rates by farmers (in many cases only when direct incentives were provided) because such practices do not always result in direct short-term benefits for the farmer (Hellin and Schrader 2003). Conservation agriculture has the win–win combination of being a soil and water conservation technology that also increases productivity in most cases.

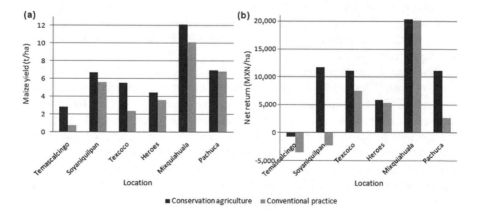

*Figure 25.3* (a) Average maize yields (t/ha) and (b) comparative net returns (MXN/ ha) for farmers' plots under conservation agriculture and the conventional practice at six different locations in the Central Mexican Highlands in 2009 (source: CIMMYT's survey data).

### Conservation agriculture adoption and the need of new extension approaches: the efforts in Mexico

The CA adoption process is complex. It is unlikely that complex, multi-component technologies, such as CA, can be successfully scaled out through traditional linear models of research and extension. Instead, this effort requires the development of innovation systems to adapt technologies to local conditions that includes functioning networks of farmer groups, machinery developers, extension workers, local business and researchers (Hall *et al.* 2005). To this end, decentralized learning hubs within different farming systems and in key agro-ecological zones are being developed (Sayre and Govaerts 2009). At the hubs, intensive contact and exchange of information is organized between the different partners in the research and extension process with a large feedback component. Because of the multi-faceted nature of CA technology development and extension, activities should be concentrated in a few defined locations representative of distinct farming systems, rather than making lower-intensity efforts on a wide scale. Through research and training, regional CA networks are established to facilitate and foment research and the extension of innovation systems and technologies. Research at the hubs also provides an example of functioning CA systems, helping to break down the culture of the plough with locally-adapted successful examples. Furthermore, the hubs act as strategic science platforms operated by international centres and national research institutes, enabling the synthesis of a global understanding of CA, and its adaptability to different environments, cropping systems and farmers' circumstances.

## Conclusions and recommendations

Climate models predict an increase in annual average temperature by the 2050s ranging from 1.6–2.0 °C on the Yucatán peninsula to 3.1–3.5 °C in northern Mexico. Annual precipitation is predicted to decrease in most of Mexico. The predicted changes in climate could severely affect agricultural production in Mexico if adequate measures are not taken.

In the highlands of Mexico, conservation agriculture (zero tillage with at least partial residue retention and crop rotation) has the potential to mitigate and adapt to climate change. Conservation agriculture results in high physical, chemical and biological soil quality that favours larger yields and reduces the net greenhouse gas flux compared to the traditional agricultural system (tillage, residue removal and monoculture of maize). The high physical soil quality ensures that the cropping system is optimized to cope with both heavy rainfall events and prolonged drought, events that are likely to increase in frequency due to climate change. Residue retention under zero and conventional tillage induces greater microbial diversity, especially higher total bacteria, fluorescent

Pseudomonas and actinomycetes both for maize and wheat cropping systems. This diversity allows an ecosystem to remain stable when facing changes in environmental conditions. The net greenhouse gas flux is near neutral in conservation agriculture, whereas systems involving conventional tillage or zero tillage contribute to global warming. Conservation agriculture requires the retention of at least part of the crop residue in the field. Therefore, research on the integration of conservation agriculture and mixed livestock systems is needed, focusing on the trade-offs for residue use.

In the irrigated, arid north-western part of Mexico, the practice of permanent beds, where all or part of the residue is retained in the field, seems to be a more sustainable option for this wheat-based cropping system than conventional tillage with incorporation of the residues. Timely planting of wheat might help to prevent negative effects of global warming. Optimizing application fertilizer rates and synchronizing them with crop development will further increase yields while reducing costs and emissions of $N_2O$. More research is needed to determine net greenhouse gas flux of different practices and to further develop conservation agriculture systems and irrigation strategies with reduced irrigation requirements.

Complex, multi-component technologies, such as conservation agriculture, can be successfully scaled out through an innovation systems approach. The case studies described in this chapter highlight the importance of key investments to establish several of those innovation systems in contrasting cropping systems in different agro-ecological environments in Mexico and worldwide to corroborate the results obtained and develop conservation agriculture practices tailored to local conditions.

# References

Aguilar, E., Peterson, T.C., Ramírez Obando, P., Frutos, R., Retana, J.A., Solera, M., Soley, J., González García, I. *et al.* (2005) Changes in precipitation and temperature extremes in Central America and northern South America 1961–2003, *J. Geophys. Res.*, vol. 110, D23107, doi:10.1029/2005JD006119.

Chocobar-Guerra, E.A. (2010) *Edafofauna como indicador de calidad en un suelo cumulic phaozem sometido a diferentes sistemas de manejos en un experimento de larga duración*, M.Sc. thesis, Colegio de Postgraduados, Montecillo, Mexico.

Dendooven, L., Patiño-Zúñiga, L., Verhulst N., Luna-Guido M., Marsch, R. and Govaerts, B. (2011) Global warming potential of agricultural systems with contrasting tillage and residue management in the central highlands of Mexico, submitted to *Agriculture, Ecosystems & Environment*.

Govaerts, B., Fuentes, M., Mezzalama, M., Nicol, J.M., Deckers, J., Etchevers, J.D., Figueroa-Sandoval, B. and Sayre, K.D. (2007a) Infiltration, soil moisture, root rot and nematode populations after 12 years of different tillage, residue and crop rotation managements, *Soil & Tillage Research*, vol. 94, pp. 209–219.

Govaerts, B., Mezzalama, M., Sayre, K.D., Crossa, J., Lichter, K., Troch, V.,

Vanherck, K., De Corte, P. and Deckers, J. (2008) Long-term consequences of tillage, residue management, and crop rotation on selected soil micro-flora groups in the subtropical highlands, *Applied Soil Ecology*, vol. 38, pp. 197–210.

Govaerts, B., Mezzalama, M., Sayre, K.D., Crossa, J., Nicol, J.M. and Deckers, J. (2006b) Long-term consequences of tillage, residue management, and crop rotation on maize/wheat root rot and nematode populations in subtropical highlands, *Applied Soil Ecology*, vol. 32, pp. 305–315.

Govaerts, B., Mezzalama, M., Unno, Y., Sayre, K.D., Luna-Guido, M., Vanherck, K., Dendooven, L. and Deckers, J. (2007b) Influence of tillage, residue management, and crop rotation on soil microbial biomass and catabolic diversity, *Applied Soil Ecology*, vol. 37, pp. 18–30.

Govaerts, B., Sayre, K.D. and Deckers, J. (2005) Stable high yields with zero tillage and permanent bed planting?, *Field Crops Research*, vol. 94, pp. 33–42.

Govaerts, B., Sayre, K.D. and Deckers, J. (2006a) A minimum data set for soil quality assessment of wheat and maize cropping in the highlands of Mexico, *Soil Till. Res. Soil & Tillage Research*, vol. 87, pp. 163–174.

Govaerts, B., Sayre, K.D., Ceballos-Ramirez, J.M., Luna-Guido, M.L., Limon-Ortega, A., Deckers, J. and Dendooven, L. (2006c) Conventionally tilled and permanent raised beds with different crop residue management: Effects on soil C and N dynamics, *Plant and Soil*, vol. 280, pp. 143–155.

Govaerts, B., Sayre, K.D., Goudeseune, B., De Corte, P., Lichter, K., Dendooven, L. and Deckers, J. (2009) Conservation agriculture as a sustainable option for the central Mexican highlands, *Soil & Tillage Research*, vol. 103, pp. 222–230.

Groisman, P.Y., Knight, R.W., Easterling, D.R., Karl, T.R., Hegerl G.C. and Razuvaev, V.N. (2005) Trends in intense precipitation in the climate record, *J. Climate*, vol. 18, pp. 1326–1350.

Hall, A., Mytelka, L. and Oyeyinka, B. (2005) 'Innovation systems: Implications for agricultural policy and practice', *Institutional Learning and Change (ILAC) Brief – Issue 2*. International Plant Genetic Resources Institute (IPGRI), Rome, Italy.

Hellin, J. and Schrader, K. (2003) The case of direct incentives and the search for alternative approaches to better land management in Central America, *Agr. Ecosys. Environ.*, vol. 99, pp. 61–81.

Lavelle, P. (1997) 'Faunal activities and soil processes: Adaptive strategies that determine ecosystem function', in: Begon, M. and Fitter, A.H. (eds) *Advances in Ecological Research*. Academic Press, New York, pp. 93–132.

Magrin, G., Gay García, C., Cruz Choque, D., Giménez, J.C., Moreno, A.R., Nagy, G.J., Nobre, C. and Villamizar, A. (2007) 'Latin America. climate change 2007: Impacts, adaptation and vulnerability. Contribution of Working Group II to the fourth assessment report of the intergovernmental panel on climate change', Parry, M.L., Canziani, O.F., Palutikof, J.P., van der Linden, P.J. and Hanson, C.E. (eds). Cambridge University Press, Cambridge, UK, pp. 581–615.

Ortiz-Monasterio, I., Wassmann, R., Govaerts, B., Hosen, Y., Katayanagi, N. and Verhulst, N. (2010) 'Greenhouse gas mitigation in the main cereal systems: rice, wheat and maize', in: Reynolds, M.P. (ed.) *Climate Change and Crop Production*. CABI, Oxfordshire, UK, pp. 151–176.

Parry, M.L., Rosenzweig, C., Iglesias, A., Livermore, M. and Fischer, G. (2004) Effects of climate change on global food production under SRES emissions and socio-economic scenarios, *Global Environ. Chang.*, vol. 14, pp. 53–67.

Raun, W.R., Ortiz-Monasterio, I. and Solie, J.B. (2009) 'Temporally and spatially dependent nitrogen management in diverse environments', in: Carver, B.F. (ed.) *Wheat: Science and Trade.* Wiley – Blackwell, Ames, IA, pp. 203–214.

Sayre, K. and Govaerts, B. (2009) 'Conserving soil while adding value to wheat germplasm', in: Dixon, J., Braun, H.-J. and Kosina, P. (eds) *Wheat Facts and Future.* CIMMYT, Mexico D.F.

Sayre, K.D., Limon-Ortega, A. and Govaerts, B. (2005) 'Experiences with permanent bed planting systems CIMMYT/Mexico', in: Roth, C.H., Fischer, R.A. and Meisner, C.A. (eds) *Evaluation and Performance of Permanent Raised Bed Cropping Systems in Asia, Australia and Mexico.* Proceedings of a workshop held in Griffith, Australia. ACIAR Proceedings 121, ACIAR, Griffith, Australia, pp. 12–25.

Verhulst, N., Kienle, F., Sayre, K.D., Deckers, J., Raes, D., Limon-Ortega, A., Tijerina-Chavez, L. and Govaerts, B. (2011b) Soil quality as affected by tillage-residue management in a wheat-maize irrigated bed planting system, *Plant and Soil,* vol. 340, pp. 453–466.

Verhulst, N., Nelissen, V., Jespers, N., Haven, H., Sayre, K.D., Raes, D., Deckers, J. and Govaerts, B. (2011a) 'Soil water content, rainfall use efficiency and maize yield and its stability as affected by tillage practice and crop residue management in the rainfed semi-arid highlands of Mexico'. *Plant and Soil,* DOI 10.1007/s11104–011–0728–8.

Wang, K.H. and McSorley, R. (2005) *Effects of soil ecosystem management on nematode pests, nutrient cycling, and plant health* [Online]. Available at: www.apsnet.org/publications/ apsnetfeatures/Pages/SoilEcosystemManagement.aspx (Accessed: 24 November 2010). APSnet.

West, T.O. and Marland, G. (2002) 'A synthesis of carbon sequestration, carbon emissions, and net carbon flux in agriculture: Comparing tillage practices in the United States', *Agriculture, Ecosystems & Environment,* vol. 91, pp. 217–232.

# 26

# POTENTIAL OF AGRICULTURAL CARBON SEQUESTRATION IN SUB-SAHARAN AFRICA

*Timm Tennigkeit, Fredrich Kahrl, Johannes Wölcke, and*
*Ken Newcombe*

Agricultural carbon sequestration has the potential to play a significant role in global greenhouse gas (GHG) mitigation efforts, with important co-benefits for soil fertility and agricultural productivity, food security, rural development, water use efficiency, and climate adaptation. Particularly because of the importance of these co-benefits, sub-Saharan Africa (SSA) will be a key region in global efforts to sequester carbon in agriculture.

There is broad consensus that soil fertility is declining in many parts of SSA (Pender *et al.* 2006; FAO 2004; Sanchez 2002; Nandwaa and Bekundab 1998; Dregne 1990). Soil degradation and declining crop yields have far-reaching consequences for food security and socio-economic development in sub-Saharan African countries (Pender *et al.* 2006; CIAT, TSBF, and ICRAF 2002; Lal and Singh 1998; Enters 1997; Bojö 1996). Climate variability and change are further exacerbating these challenges. Restoring soil carbon is an important part of efforts to reverse these declines in soil fertility, increase agricultural productivity and provide for a more climate-resilient agricultural sector. Agricultural carbon sequestration in SSA could thus provide a rare triple win that offers GHG mitigation, food security, and socio-economic development benefits (Antle and Stoorvogel 2008; Batjes 2004; Lipper and Cavatassi 2004; Antle and Diagna 2003).

This chapter situates agricultural carbon sequestration in the context of demographic, socio-economic, and agronomic trends in SSA. We describe four important themes in agricultural sequestration projects: management practices, risk, permanence, and emissions attribution. This chapter supplements the chapter on the economics of agricultural carbon sequestration in this book (see Tennigkeit *et al.* (a)).

# Potential for agricultural carbon sequestration in sub-Saharan Africa

Greenhouse gas mitigation in agriculture covers a wide range of practices, from reduced tillage that sequesters carbon in soils to changes in nutrient management practices, including green manure and leguminous cover crops, and changes in feeding practices and manure management that reduce methane emissions from livestock. Both at a socio-economic and a biophysical level these practices are interactive and their effects interlinked. Because our focus in this analysis is on carbon sequestration, we distinguish between agricultural activities that sequester carbon and those that reduce emissions, while recognizing that in practice this division may not always hold.

Within 'agricultural carbon sequestration,' we consider carbon additions to both soil and plant biomass that result from changes in management practices. Agroforestry in tandem with changes in residue management, for instance, could increase carbon stored both in the soil and in tree biomass. However, because soils are by far the dominant source of agricultural carbon sequestration (Smith *et al.* 2008), in examining the potential for agricultural carbon sequestration in SSA we focus initially and predominantly on soil carbon.

At approximately 1,500–1,600 Gt C, soils are the earth's largest terrestrial store of carbon (Lal 2004a; Schlesinger 1997).[1] Since the Industrial Revolution, soil carbon emissions (78 ± 12 Gt C) from land-use change and agricultural activities have accounted for about 19% of total atmospheric carbon emissions (406 ± 63 GtC), with cumulative losses of as much as 30–40 t C/ha on cultivated soils (Lal 2004b).

A significant portion of these losses are recoverable. Lal (2004a) reports attainable sequestration rates between 0.4–1.2 Gt C/yr,[2] equivalent to 3–9% of global GHG emissions in 2004 (49.0 GtCO$_2$e) (IPCC 2007). Changes in land-use and management practices could add carbon to the soil for 15–60 years (Lewandrowski *et al.* 2004; Lal 2004a), depending on site-specific conditions, after which soil carbon levels would reach a new steady state and sequestration rates would fall to zero. Although soil carbon sequestration is limited both in space and time, it could be a cost-effective, nearer-term option for reducing atmospheric CO$_2$ concentrations and is unequivocally an important investment in longer-term soil productivity (see also chapter by Verhulst *et al.* on conservation agriculture).

Existing estimates provide a useful sense of the economic mitigation potential for agricultural carbon sequestration in sub-Saharan Africa. Lal's (2004a) estimated sequestration potential in world cropland soils (0.4–0.8 Gt C/yr) scaled by Africa's share of global cropland (11%) leads to a technical potential of 44–88 Mt C/yr. For the full suite of agricultural mitigation activities, Smith *et al.* (2008) estimate an economic (US$0–20/t

$CO_2e$) mitigation potential for Africa of 72 Mt Ce/yr, of which cropland management, grazing land management, and restoration of degraded land account for 46 Mt Ce/yr.[3] Excluding "Northern Africa," Smith *et al.*'s estimates are 65 Mt Ce/yr and 41 Mt Ce/yr, respectively. Based on these ranges, we argue that 40 Mt C/yr is a reasonable anchor point for attainable agricultural carbon sequestration in sub-Saharan Africa. As per Smith *et al.* (2008), we assume that roughly half of this potential is on croplands and half on grasslands.

At 40 Mt C/yr (147 Mt $CO_2$/yr) and a discounted carbon price of US$10/t $CO_2$ (see *Permanence*, below), soil carbon sequestration would generate annual revenues of US$1.47 billion. Although small as a percentage of total Official Development Assistance (ODA) to SSA (US$36.6 billion in 2008), US$1.47 billion is almost double the total ODA directed to agriculture in SSA (US$1.68 billion in 2007, 4.6% of total ODA),[4] and more than 40% larger than the total annual public spending on agricultural R&D in SSA in 1999 (US$1.09 billion) (World Bank 2007). In aggregate, and even at lower carbon prices (e.g., US$5/t $CO_2$), revenues from agricultural sequestration would represent a major infusion of funds into sub-Saharan African agriculture.

Sequestering 40 Mt C/yr in SSA would require considerable scale. At 20 Mt C/yr (73 Mt $CO_2$/yr) of cropland potential and an average carbon sequestration rate of 1.5 t $CO_2$/ha/yr, the total area of farms involved across the region would be 49 Mha, or 25% of the region's total arable and permanent cropland in 2003.[5] Assuming an average farm size of 1.5 ha across the region, the number of households involved would be more than 30 million.

## Population growth, soil degradation, food security, and poverty in sub-Saharan Africa, with a focus on East Africa

We examine the demographic, socio-economic, and agronomic trends for the three largest members of the East African Community—Kenya, Tanzania, and Uganda. The three countries illustrate the nature and magnitude of challenges at the nexus of agriculture, food security, and economic development in SSA, as well as the differences in the scale of these challenges across countries. Eastern Africa also has the highest agricultural sequestration potential among African regions, accounting for more than 40% of Smith *et al.*'s (2008) estimated agricultural mitigation potential for Africa.

### Population growth

Kenya, Tanzania, and Uganda have all experienced rapid population growth over the past three decades, at net growth rates that have, at times, exceeded 3% and have historically exceeded the average for SSA (Figure 26.1). In all

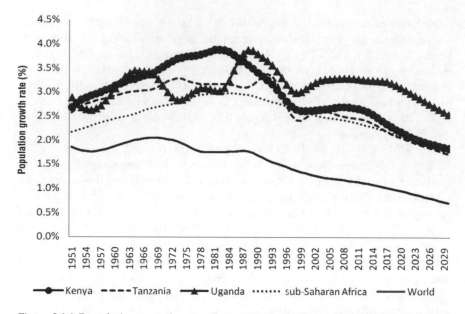

*Figure 26.1* Population growth rates, Kenya, Tanzania, Uganda, sub-Saharan Africa, and world, 1951–2030 (source: UN 2007).

three countries, populations more than doubled between 1980 and 2005, reaching an estimated 35.6 million, 38.5 million, and 28.9 million in Kenya, Tanzania, and Uganda, respectively, in 2005 (UN 2007). The UN projects that each country will add more than one million people per year until at least 2040 (UN 2007).

Population growth is itself not a cause of land degradation. However, the combination of population pressures and rigidities in off-farm labor markets has steadily reduced available cropland per farmer over the last four decades (Figure 26.2). This pattern is part of a global trend, but is particularly important for East African countries because cropland–farmer ratios are already significantly below the world average. Although growth in the agricultural labor force has slowed significantly since the beginning of the 1990s in Kenya and Tanzania,[6] continued growth in the agricultural labor force either requires households to further subdivide their land or convert pasture to cropland.

### Soil degradation

To meet minimum food and cash income needs, small farmers are forced to cultivate their land more intensively. As a result, farms in many parts of East Africa have undergone a shift from traditional fallow-based to continuously farmed systems, leading to a net loss of soil organic matter and nutrients (Fermont *et al.* 2008; Dreschel *et al.* 2001; Lal and Singh 1998).

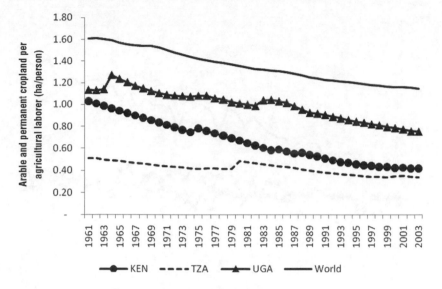

*Figure 26.2* Arable and permanent cropland per agricultural laborer, Kenya, Tanzania, Uganda, and world, 1961–2004 (source: all data are from FAOSTAT).

Nutrient depletion has long been recognized as a problem in East Africa, and more generally throughout SSA (FAO 2004; Stoorvogel and Smaling 1990). From 1945–90, the productivity loss from soil degradation alone has been estimated at 25% for cropland and 8–14% for cropland and pasture combined in SSA.[7]

### Food security

Nutrient depletion is one factor behind a trend of flat or declining crop yields in East Africa that began in the mid-1980s (Figure 26.3). Between 1980 and 2005, more than 90% of the near doubling of total maize output in East Africa was the result of bringing more land under maize cultivation; higher yields accounted for less than 10% of output increases over this time frame,[8] though differences among countries suggest caution in interpreting this trend. Kenyan maize yields appeared to be on an upward trend beginning in the late 1990s (Figure 26.3), but more recent trends for Tanzania and Uganda are less certain.

More importantly, cereal production in East Africa has largely failed to keep pace with population growth (Figure 26.4), as per capita domestic cereal production has declined in all three countries over the past four decades, but particularly since the mid-1980s. All three countries have become significantly larger cereal importers over the past decade. From 1985–1995 Kenyan cereal imports were equivalent to 13% of domestic production; from 1995–2004 imports rose to an average of 31% of domestic production.

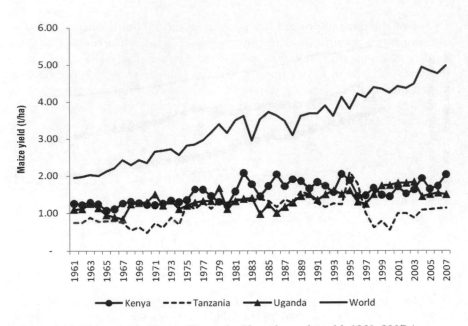

*Figure 26.3* Maize yields, Kenya, Tanzania, Uganda, and world, 1961–2007 (source: all data are from FAOSTAT).

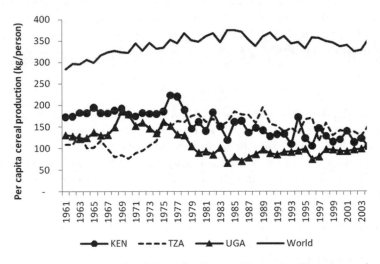

*Figure 26.4* Per capita cereal production, Kenya, Tanzania, Uganda, and world, 1961–2005 (sources: cereal production data are from FAOSTAT; population data are from UN (2007)).

One of the most important investments in sub-Saharan African agriculture will be in restoring soil fertility, because of its bidirectional links to poverty (Nkonya *et al.* 2005), basic food provision and food security (Sanchez 2002), and environmental services (Lipper and Cavatassi 2004). Agricultural carbon sequestration in SSA thus has a multifaceted role, as a strategy for reducing net GHG emissions and as an important investment in soil productivity.

## An overview of management practices, risk, permanence, and emissions attribution

Regardless of whether they are publicly or privately run, projects to sequester agricultural carbon will typically involve four primary actors: Farmers, aggregators, project developers, and carbon buyers. This formulation over-simplifies the institutional and operational complexity that will likely characterize agricultural sequestration projects. Four areas of potential complexity—management practices, risk, permanence, and emissions attribution—are particularly relevant and will arise repeatedly throughout this analysis. We describe each area in greater detail below.

### *Management practices*

Carbon can be sequestered in soils by changing land management practices or land use. In many parts of sub-Saharan Africa, population pressure and food security considerations do not allow land to be diverted from agriculture, and thus at the smallholder farm level additions to the carbon stock will have to be achieved by incentivizing changes in land management practices, or what is known in the environmental services literature as 'working land' programs (Zilberman *et al.* 2008).

For improved management practices, Nkonya *et al.* (2008) demonstrate that soil and water conservation techniques are often most effective when adopted as a "package" of practices. For agricultural practices that sequester carbon in soils and above- and below-ground biomass, these practices might include some combination of:

- improved agronomic practices;
- nutrient management;
- reduced tillage and residue management; and
- agroforestry.[9]

Sequestration rates for individual management practices that sequester carbon vary by as much as two orders of magnitude on a per area basis, with estimates ranging from tens of kilograms of carbon sequestered annually per hectare to tons of carbon per hectare. The significant range in these estimates is driven by both site-specific conditions and the level of

Table 26.1 Estimated mitigation potential for different practices by climate region (t $CO_2$ ha/yr)

| | Cool–dry | | | Cool–moist | | | Warm–dry | | | Warm–moist | | |
|---|---|---|---|---|---|---|---|---|---|---|---|---|
| | Mean | Low | High | Mean | Low | High | Mean | Low | High | Mean | Low | High |
| Agronomy | 0.29 | 0.07 | 0.51 | 0.88 | 0.51 | 1.25 | 0.29 | 0.07 | 0.51 | 0.88 | 0.51 | 1.25 |
| Nutrient management | 0.26 | −0.22 | 0.73 | 0.55 | 0.01 | 1.10 | 0.26 | −0.22 | 0.73 | 0.55 | 0.01 | 1.10 |
| Tillage and residue management | 0.15 | −0.48 | 0.77 | 0.51 | 0.00 | 1.03 | 0.33 | −0.73 | 1.39 | 0.70 | −0.40 | 1.80 |
| Agroforestry | 0.15 | −0.48 | 0.77 | 0.51 | 0.00 | 1.03 | 0.33 | −0.73 | 1.39 | 0.70 | −0.40 | 1.80 |

Source: The table is drawn from IPCC (2007).

Note
Mitigation potentials here include both practices that reduce emissions and those that sequester carbon.

biomass additions associated with different management practices. Table 26.1 illustrates the considerable range in mitigation potential across agricultural mitigation practices in different climatic regions. There is also a significant range in different researchers' estimates of mitigation potential from the same practices, often as a result of what is or is not included in estimates.

### Risk

Risk management is central to the economic feasibility and distributional implications of carbon sequestration (see also chapters by Shames *et al.*, Havemann, and Baroudy and Hooda). Indeed, one of the central roles of project developers in agricultural sequestration projects is to take on and manage risk. Although other actors and other risks are important for risk management in agricultural sequestration projects in SSA, we focus here on four broad categories of risk that will be faced by project developers and farmers (Table 26.2). Risk-hedging strategies are considered in the proposed operational models.

The sequestration potential and effectiveness of agricultural management practices varies as outlined above. A second source of physical uncertainty lies in the net attainable sequestration with management practices themselves. For instance, there has been a vigorous debate about the net carbon dioxide-nitrous oxide $(CO_2\text{-}N_2O)$ trade-offs with low- or no-tillage agriculture. More recent assessments suggest that soil aeration, which is also site and practice specific, plays a key role in determining net GHG emissions from no-till agriculture (Rochette 2008).

Crop response to changes in soil management practices is also site specific and highly uncertain. For instance, Lal (2006) reports that a 1%

*Table 26.2* Risks and risk bearers in agricultural sequestration projects

| Source of uncertainty | Fundamental question | Main affected party |
|---|---|---|
| Sequestration rates | Is sequestration lower than expected? | Farmers and project developers |
| Crop response | Are yield changes associated with prescribed practices lower than expected? | Farmers |
| Adoption and disadoption | Have farmers actually adopted and do they continue to use specified management practices? | Project developers |
| Price risk | Are prices for inputs and outputs as expected? | Farmers and project developers |

increase in root zone soil organic carbon could increase maize yields by 30–300 kg/ha, or an order of magnitude difference, which is not even considering other factors and repercussions that contribute to increasing crop response, e.g., improved varieties or fertilizer use. Projects that include seeds and fertilizers may have less variance in crop response, but actual yield improvements will depend on the quality of extension services. If yield benefits are lower than expected, farmers', and more indirectly project developers' (e.g., through adoption and disadoption rates), expected revenues may not materialize. While project developers can diversify away some of this risk, for farmers yield risks are substantial and are likely to factor into adoption decisions.

Price risk is substantial for both farmers and project developers. A fall in maize prices would affect farmer incomes, while at the same time increasing the risk of disadoption for project developers. Similarly, a rise in input prices would create a substantial burden for project developers that supply inputs for farmers. The notion of price risk in agricultural sequestration projects is largely unexplored, and should be an important consideration in program design.

### Permanence

Carbon sequestration differs from a permanent GHG emission reduction (e.g., replacing a kilowatt-hour of coal-fired electricity generation with a kilowatt-hour of wind-powered generation):

- The biosphere's capacity to sequester carbon declines over time, eventually reaching a maximum value as the system finds a new equilibrium;
- Sequestered carbon can be released back into the atmosphere at any point; and
- Humans operate on time scales that are shorter than typical residence times in the carbon cycle; contracts for sequestered carbon are likely to be shorter (e.g., 20 years) than the average residence time of an equivalent unit of $CO_2$ emitted into the atmosphere (e.g., 100 years) (Kim *et al.* 2008).

There is an extensive literature on possible approaches to permanence discounting (Kim *et al.* 2008; Maréchal and Hecq 2006; Lewandrowski *et al.* 2004; Marland *et al.* 2001; Moura-Costa and Wilson 2000). The Verified Carbon Standard (VCS) has introduced a buffer concept to consider permanence. Depending on the perceived permanence risk, a certain proportion of the generated carbon credits are kept in a buffer to cope with potential non-permanence of certain projects.

In theory, sequestered carbon could be stored indefinitely, long after carbon sequestration rates have slowed to zero. However, if storage is to occur beyond a maximum attainable value (i.e., when payments for sequestered carbon have declined to zero), farmers will have to be given an incentive to continue to store carbon, or what is typically referred to as a 'maintenance cost'. Whether maintenance costs are required in SSA will depend, among other things, on how economic forces in rural areas evolve over the next three to five decades.

### Emissions attribution

Depending on management practices and where project boundaries are drawn, agricultural carbon sequestration projects may also lead to an increase in GHG emissions. In designing the rules that govern how sequestration projects are integrated into international climate policies and carbon markets, these emissions should be accounted for. However, designs should also consider the dual role of sequestration projects as achieving mitigation and development objectives. When there are trade-offs between these two objectives, the challenge will be to enable development without significantly increasing net GHG emissions.

Fertilizer use in particular presents a problem for agricultural mitigation projects in sub-Saharan Africa. Use of synthetic fertilizers in many parts of SSA is reportedly below minimum recommended requirements to sustain nutrient levels.[10] If agricultural sequestration projects are being designed to improve soil fertility and increase yields, some amount of increase in nitrogen-based fertilizers will likely be necessary. Although the embodied carbon emissions from producing and transporting nitrogen fertilizers may be offset by the increase in soil carbon and biomass associated with their use (Alvarez 2005), $N_2O$ emissions associated with nitrogen fertilizer use are likely to be positive and could be significant if the soil absorption capacity is low. Because of sub-Saharan Africa's historically low fertilizer use, $N_2O$ emissions associated with nitrogen fertilizer are expected to be low. Accounting for indirect $N_2O$ emissions could raise $N_2O$ emissions 2–5 fold, depending on what range of indirect $N_2O$ emission factor estimates are used (IPCC 2006; Crutzen et al. 2008). In summary, research on this question is quite crucial.

### Conclusions

•  Agricultural carbon sequestration can be a near-term, cost-effective strategy for reducing atmospheric GHG concentrations, with important co-benefits for soil fertility and agricultural productivity, food security, rural development, water use efficiency, and climate adaptation. For SSA, the co-benefits of agricultural sequestration are particularly important because of the region's high population growth rates, historically low and even declining agricultural productivity, and vulnerability to climate change. Large-scale efforts to sequester carbon in agriculture could have transformative effects for agro-ecosystems and economies in the region.

• Over the longer term, revenues from agricultural sequestration could be substantial for the agricultural sector in sub-Saharan African countries. At a total estimated economic potential of 40 million tons of carbon (147 million tons of $CO_2$) per year and a conservative estimate of US$10 per ton $CO_2$, soil carbon sequestration would generate annual revenues of US$1.47 billion, larger than both total ODA to agriculture in SSA (2002) and total annual public spending on agricultural R&D in SSA (1999).

## Acknowledgement

The initial investigation for this chapter was financed by the BioCarbon Fund of the World Bank in early 2009. Views expressed in this article are the sole responsibility of the authors, not the associated institutions.

## Notes

1 The estimated 4,000 Gt C of carbon in fossil fuels (Schlesinger 1997) is not included in the definition of 'terrestrial' here.
2 The IPCC's Second Assessment Report estimates a narrower range of 0.4–0.8 Gt C/yr (IPCC 1996).
3 Smith *et al.* (2008) use five categories of mitigation activities: cropland management, restoration of cultivated organic soils, restoration of degraded land, grazing land management, and rest.
4 Preliminary data are from the OECD's Development Assistance Committee.
5 Data are from FAOSTAT.
6 Additions to the agricultural labor force in Kenya and Tanzania decreased by 45% and 35%, respectively, between 1990 and 2004. In Uganda, alternatively, the agricultural labor force grew by 10% between 1990 and 2004. Data are from FAOSTAT.
7 From the widely referenced Global Assessment of Soil Degradation (GLASOD) study, the first serious global attempt at an estimation of land degradation.
8 A simple decomposition demonstrates this. Total maize output (Y) is equal to land under maize cultivation (L) times per hectare yield (y), which implies that $(Y_t - Y_0) = (L_t - L_0)y_t + (y_t - y_0)L_0$, and where the first term captures the impact of land and the second captures the impact of yields. For Kenya, higher yields (the second term) contributed a significant share (46%) to higher output, but for Uganda and Tanzania the effect of higher yields was much smaller (11% and 12%, respectively), and in the case of Tanzania was actually negative. Across the three countries, higher maize yields accounted for only 9.8% of higher maize output. All data are from FAOSTAT.
9 We do not include grassland management practices in this list, although these would be important in a broader agricultural sequestration program. See Smith *et al.* (2008) for a detailed enumeration of grassland management practices.
10 Most of the data in the literature on nutrient balances in SSA appear to be from the 1990s. With more recent increases in fertilizer use in, for instance, Kenya and Tanzania, the extent of current nutrient shortages is unclear. Vis-à-vis other countries, fertilizer use is quite low. For instance, per-hectare fertilizer

use in Kenya (33.4 kg/ha), Tanzania (8.6 kg/ha), and Uganda (1.3 kg/ha) is significantly lower in the U.S. (146.8 kg/ha) or China (258.3 kg/ha). Data are from FAOSTAT.

## References

Alvarez, R. (2005) A review of nitrogen fertilizer and conservation tillage effects on soil organic carbon storage, *Soil Use and Management*, vol. 21, pp. 38–52.

Antle, J.M. and Diagna, B. (2003) Creating incentives for the adoption of sustainable agricultural practices in developing countries: The role of soil carbon sequestration, *American Journal of Agricultural Economics*, vol. 85, no. 5, pp. 1178–1184.

Antle, J.M. and Stoorvogel, J.J. (2008) Agricultural carbon sequestration, poverty, and sustainability, *Environment and Development Economics*, vol. 13, pp. 327–352.

Batjes, N. (2004) Estimation of soil carbon gains upon improved management within croplands and grasslands of Africa, *Environment, Development and Sustainability*, vol. 6, pp. 133–143.

Bojö, J. (1996) The cost of land degradation in Sub-Saharan Africa, *Ecological Economics*, vol. 16, pp. 161–173.

International Center for Tropical Agriculture (CIAT), Tropical Soil Biology and Fertility Institute (TSBF), and World Agroforestry Centre (ICRAF) (2002) 'Soil fertility degradation in Sub-Saharan Africa: Leveraging lasting solutions to a long-term problem'. Workshop Conclusions, Rockefeller Foundation Bellagio Study and Conference Center, 4–8 March.

Crutzen, P.J., Mosier, A.R., Smith, K.A and Winiwarter, W. (2008) $N_2O$ release from agrobiofuel production negates global warming reduction by replacing fossil fuels, *Atmospheric Chemistry and Physics*, vol. 8, pp. 389–395.

Dreschel, P., Gyiele, L., Kunze, D. and Cofie, O. (2001) Population density, soil nutrient depletion, and economic growth in sub-Saharan Africa, *Ecological Economics*, vol. 38, no. 2, pp. 251–258.

Dregne, H. (1990) Erosion and soil productivity in Africa, *Journal of Soil and Water Conservation*, vol. 45, pp. 432–436.

Enters, T. (1997) 'Methods for the economic assessment of the on- and off-site impacts of soil erosion', *Issues in Sustainable Land Management No. 2*. International Board for Soil Research and Management, Bangkok.

FAO 2010. *FAOSTAT* [Online]. Available at: http://faostat.fao.org/.

Fermont, A.M., Van Asten, P.J.A. and Giller, K.E. (2008) Increasing land pressure in East Africa: The changing role of cassava and consequences for sustainability of farming systems, *Agriculture, Ecosystems & Environment*, vol. 128, pp. 239–250.

Food and Agricultural Organization of the United Nations (FAO) (2004) *Scaling Soil Nutrient Balances: Enabling Mesolevel Applications for African realities*. FAO, Rome, Italy.

Intergovernmental Panel on Climate Change (IPCC) (2006) 'Chapter 11: $N_2O$ emissions from managed soils, and $CO_2$ emissions from lime and urea application', *2006 IPCC Guidelines for National Greenhouse Gas Inventories, Volume 4: Agriculture, Forestry and Other Land Use*. IPCC, Geneva, Switzerland.

Intergovernmental Panel on Climate Change (IPCC) (2007) *Climate Change 2007: Fourth Assessment Report*. IPCC, Geneva, Switzerland.

Kim, M.K., McCarl, B.A. and Murray, B.C. (2008) Permanence discounting for land-based carbon sequestration, *Ecological Economics*, vol. 64, pp. 763–769.

Lal, R. (2004a) Soil carbon sequestration impacts on global climate change and food security, *Science*, vol. 304, pp. 1623–1626.

Lal, R. (2004b) Soil carbon sequestration to mitigate climate change, *Geoderma*, vol. 123, pp. 1–22.

Lal, R. (2006) Enhancing crop yields in the developing countries through restoration of the soil organic carbon pool in agricultural lands, *Land Degradation & Development*, vol. 17, pp. 197–209.

Lal, R. and Singh, B.R. (1998) Effects of soil degradation on crop productivity in East Africa, *Journal of Sustainable Agriculture*, vol. 13, no. 1, pp. 15–36.

Lewandrowski, J., Peters, M., Jones, C., House, R., Sperow, M., Eve, M. and Paustian, K. (2004) 'Economics of sequestering carbon in the U.S. agricultural sector', *USDA Technical Bulletin Number 1909*.

Lipper, L. and Cavatassi, R. (2004) Land-use change, carbon sequestration, and poverty alleviation, *Environmental Management*, vol. 33, pp. S374–S387.

Maréchal, K. and Hecq, W. (2006) Temporary credits: A solution to the potential non-permanence of carbon sequestration in forests?, *Ecological Economics*, vol. 58, pp. 699–716.

Marland, G., Fruit, K. and Sedjo, R. (2001) Accounting for sequestered carbon: The question of permanence, *Environmental Science & Policy*, vol. 4, no. 6, pp. 259–268.

Moura-Costa, P. and Wilson, C. (2000) An equivalence factor between $CO_2$ avoided emissions and sequestration—descriptions and applications in forestry, *Mitigation and Adaptation Strategies for Global Change*, vol. 5, pp. 51–60.

Nandwaa, S.M. and Bekundab, M.A. (1998) Research on nutrient flows and balances in East and Southern Africa: State-of-the-art, *Agriculture, Ecosystems & Environment*, vol. 71, pp. 5–18.

Nkonya, E., Pender, J., Kaizzi, C., Edward, K. and Mugarura, S. (2005) 'Policy options for increasing crop productivity and reducing soil nutrient depletion and poverty in Uganda', *IFPRI EPT Discussion Paper 134*.

Nkonya, E., Gicheru, P., Wölcke, J., Okoba, B., Kilambya, D. and Gachimbi, L. (2008) 'On-site and off-site long-term economic impacts of soil fertility management practices: The case of maize-based cropping systems in Kenya', *IFPRI Discussion Paper 00778*.

Pender, J., Place, F., and Ehui, S. (eds) (2006) *Strategies for Sustainable Land Management in the East African Highlands*, IFPRI, Washington, DC.

Rochette, P. (2008) No-till only increases $N_2O$ emissions in poorly-aerated soils, *Soil & Tillage Research*, vol. 101, pp. 97–100.

Sanchez, P.A. (2002) Soil fertility and hunger in Africa, *Science*, vol. 295, pp. 2019–2020.

Schlesinger, W.H. (1997) *Biogeochemistry: An Analysis of Global Change*, Academic Press, San Diego, CA.

Smith, P. Martino, D., Cai, Z., Gwary, D., Janzen, H., Kumar, P., McCarl, B., Ogle, S., O'Mara, F., Rice, C., Scholes, B., Sirotenko, O., Howden, M., McAllister, T., Pan, G., Romanenkov, V., Schneider U., Towprayoon S., Wattenbach, M. and Smith, J. (2008) Greenhouse gas mitigation in agriculture, *Philosophical Transactions of the Royal Society*, vol. 363, pp. 789–813.

Stoorvogel, J.J. and Smaling, E.M.A. (1990) 'Assessment of soil nutrient depletion in Sub-Saharan Africa: 1983–2000', *Winand Staring Centre Report 28*.

United Nations Secretariat, Population Division of the Department of Economic and Social Affairs (UN) (2007) *World Population Prospects: The 2006 Revision.* United Nations, New York, NY.

World Bank (2007) *Agriculture for Development: World Development Report 2008.* World Bank, Washington, DC.

Zilberman, D., Lipper, L. and McCarthy, N. (2008) When could payments for environmental services benefit the poor? *Environment and Development Economics*, vol. 13, pp. 255–278.

# 27

# LIVESTOCK AND GREENHOUSE GAS EMISSIONS

## Mitigation options and trade-offs

*Mario Herrero, Philip K. Thornton, Petr Havlík, and Mariana Rufino*

## Introduction

Livestock systems, especially in developing countries, are changing rapidly in response to a variety of drivers. Globally, human population is expected to increase from around 6.5 billion today to 9.2 billion by 2050. More than 1 billion of this increase will occur in Africa. Rapid urbanisation is taking place and incomes in developing countries are increasing significantly, and they are largely responsible for the large increases in demand for livestock products projected for the next 40 years.

Considering that the demand for meat and milk is increasing, and that in the future livestock systems are likely to have to operate in carbon-constrained markets, mitigation options for the livestock sector will become an essential component for supplying livestock products within equitably negotiated and sustainable greenhouse gas (GHG) emission targets. Emissions from livestock systems can be reduced significantly through technologies and policies, and the provision of adequate incentives for their implementation (Steinfeld and Gerber 2010). The objective of this chapter is to highlight options to mitigate GHG from livestock systems and their potential consequences for the use of natural resources and livelihoods of farmers.

## Importance of the global livestock industry

Several authors have recently reviewed the economic, social and environmental importance of the global livestock industry (FAO 2006, Herrero *et al.* 2009, Thornton 2010). Livestock systems occupy 30% of the ice-free global surface area and are a significant global asset with a value of at least

US$1.4 trillion. Apart from the value of animals, livestock products have some of the highest value of production globally (FAOSTAT 2010). Livestock industries are a significant source of livelihoods globally. They are organised in long market chains that employ at least 1.3 billion people globally and directly support the livelihoods of 800 million poor smallholder farmers in the developing world. Keeping livestock is also an important risk reduction strategy for vulnerable communities. At the same time they are important providers of nutrients and traction for growing crops in smallholder systems. Livestock are also an important source of nourishment. Livestock products contribute to 17% of the global kilocalorie consumption and 33% of the protein consumption globally.

### The demand for livestock products is rising in the developing world

Vast differences in the level of consumption of livestock products exist between rich and poor countries. The level of consumption of milk and meat in the developed world is at least five times higher than in the developing world. However, in developing countries the demand for livestock products is rising rapidly, mainly as a consequence of increased human population and rapidly increasing incomes (Figure 27.1). This increased consumption of livestock products is having positive impacts on mortality and cognitive development of children in these countries and on the overall food security of the developing world, but is also the source of health problems in parts of the developed world.

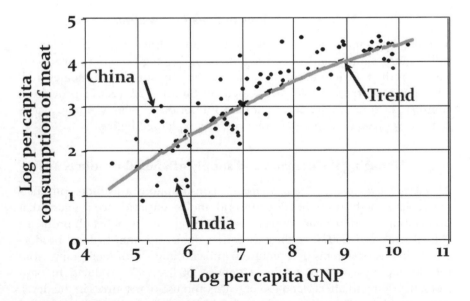

*Figure 27.1* The relationship between per capita income and meat consumption (Delgado *et al.* 1999).

## Livestock and greenhouse gas emissions

Livestock are a significant source of greenhouse gases (GHGs) globally. Accepted global GHG emissions estimates from the 17 billion domestic food-producing animals in the world (FAO 2009) vary from 8 to 18% of global anthropogenic emissions (Olivier and Berdowski 2001, Baumert *et al.* 2005, US EPA 2006, FAO 2006, O'Mara 2011), with the range reflecting methodological differences (inventories vs. life cycle analysis) and attribution of carbon dioxide ($CO_2$) from land-use changes (livestock rearing, feed production). The main sources and types of GHGs from livestock systems include methane ($CH_4$) production from enteric fermentation and animal manures, $CO_2$ from land use and its changes, and nitrous oxide ($N_2O$) from manure and slurry management (FAO 2006). These emissions represent 9%, 37% and 65% of the anthropogenic cmissions of $CO_2$, $CH_4$ and $N_2O$, respectively (FAO 2006).

Large differences in GHG emissions from livestock exist between different regions and countries. In general terms, livestock emissions of methane and nitrous oxide represent a significant proportion of the total emissions of agriculture-based countries (i.e., Argentina, Uruguay, sub-Saharan Africa, up to 60–70% of GHG). As countries become more industrialised, emissions from other sectors (energy, industry) become more important in proportional terms. For example, in the US, livestock methane and nitrous oxide emissions account for 2.8% of total GHG emissions (Pitesky *et al.* 2009).

## Mitigation options in livestock systems

Meeting the demand for livestock products in carbon-constrained markets will require simple and effective mitigation strategies. There are many ways to reduce GHG emissions from livestock through technologies, policies and incentives. Table 27.1, based on Smith *et al.* (2007) provides a summary of some mitigation options. Some of these, which in many cases can also represent adaptation practices, are explained below.

## Managing the demand and supply of animal products

*Managing the demand for livestock products:* consumption of livestock products per capita is high in the developed world and recent evidence suggests that over-consumption in some countries is increasing the risk of health problems (McMichael *et al.* 2007). In these countries demand is met by local production in intensive systems that require significant imports of feed grains from the developing world, or by direct imports of livestock products. In both cases, this demand affects land-use practices and use of resources in the developing world which lead to increases in GHG emissions. Reducing demand for livestock products in the developed world could lead to healthier people

318

Table 27.1 Some livestock-related options for mitigating GHG emissions, their mitigative effect on individual gases, and where they may have potential

| Options | Mitigative effect | | | Scope for application | |
|---|---|---|---|---|---|
| | $CO_2$ | $CH_4$ | $N_2O$ | in developed countries | in developing countries |
| Changing the diets of animals: improved feeding practices, use of legumes, etc. | + | + | +/– | Mixed systems, dairy and beef production | Mixed rainfed systems<br>Mixed irrigated systems<br>Intensive systems where efficiency gains are needed |
| Species shifts from ruminants to monogastrics | +/– | + | +/– | | |
| Control of animal numbers | | + | +/– | | Rangeland systems<br>Mixed extensive systems |
| Modification of rumen microbial flora and microflora | | + | | Intensive livestock systems | |
| Manure management: improved storage and handling of manure, more efficient use of manure as source of nutrients | + | + | +/– | Intensive and industrial livestock systems | Intensive and industrial livestock systems, mixed crop–livestock system |
| Managing the demand for livestock products | + | +/– | +/– | Reduce over-consumption of meat<br>Reduce imports of livestock products from places where livestock rearing has significant impacts on the environment | |

Adapted from Smith *et al.* (2007).

*Note*
1 "+" reduced emissions or enhanced removal (positive mitigative effect); "–" increased emissions or suppressed removal (negative mitigative effect); "+/–" uncertain or variable response.

and also to reduce pressures on land use and natural resources in developing countries. Stehfest *et al.* (2009) (Table 27.2) studied the impacts of alternative scenarios of global dietary changes on GHG emissions from land-use change. They found that reductions in red meat consumption were the single most important path for reducing $CO_2$ emissions from land use. Additional measures such as reduced consumption of white meats or other animal products (milk and eggs) reduced emissions further, but only modestly. The main reductions were mediated via significant reductions in $CO_2$ from reduced land-use changes for feed production and grazing land conversion, and also from reduced methane emissions from ruminants, due to lower animal numbers required to meet meat demands. While these are important results to understand the magnitude and scope for change, they need to be analysed with caution, as the authors did not study the socio-economic, cultural and livelihood implications of the scenarios analysed. These are essential aspects to consider when assessing the impacts of mitigation in livestock systems, and methodologies for their study warrant further research.

*Managing the supply of livestock products:* While changes in livestock product demand could have profound implications for reducing GHG emissions, managing their supply could also play a significant role. Two key ways in which we can alter the supply of livestock products is by making choices on the types of animal products we consume, and on the types and intensity of production systems in which we produce them.

The results of Stehfest *et al.* (2009) are also partially explained by the differences in GHG emission intensities of different animal products (Figure 27.2, de Vries and de Boer (2009)). Monogastrics, like chickens and pigs, which consume the highest quality diets and have high feed conversion ratios, have also the highest GHG efficiencies per unit of animal protein. Of the livestock products from ruminants, milk is almost as GHG efficient as the products from monogastrics, while red meat the most inefficient. Nevertheless, for all products, there is at least a twofold difference in the range of GHG intensities. This is caused mostly by differences in feeding and manure management, production systems, and breeds, for example. This suggests that there is a significant room for improvement of the efficiency of GHG emissions per unit of animal products.

*Table 27.2* Land-use emissions in 2000 and 2050 for the reference scenario and four dietary variants (Stehfest *et al.* 2009)

| Scenario | Gt Ce |
|---|---|
| 2000 | 3.0 |
| 2050 – Reference | 3.3 |
| 2050 – No red meat | 1.7 |
| 2050 – No meat | 1.5 |
| 2050 – No animal products | 1.1 |
| 2050 – Healthy diet | 2.1 |

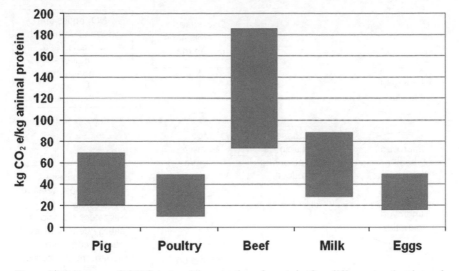

*Figure 27.2* Range of GHG intensities per kg of protein for different animal products in OECD countries (de Vries and de Boer 2009).

Havlik *et al.* (2009) demonstrated that structural changes in the supply of livestock products, represented as shifts from grassland-based or mixed extensive to mixed intensive systems, could have significant reductions on GHG emissions from livestock, while still meeting the projected demand for livestock products to 2030. The most important reductions in their study were effected via reductions in $CO_2$ emissions due to diminished livestock-induced deforestation from grassland-based systems. Total emissions of methane and nitrous oxide changed relatively little, but emissions per unit of livestock product were reduced significantly. The transition from more extensive systems to more intensive mixed-systems led to higher yielding and better fed animals.

## Technological options

Gill *et al.* (2010) proposed a framework for assessing the mitigation impacts of technologies and management practices designed for improving livestock feeding (Figure 27.3). A transition from large numbers of poorly fed animals to fewer but better fed animals can be achieved through technology. Some of the strategies are described below.

*Intensification of the diets of animals*: It is well established that feeding better quality diets to animals reduces the amount of GHG produced per unit of animal product (Blaxter and Clapperton 1965). An example is provided for a range of dairy production systems globally (Herrero *et al.* 2011, Figure 27.4). This increased efficiency can be achieved through improved supplementation practices or through land-use management with practices like improved pasture management (grazing rotation, fertilizer

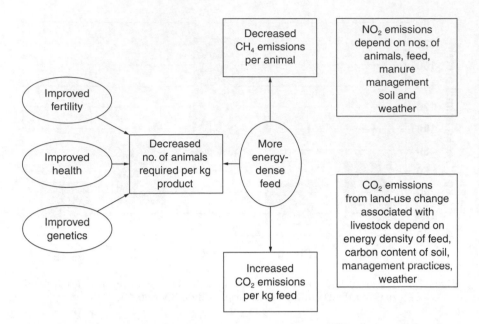

*Figure 27.3* Routes for impact of management and technology interventions designed to improve productivity on GHG emissions (Gill *et al.* 2010).

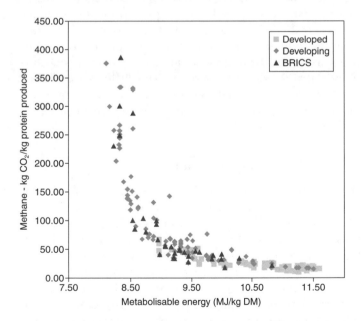

*Figure 27.4* Efficiency of methane production ($CO_2e$) per kg protein produced for dairy ruminant diets of different metabolisable energy densities (MJ/kg DM) in developed, developing and the BRIC countries (data from Herrero *et al.* forthcoming).

applications, development of fodder banks, improved pasture species, use of legumes and others) and the use of improved crop-by-products and others. When done through increased feeding of grains, transitions to improved diets shift the contributions of different GHGs to the total emissions. $CO_2$ emissions from land-use change increase while methane emissions per unit of output decrease (Gill *et al.* 2010).

Of all the technical mitigation options, these are probably the easiest to implement at the farm level. A prerequisite for these options to work is that systems need to be geared towards market-orientated production, as otherwise there is little incentive to improve feeding systems. Examples of where this option could be applicable are smallholder dairy systems in Africa and Asia, dual-purpose and dairy production in Latin America and beef cattle operations. Other options include manipulation of rumen microflora, breeding for lower methane production, and the use of feed additives (McCrabb 2002, Gill *et al.* 2010).

The largest GHG efficiency gaps are observed in systems where the quality of the diet is the poorest (i.e., grassland-based and some arid and humid mixed systems in the developing world). The highest marginal gains of improving diets through simple feeding practices, both biologically and economically, are in these systems.

*Control of animal numbers and shifts in breeds*: increases in animal numbers are one of the biggest factors contributing directly to GHG emissions (US EPA 2006, Thornton and Herrero 2010). In the developing world, many low-producing animals could be replaced by fewer but better fed animals of a higher potential to be able to reduce total emissions while maintaining or increasing the supply of livestock products. These kinds of efficiency gains will be essential in carbon-constrained markets.

*Shifts in livestock species*: switching species to better suit particular environments is a strategy that could yield higher productivity per animal for the resources available. At the same time switches from ruminants to monogastrics could lead to reduced methane emissions and higher efficiency gains, though this could increase the price of grains as competition with human consumption increases.

## Managing nitrous oxide emissions from manure

In the developing world, large amounts of nutrients are lost due to poor manure management. The opportunistic nature of many feeding systems means that large amounts of nutrients and carbon are lost before manure is stored (Rufino *et al.* 2006). In many places in Africa and Latin America, pig manure is not recycled; considered a waste, it is often discharged to water bodies or left to accumulate unused. This creates serious problems, especially in urban and peri-urban systems with alarmingly high water pollution and air emissions levels (Diogo *et al.* 2010). Research in intensive African ruminant livestock systems has shown that up to 70% of the manure nitrogen (N) can be lost within 6 months of excretion when manure is poorly managed (Rufino et al. 2007, Tittonell et al. 2010).

Excreta of livestock fed protein-rich diets are rich in mineral N and prone to large ammonia losses (Murwira 1995, Delve et al. 2001, Van der Stelt et al. 2008). When there are large ammonia losses after excretion, direct emissions of nitrous oxide during storage are relatively small due to lack of substrate (Chadwick 2005, Predotova et al. 2010), but indirectly emissions after N deposition might be large, and are difficult to account for in N emissions assessments (Oenema et al. 2007, Bouwman et al. 2010). In extensive livestock systems (e.g., African savannas, Mongolian steppe), where livestock diets are relatively poor in protein content, the largest N emissions occur in the places where livestock overnight and manure accumulates (Holst et al. 2007, Augustine 2003).

Options to manage emissions are not easy to design because they require systems thinking and awareness of key driving factors in different livestock systems. Reducing N emissions starts with feeding livestock balanced diets so that excreta are not rich in labile N, which is easily lost as ammonia and enters the N cascade (Galloway et al. 2003). In intensive systems, mineral N can be captured effectively using bedding material, which has been increasingly excluded from livestock facilities to reduce operational costs. Capturing mineral N from excreta in organic material reduces both direct ammonia and nitrous oxide emissions (Monteny et al. 2006). In intensive livestock systems, manure is increasingly handled as slurry in tanks or anaerobic lagoons, which may reduce direct nitrous oxide emissions during storage but can increase the risk of emission during land spreading (Huijsmans et al. 2003, Velthof and Mosquera 2011). In extensive systems, emissions of ammonia and nitrous oxide can be managed by shifting spatially livestock pens or the facilities where they overnight.

Recent research shows that direct nitrous oxide emissions after application to soils can be reduced by managing organic matter content in soils. Nitrogen emissions from applications of manure and fertilizer to soils poor in carbon (C) can be reduced by adding organic residues (Dick et al. 2008, Sanchez-Martin et al. 2010, Dalal et al. 2010). Synchronizing nutrient demand of crops with nutrient release from fertilizers remains the most sensible option to reduce nutrient losses and emissions, but so far has been difficult to implement by farmers. Future studies must measure all pathways for N emissions because techniques that reduce direct nitrous oxide emissions may enhance ammonia emissions and therefore indirect nitrous oxide emission, or leaching of N organic and inorganic compounds, and shuffling the pollution problem somewhere else.

## Land use and $CO_2$ mitigation

Some of the carbon lost from agricultural ecosystems through time can be recovered via improved management, thereby withdrawing atmospheric

*Table 27.3* Potential for carbon sequestration (Tg C/yr) in global rangelands of different overgrazing severity, by continent (Conant and Paustian 2002)

|  | *Light* | *Moderate* | *Strong* | *Extreme* | *Total* |
|---|---|---|---|---|---|
| Africa | 1.9 | 8.6 | 6.1 | 0.1 | 16.7 |
| Australia/Pacific | 4.5 | −0.1 | 0.0 | 0.3 | 4.4 |
| Eurasia | 0.8 | 3.2 | 0.6 | 0.4 | 4.3 |
| North America | 0 | 1.6 | 0.7 |  | 2.2 |
| South America | 6.1 | 11.3 | 7.4 |  | 18.1 |
| Total | 13.3 | 24.4 |  |  | 45.7 |

$CO_2$. Any practice that increases the photosynthetic input of carbon or slows the return of stored carbon to $CO_2$ via respiration, fire or erosion will increase carbon reserves, thereby sequestering carbon (Smith *et al.* 2007). Significant amounts of soil carbon could be stored in rangelands or in silvopastoral systems through a range of practices suited to local conditions (Table 27.3) (see chapter by Neely and De Leeuw). This not only improves carbon sequestration but could also turn into an important diversification option for sustaining livelihoods of smallholders and pastoralists through payments for ecosystems services.

## Mitigation through emissions offsets

Crops and residues from agricultural lands can be used as a source of fuel, either directly or after conversion to fuels such as ethanol or diesel. While these bio-energy feedstocks still release $CO_2$ upon combustion, the carbon is of recent atmospheric origin (via photosynthesis), rather than from fossil carbon. The net benefit of these bio-energy sources to the atmosphere is equal to the fossil-derived emissions displaced, less any emissions from producing, transporting, and processing. Numerous life cycle assessments focused on GHG emission balances of conventional biofuels like sugarcane and corn ethanol or rapeseed and soybean biodiesel (for a review, see e.g., OECD (2008)). Although the ranges of the GHG savings estimates are large, the savings tend to be positive. But these assessments did not include emissions caused by land-use changes. Recent studies showed that the overall balance can be negative for several decades if land-use change is accounted for (Fargione *et al.* 2008; Searchinger *et al.* 2008; Havlík *et al.* 2010). From this perspective, significant intensification of the livestock sector could lead not only to avoiding $CO_2$ emissions from expansion of pasture and feed crop production into natural lands, but also to freeing land for bioenergy crops production and hence to improving the carbon balance of the latter.

Land-sparing measures in the livestock sector seem also necessary for large-scale afforestation and reforestation (A/R) projects which would substantially increase carbon sinks in the landscape compared to uses as cropland and grassland. Energy use of biomass from such plantations, either for direct combustion or, once the technological development allows, for second generation biofuels, represents the most GHG-efficient land management for many areas (Marland *et al.* 2007).

## Trade-offs and synergies with mitigation

Mitigating GHGs in livestock systems can lead to trade-offs and synergies in the use of other resources. Given that almost all human activity is associated with GHG emissions, the complex balancing act of efficient resource use, GHG emissions and sustainable livelihoods requires clear understanding and needs to be taken into account when discussing the contribution of livestock to global GHG. Especially for the developing world, this complex issue needs to move beyond a commodity-based approach towards a livelihoods approach, so that trade-offs between the use of different natural resources and the production objectives of farmers are adequately represented when designing mitigation strategies.

In smallholder crop-livestock and agro-pastoral and pastoral livestock systems, livestock are one of a limited number of broad-based options to increase incomes and sustain the livelihoods of people who have a limited environmental footprint. Livestock are particularly important for increasing the resilience of vulnerable poor people, subject to climatic, market and disease shocks through diversifying risk and increasing assets. Under these circumstances, livestock rearing is not driven by productivity-orientated objectives. Overall emissions are low per animal as feed intakes are low due to lack of biomass or its quality, but GHG efficiencies per unit of livestock product are some of the lowest found. These systems are, however, efficient at balancing the trade-offs in the use of limiting resources under considerable variability. We might need to accept GHG inefficiencies as inevitable in systems that are not managed with productivity as a key objective, unless it would be possible to compensate methane emissions especially, through carbon sequestration practices. However, in some cases these practices could lead to fewer animals per person, thus reducing the amount of meat and milk available to subsistence communities. The feasibility of a win–win solution would largely depend on carbon prices and the accessibility of food to these communities.

On the other hand, in mixed crop-livestock systems, trade-offs in land use can occur when mitigating GHG emissions. For example, decisions on land allocation between food-feed crops (i.e., maize) and improved forages (i.e., Napier grass) to raise productivity can have consequences on incomes, food security and soil fertility. The outcomes of these decisions on farmers' livelihoods largely depend on the relative prices of labour, crop and livestock products (Herrero *et al.* 1999, Waithaka *et al.* 2007).

Technological strategies that improve productivity such as improving animal feeding can often lead to efficiency synergies in the use of resources. For example, van Breugel *et al.* (2010) using data from Herrero *et al.* (2008) to compute the livestock water productivity of meat and milk in different production systems in the Nile basin, found that the systems that had the highest GHG efficiencies per unit of output also had higher water productivities per kg of meat or milk. These were often mixed crop livestock systems instead of grazing systems.

Economic trade-offs also arise when mitigating GHG from livestock, and these are dependent on the marginal gains in GHG reduction versus increased livestock production of the strategies implemented. In smallholder systems with low productivity, improving the diet quality of animals often leads to increases in income, but in the developed world, with high-producing animals, the adoption of sophisticated improved feeding practices requires close inspection to ensure economic benefits (US EPA 2006).

## Incentives for mitigation and the role of regulation

While technical options for mitigating emissions from livestock in developing countries do exist, there are various problems to be overcome, related to incentive systems, institutional linkages, policy reforms, monitoring techniques for carbon stocks, and appropriate verification protocols. For the pastoral lands, Reid *et al.* (2004) conclude that mitigation activities have the greatest chance of success if they build on traditional pastoral institutions and knowledge, while providing pastoralists with food security benefits at the same time (see also Neely and De Leeuw). On the other hand, promoting sustainable intensification of smallholder systems requires the development of markets for livestock products, services and inputs; and this in turn needs investment in infrastructure and agricultural development in general (roads, storage systems, etc.) (Herrero *et al.* 2010). Without product outlets, efficiency of production will not increase.

In some commercial dairy systems in the developed and the developing world, incentives for GHG reduction are being developed as part of payment schemes for animal products. Costa Rica, for example, is developing farm-level monitoring and milk payment systems for rewarding more environmentally efficient dairy farmers (H. Leon, Dos Pinos Dairy Cooperative, personal communication).

Some of the best examples of environmental regulation in the livestock sector come from the developed world. A mixture of technology and regulation has had significant impact in the reduction of GHG emissions in intensive systems in industrialised countries. For example, in Europe (EU-12) livestock production increased slightly between 1990 and 2002, while the emissions of $CH_4$ and $N_2O$ decreased 8–9% over the same period (EEA 2009), with some countries like Denmark achieving GHG reductions of more than 20%. There is a long way to go to regulate intensive systems in the developing world, which adopted intensive production practices (breeds, feeds, management), primarily for monogastrics, but not the regulations and policies for managing environmental loads, especially from manure (FAO 2006).

## Conclusions

1   There is potential for mitigating GHG gases from some livestock systems. Reduction in $CO_2$ emissions from land-use change through demand management and structural change adjustments in the supply of animal products has significant potential for reducing emissions. Technological options can also play a large role in methane and nitrous oxide mitigation.

2   The magnitude and scope for mitigation is largely dependent on the objectives of the system, the GHG efficiency gap in terms of GHG emissions per unit of product and the marginal abatement costs of the mitigation strategies selected. The marginal gains are higher in systems with low GHG efficiencies per unit of product. However, often these high marginal gains are associated with systems producing low absolute amounts of GHG, and often in systems where objectives may have little to do with efficiency or productivity. Trade-offs and synergies can occur when mitigating GHG in livestock systems. They depend on the type of system, the productive goals of the farmers and the nature and type of resource constraints they experience. Mitigation efforts do not always produce the best compromise in the use of multiple resources, hence indicators like GHG emissions per unit of output need to be part of a more comprehensive set of livelihoods and environmental indicators when evaluating production alternatives.

3   Significant synergies can occur also between mitigation and adaptation efforts, but in some cases inevitable adaptations necessary to support livelihoods (i.e., like in some marginal environments) may not lead to mitigation of GHG from livestock systems.

4   Incentives such as market development and access, investments, regulations and policies are essential for sustaining mitigation efforts.

## References

Augustine, D.J. (2003) Long-term, livestock-mediated redistribution of nitrogen and phosphorus in an East African savanna, *Journal of Applied Ecology*, vol. 40, pp. 137–149.

Baumert, K., Herzog, T. and Pershing, J. (2005) 'Navigating the numbers: Greenhouse gas data and international climate policy'. World Resources Institute, Washington, DC.

Blaxter, K.L. and Clapperton, J.L. (1965) Prediction of the amount of methane produced by ruminants, *British Journal of Nutrition*, vol. 19, pp. 511–522.

Bouwman, A.F., Stehfest, E. and van Kessel, C. (2010) 'N$_2$O emissions from the N cycle in arable agriculture: estimation and mitigation', in K. Smith (ed.) *Nitrous Oxide and Climate Change*. Earthscan, UK.

Chadwick, D.R. (2005) Emissions of ammonia, nitrous oxide and methane from cattle manure heaps: Effect of compaction and covering, *Atmospheric Pollution*, vol. 39, pp. 787–799.

Conant, R.T. and Paustian, K. (2002) Potential soil carbon sequestration in over-grazed grassland ecosystems, *Global Biogeochemical Cycles*, vol. 16, pp. 1143–1152.

Dalal, R.M., Gibson, I., Allen, D.E. and Menzies, N.W. (2010) Green waste compost reduces nitrous oxide emissions from feedlot manure applied to soil, *Agriculture Ecosystems and Environment*, vol. 136, pp. 273–281.

Delgado, C., Rosegrant, M., Steinfeld, H., Ehui, S. and Courbois, C. (1999) 'Livestock to 2020: the next food revolution', Food, Agriculture and the Environment Discussion Paper 28. IFPRI/FAO/ILRI, Washington, DC.

Delve, R.J., Cadisch, G., Tanner, J.C., Thorpe, W., Thorne, P.J. and Giller, K.E. (2001) Implications of livestock feeding management on soil fertility in the smallholder farming systems of sub-Saharan Africa, *Agriculture Ecosystems and Environment*, vol. 84, pp. 227–243.

De Vries, M and De Boer, I.J.M. (2009) Comparing environmental impacts for livestock products: A review of life cycle assessments, *Livestock Science*, vol. 128, pp. 1–11.

Dick, J., Kaya, B., Soutoura, M., Skiba, U., Smith, R., Niang, A. and Tabo, R. (2008) The contribution of agricultural practices to greenhouse gas emissions in semi-arid Mali, *Soil Use Management*, vol. 21, pp. 292–301.

Diogo, R.V.C., Buerkert, A. and Schlecht, E. (2010) Horizontal nutrient fluxes and food safety in urban and peri-urban vegetable and millet cultivation of Niamey, Niger, *Nutrient Cycling in Agroecosystems*, vol. 87, no. 1, pp. 81–102.

EEA (2009) 'Annual European community greenhouse gas inventory 1990–2007 and inventory report 2009', *Submission to the UNFCCC Secretariat*, European Environment Agency, Brussels, Belgium.

FAO (2006) 'Livestock's long shadow: Environmental issues and options'. Food and Agriculture Organization of the United Nations, Rome, Italy.

FAO (2009) 'The state of food and agriculture', *Livestock in the Balance*. Food and Agriculture Organization of the United Nations, Rome, Italy.

FAOSTAT (2010) 'FAO Statistical Databases'. Food and Agriculture Organization of the United Nations, Rome, Italy.

Fargione, J., Hill, J., Tilman, D., Polasky, S. and Hawthorne, P. (2008) Land clearing and the biofuel carbon debt, *Science*, vol. 319, pp. 1235–1238.

Galloway, J.N., Aber, J.D., Erisman, J.W., Seitzinger, S.P., Howarth, R.W., Cowling E.B. and Cosby, B.J. (2003) The nitrogen cascade, *BioScience*, vol. 53, pp. 341–356.

Gill, M., Smith, P. and Wilkinson, J.M. (2010) Mitigating climate change: The role of domestic livestock, *Animal*, vol. 4, pp. 323–333.

Havlík, P., Herrero, M., Obersteiner, M., Mosnier, A., Schmidt, E., Thornton, P.K., Kraxner, F. Fritz, S., Aoki, K., Kruska, R. and Notenbaert, A. (2009) 'Livestock, GHG and global models for integrated assessment', *Livestock Sector Evolution: Trade-offs With Food, Feed and Biofuels, and Solutions to Deforestation.* Side-event at the 15th Conference of Parties of the UNFCCC (COP15), December 2009, Copenhagen, Denmark.

Havlík, P., Schneider, A.U., Schmid, E., Böttcher, H., Fritz, S., Skalský, R., Aoki, K., de Cara, S., Kindermann, G., Kraxner, F., Leduc, S., McCallum, I., Mosnier, A, Sauer, T. and Obersteiner, M. (2010), 'Global land-use implications of first and second generation biofuel targets', *Energy Policy*, doi:10.1016/j.enpol.2010.03.030.

Herrero, M., Fawcett, R.H. and Dent, J.B. (1999) Bio-economic evaluation of dairy farm management scenarios using integrated simulation and multiple-criteria models, *Agricultural Systems*, vol. 62, pp. 149–168.

Herrero, M., Thornton, P.K., Kruska, R. and Reid, R.S. (2008) Systems dynamics and the spatial distribution of methane emissions from African domestic ruminants to 2030, *Agriculture, Ecosystems & Environment*, vol. 126, pp. 122–137.

Herrero, M., Thornton, P.K., Gerber, P. and Reid, R.S. (2009) Livestock, livelihoods and the environment: Understanding the trade-offs, *Current Opinion in Environmental Sustainability*, vol. 1, pp. 111–120.

Herrero, M., Thornton, P.K., Notenbaert, A.M., Wood, S., Msangi, S., Freeman, H.A., Bossio, D., Dixon, J., Peters, M., van de Steeg, J., Lynam, J., Parthasarathy Rao, P., Macmillan, S., Gerard, B., McDermott, J., Seré, C. and Rosegrant, M. (2010) Smart investments in sustainable food production: Revisiting mixed crop-livestock systems, *Science*, vol. 327, no. 5967, pp. 822–825.

Herrero, M., Havlík, P., Rufino, M., Notenbaert, A.M., Heike, J., Kruska, R., Thornton, P.K., Obersteiner, M., Duncan, A., Blummel, M., Wright, I. and Reid, R.S. (2011) Global livestock: biomass use, livestock products, excretions and greenhouse gas emissions, *PNAS (submitted)*.

Holst, J., Liu, C., Yao, Z., Bruggemann, N., Zheng, X., Han, X. and Butterbach-Bahl, K. (2007) Importance of point sources on regional nitrous oxide fluxes in semi-arid steppe of Inner Mongolia, China, *Plant Soil*, vol. 296, pp. 209–226.

Huijsmans, J.F.M., Hol, J.M.G. and Vermeulen, G.D. (2003) Effect of application method, manure characteristics, weather and field conditions on ammonia volatilization from manure applied to arable land, *Atmospheric Environment*, vol. 37, pp. 3669–3680.

Marland, G., Obersteiner, M. and Schlamadinger, B. (2007) The carbon benefits of fuels and forests, *Science*, vol. 318, pp. 1066.

McMichael, A.J., Powles, J.W., Butler, C.D. and Uauy, R. (2007) Food, livestock production, energy, climate change, and health, *The Lancet*, vol. 370, no. 9594, pp. 1253–1263, DOI:10.1016/S0140–6736(07)61256–2.

McCrabb, G.J. (2002) 'Nutritional options for abatement of methane emissions from beef and dairy systems', in Takahashi, J. and Young, B.A. (eds) *Greenhouse Gases and Animal Agriculture*. Elsevier Science B.V., The Netherlands, pp. 115–124.

Monteny, G.J., Bannink, A., Chadwick, D. (2006) Greenhouse gas abatement strategies for animal husbandry, *Agriculture, Ecosystems & Environment*, vol. 112, pp. 163–170.

Murwira, H.K. (1995) Ammonia losses from Zimbabwean cattle manure before and after incorporation into soil, *Tropical Agriculture*, vol. 72, pp. 269–273.

OECD (2008) *Biofuel Support Policies: An Economic Assessment*, OECD publishing.

Oenema, O., Oudendag, D. and Velthof, G.L. (2007) Nutrient losses from manure management in the European Union, *Livestock Science*, vol. 112 pp. 261–272.

Olivier, J.G.J. and Berdowski, J.J.M. (2001) 'Global emissions sources and sinks', in: Berdowski, J., Guicherit, R. and Heij, B.J. (eds) *The Climate System*, pp. 33–78. A.A. Balkema Publishers/Swets & Zeitlinger Publishers, Lisse, The Netherlands.

O'Mara, F.P. (2011) The significance of livestock as a contributor to global green-house gas emissions today and in the near future, *Animal Feed Science and Technology* (in press).

Pitesky, M., Stackhouse, K. and Mitloehner, F. (2009) Clearing the air: livestock's contribution to climate change, *Advances in Agronomy*, vol. 103, pp. 1–40.

Predotova, M., Schlecht, E. and Buerkert, A. (2010) Nitrogen and carbon losses from dung storage in urban gardens of Niamey, Niger, *Nutrient Cycling in Agroecosystems*, vol. 87, no. 1, pp. 103–114.

Reid, R.S., Thornton, P.K., McCrabb, G.J., Kruska, R.L., Atieno, F. and Jones, P.G. (2004) Is it possible to mitigate greenhouse gas emissions in pastoral ecosystems of the tropics? *Environment Development and Sustainability*, vol. 6, pp. 91–109.

Rufino, M.C., Rowe, E.C., Delve, R.J. and Giller, K.E. (2006) Nitrogen cycling efficiencies through resource-poor African crop-livestock systems: A review, *Agriculture, Ecosystems & Environment*, vol. 112, pp. 261–282.

Rufino, M.C., Tittonell P., van Wijk, M.T., Castellanos-Navarrete, A., Delve, R.J., de Ridder, N. and Giller, K.E. (2007) Manure as a key resource within smallholder farming systems: Analysing farm-scale nutrient cycling efficiencies with the NUANCES framework, *Livestock Science*, vol. 112, pp. 273–287.

Sanchez-Martin, L., Dick, J., Bocary, K., Vallejo, A. and Skiba, U.M. (2010) Residual effect of organic carbon as a tool for mitigating nitrous oxides emissions in semi-arid climate, *Plant and Soil*, vol. 326, pp. 137–145.

Searchinger, T., Heimlich, R., Houghton, R.A., Dong, F., Elobeid, A., Fabiosa, J., Tokgoz, S., Hayes, D. and Yu, T.H. (2008) Use of U.S. croplands for biofuels increases greenhouse gases through emissions from land-use change, *Science*, vol 319, pp. 1238–1240.

Smith, P., Martino, D., Cai, Z., Gwary, D., Janzen, H., Kumar, P., McCarl, B., Ogle, S., O'Mara, F., Rice, C., Scholes, B. and Sirotenko, O. (2007) 'Agriculture', in *Climate Change 2007: Mitigation, Contribution of Working Group III to the Fourth Assessment Report of the Intergovernmental Panel on Climate Change*, Edited by Metz, B., Davidson, O.R., Bosch, P.R., Dave, R. and Meyer, R.L. Cambridge University Press, Cambridge, UK.

Stehfest, E., Bouman, L., van Vuuren, D., den Elzen, M., Eickhout, B. and Kabat, P. (2009) Climate benefits of changing diet, *Climatic Change*, vol. 95, pp. 83–102.

Steinfeld, H. and Gerber, P. (2010) Livestock production and the global environment: Consume less or produce better, *PNAS*, vol. 107, pp. 18237–18238.

Tittonell, P., Rufino, M.C., Janssen, B.H. and Giller, K.E. (2010) Carbon and nutrient losses during manure storage under traditional and improved practices in small-holder crop-livestock systems-evidence from Kenya, *Plant Soil*, vol. 328, pp. 253–269.

Thornton, P.K. and Herrero, M. (2010) The potential for reduced methane and carbon dioxide emissions from livestock and pasture management in the tropics, *PNAS*, vol. 107, pp. 19667–19672.

US EPA (2006) 'Global anthropogenic non-$CO_2$ emissions: 1990–2020'. United States Environmental Protection Agency, Washington, DC.

Velthof, G.L. and Mosquera, J. (2011) The impact of slurry application technique on nitrous oxide emission from agricultural soils, *Agriculture, Ecosystems & Environment*, vol. 140, pp. 298–308.

Van Breugel, P., Herrero, M., Van de Steeg, J. and Peden, D. (2010) Livestock water use and productivity in the Nile Basin, *Ecosystems*, vol. 13, pp. 205–221.

Van der Stelt, B., Van Vliet, P.C.J., Reijs, J.W., Temminghoff, E.J.M. and Van

Riemsdijk, W.H. (2008) Effects of dietary protein and energy levels on cow manure excretion and ammonia volatilization, *Journal of Dairy Science*, vol. 91, pp. 4811–4821.

Waithaka, M.M., Thornton, P.K., Shepherd, K.D. and Herrero, M. (2007) Bio-economic evaluation of farmers' perceptions of sustainable farms in Western Kenya, *Agricultural Systems*, vol. 90, pp. 243–271.

# 28

# HOME ON THE RANGE

## The contribution of rangeland management to climate change mitigation

*Constance L. Neely and Jan De Leeuw*

### State of knowledge

#### *Global significance of extensive grazing systems*

The world's 34 million km² of rangeland[1] make up 24% of its land area and support 50% of its livestock (UNCCD 2010). In the developing world livestock is the main output from rangelands and socially and economically critical to rural livelihoods. Livestock also composes a fast-growing agricultural sub-sector, accounting for as much as 50% of GDP in countries with significant areas of rangeland (World Bank 2007). The economic importance of rangeland might increase when demand for meat in the developing world increases as forecasted (FAO 2009).

Large areas of rangeland are under pastoralism, which supports around 200 million households worldwide (Nori *et al.* 2005). Poverty is widespread among African pastoralists; in 2005 40% of the 314 million African poor were in pastoral areas[2]. Yet, poverty is a complex issue in a pastoral context (Little *et al.* 2008) and pastoralism is considered the most appropriate strategy to maintain human well-being in rangelands, because it provides secure livelihoods, conserves ecosystem services, promotes wildlife conservation and honours cultural values and traditions (ILRI 2006; IUCN 2006). Rangelands also store significant amounts of carbon with significant potential for carbon storage-based climate change mitigation. Some 36% of the global terrestrial carbon stock[3] and 59% of the African stock is in drylands (Campbell *et al.* 2008; UNEP 2008). Because of this it has been suggested (Reid *et al.* 2004; Gerber and Steinfeld 2008) that carbon sequestration could offer potential to reduce poverty among pastoralists.

This chapter reviews the interaction between rangelands and climate change and the potential for these systems to sequester carbon for climate change mitigation.

*Impacts of livestock grazing systems on greenhouse gases*

## Livestock

Livestock directly contributes 18% to emissions of global anthropogenic greenhouse gases (GHGs): 37% of global methane ($CH_4$) emissions and 65% of nitrous oxide ($N_2O$) emissions (FAO 2006; Steinfeld *et al.* 2006, see also chapter by Herrero *et al.*). Of the total emissions from livestock production systems, extensive systems contribute more than two times that of intensive systems, 70% and 30% respectively. (Steinfeld *et al.* 2006).

## Land degradation

Drylands are sensitive to desertification; globally 10–20% is considered degraded (Millennium Ecosystem Assessment 2005) and 31% of the African rangeland soils (Oldeman 1994). Grazing is an important driver of this degradation and grazing-induced desertification is estimated to emit 8.2 Mt C/yr (Steinfeld *et al.* 2006), with 0.27 Gt C/yr emitted as a result of desertification.

## Burning

The global carbon released from grassland by predominately human-induced burning is 1.6 Gt C/yr (Andreae 1991; Andreae and Warneck 1994; Andreae and Merlet 2001). Savannahs contribute 42% of the total carbon released from biomass burning (Levine *et al.* 1999; Andreae 1991). Biomass burning reduces the replenishment of soil organic carbon (SOC), with consequences for the water infiltration and retention capacity and soil health of the upper soil horizon (Vagen *et al.* 2005).

## Land conversion

Conversion of rangelands to cropland, including biofuel production, is a major cause of emissions, resulting in a loss of 95% above-ground and up to 60% below-ground carbon (Reid *et al.* 2004; Guo and Gifford 2002). Degradation of above-ground vegetation can lose 6 t C/ha and soil degradation leads to a loss of 13 t C/ha (Woomer *et al.* 2004). Almost all Latin and North American grasslands have been converted to cropland, while significant areas of the original grassland biomes remain under grassland in Africa, Central Asia and Australia (Fig. 1; White *et al.* 2000).

## Potential for sequestration and avoidance of emissions

Apart from their large extent, rangelands have significant potential to mitigate climate change through carbon management for two reasons. First,

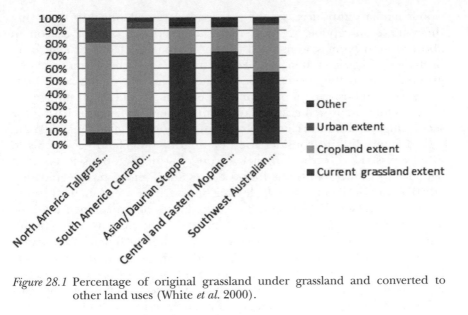

*Figure 28.1* Percentage of original grassland under grassland and converted to other land uses (White *et al.* 2000).

rangelands are under-saturated in carbon because of the above reported historical losses due to land-use change, burning and desertification. It is this human-induced under-saturation that offers potential to sequester carbon while restoring depleted soil and above-ground carbon stocks. Besides, rangelands also have potential to avoid emissions of carbon from land-use change and erosion.

Most studies have focussed on the potential to sequester carbon in rangeland soils. Lal (2004) estimated that worldwide dryland soils could sequester 1 Gt C/yr, while Smith *et al.* (2007) estimated that improved rangeland management would globally sequester 0.35–0.55 Gt C/yr up to 2030. Batjes (2004) estimated that improved management of 10% of the African grazing lands could increase soil carbon stocks by 13–28 Mt C/yr. Natural or improved fallow systems, under agroforestry and managed for resting of land, have potential sequestration rates of 0.1–5.3 t C/ha/yr (Vagen *et al.* 2005). Dryland soils may continue to sequester carbon up to 20 to 50 years, upon which the soils saturate and the system reaches steady state (Lal *et al.* 1998; Conant *et al.* 2001).

Far less attention has been given to the potential for above-ground carbon storage. There is potential, as the development of woody biomass in many rangelands is suppressed by overgrazing, fuelwood collection and control of bush encroachment. Sequestering carbon in above-ground biomass is surrounded by a few complexities. Increased woody biomass creates trade-offs with existing land uses, such as livestock production and fuelwood collection, which need to be considered when implementing sequestration of above-ground carbon in rangelands. Accumulation of woody biomass also increases the fuel load in rangelands thus rendering them more susceptible to bush fires. Long-term storage of carbon in above-ground biomass thus needs to be accompanied by fire control to overcome risks of carbon loss. Sequestration of carbon in the above-ground biomass thus needs to be considered in view of the restrictions that recurrent bush fire imposes on the permanence of carbon storage.

Altogether, improved grazing land management is viewed as having the second highest technical climate change mitigation potential (Figure 28.2).

It is interesting to consider the potential of rangeland ecosystems to avoid emissions of carbon, in view of ongoing deliberations to broaden agendas to reduce emissions from deforestation and forest degradation (REDD) to also include avoidable emissions driven by other than forest-related processes. We suggest that avoidance of emissions could be achieved when reducing land cover change at the arid frontier (Bruins and Lithwick 1998). One might consider developing financial incentives to reward carbon emission-avoiding land management, to reduce the largely poverty-driven conversion at the arid frontier of productive range-land to marginal croplands. Further avoidance of emissions appears achievable through reduction of bush fire and erosion control.

Achieving climate change mitigation in rangelands needs improved understanding of the effects of current and alternative grazing land management practices on soil carbon sequestration (Schuman *et al.* 2002). Conant *et al.* (2001), IPCC (2007) and FAO (2010) suggest the following practices to sequester carbon in grasslands: improved grazing management,

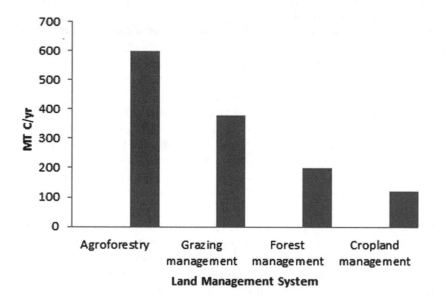

*Figure 28.2* Carbon sequestration potential of four land-use systems (adapted from IPCC 2000; Swaminathan 2009).

sowing improved species, direct inputs of water, fertilizer or organic matter (which can enhance water and nitrogen balances), managing fire and restoring degraded lands.

Conant *et al.* (2001) considered that improved grazing management may sequester 0.54t C/ha/yr. It is estimated that mitigation from rangelands and restoration of degraded soils together have the potential to sequester around 0.53 Gt C/yr until 2030 (Smith *et al.* 2008). Improved pastures offer significant sequestration within the context of the livestock sector (FAO 2006).

Cessation of overgrazing may sequester 45.7 Mt C/yr globally (Conant and Paustian 2002). The relation with grazing intensity is however not clear cut. Reeder and Schuman (2002) reported higher soil carbon in grazed pastures, while Milchunas and Lauenroth (1993), reviewing 34 studies comparing grazed and ungrazed sites, reported that soil carbon increased in 60% and decreased in 40% of all cases.

Cases where degraded rangelands have been regenerated include the Shinyanga in Tanzania, where agro-silvopastoral practices on some 500,000 ha, incorporating a combination of woodlots, fodder banks, alley and mixed cropping, boundary and tree plantings and natural revegetation resulted in 1.7–2.4 t/ha/yr C sequestration (Rubanza *et al.* 2009; Pye-Smith 2010). Another range management intervention with potential for carbon sequestration is Holistic Planned Grazing (Savory 1999), based on managing timing of grazing rather than number of animals, implemented by private ranchers and pastoralists on over 12 million ha.

Apart from its climate change mitigation potential, soil C sequestration has co-benefits as it increases adaptation capacity through improving soil physically (e.g., improved structural stability, erosion resistance, water-holding capacity and aeration), chemical (e.g., enhanced availability of micronutrients) and biological (e.g., enhanced faunal activity and species diversity) attributes (Lal 2004; FAO 1995), along with productivity.

## Knowledge gaps and issues

While a number of the knowledge gaps and issues identified below apply to contexts other than extensive grazing lands, addressing them is critical for capitalizing on the mitigation potential that can be achieved through grazing lands management.

### Institutional and policy dimension

#### Land tenure and governance issues

Tenure and governance of commonly managed land must be addressed for pastoral people to benefit from mitigation of GHG emissions and sequestered carbon on grazing lands. Carbon credits require a legal entity delivering the service; this is easier with privately owned property than with communally owned lands. This issue is particularly complicated under mobility, where those managing the land may reside there only occasionally. Experience gained with benefits from wildlife conservation reveal significant leakages of benefits towards pastoral elites; such elite capture reduces the incentive for pastoral land managers to manage their land for carbon. Equity issues thus need to be resolved before credit schemes may stimulate the de facto land manager to sequester carbon.

### Recognition of the contributions made by pastoralist populations

Notwithstanding the potential for livestock keepers to manage extensive grasslands with minimal external inputs while providing beneficial public goods, marginalization of pastoralist populations is still the norm. Unless pastoralists are recognized by policy makers for their potential to manage ecosystem resources, the risk is run of losing 'mitigation managers'.

### Lack of agreement on measuring and monitoring carbon stock changes

With the promise of carbon credit schemes for agricultural land management, there is a clear need for consistent methods to measure, monitor and report below- and above-ground carbon stocks (see chapter by Olander). These methods need to be objective and agreeable to both service providers (pastoral land managers) and investors who are often at a distance from one another (Reid *et al.* 2004). Further, regular monitoring at fine resolution might consume significant fractions of the carbon credits, leaving little as an incentive to landowners to develop carbon storage-oriented land management. An option to reduce monitoring costs would be to make payments on land practices that stimulate carbon storage among other co-benefits. Remote sensing techniques have potential to monitor land-use change, fire and exposure of bare soil, and could thus support carbon sequestration (De Leeuw *et al.* 2010), when making carbon credits available on the condition that the land is managed to store carbon, e.g., not burned, not converted to cropland and retains a protective vegetation cover.

### Incentive packages

There is too little insight in the complexities of organizing payment for service delivery in communal lands, and the institutional arrangements required to ensure that service providers tap into carbon markets. Beyond carbon credits as a means of incentivizing good grazing management strategies, other innovations in incentives that can promote sustainability should be aggressively explored. For example, improved access to services (health, education, extension, markets) should be considered along with financial credit schemes such as the Equity Bank of Kenya whose mission is to offer inclusive, customer-focused financial services that socially and economically empower clients and other stakeholders. Given their current low value, carbon credits would serve only as "icing." Increased productivity can be the incentive of choice with appropriate capacity development and available livelihood services. Further, there is a need to ensure that awaiting working carbon credit schemes does not provide a disincentive for improving land management strategies now.

## Implementation

### Impermanence and saturation

While there is great potential for carbon to be accumulated through good grazing management strategies, sequestered carbon can be lost readily (FAO 2010) through fire, tillage, and land conversion. Soil carbon also can reach a saturation level after 25–50 years. Despite these limitations, the cost of not improving land health and productivity in grazing lands is immense, and 25–50 years affords a time frame for finding additional mitigation solutions.

### Stocktaking of good practices

There are numerous innovations in regenerating degraded land and improving grazing lands (pastoral and silvopastoral) management. Given the potential for mitigation of this land use, a descriptive cataloguing of good practices should be carried out. This can be an important addition to the World Overview of Conservation Technologies and Approaches (WOCAT 2009), which scores the multiple benefits of different sustainable land management technologies.

### Lack of awareness and capacity development

For long-term productive landscapes, carbon accumulation must be accompanied by the enhancement of each of the ecosystem processes. Pastoralists can manage for effective water and nutrient cycles, enhanced biological diversity and enhanced capture of solar energy. When management practices address these different processes, the underpinning natural resource foundation can be sustained. Capacity development associated with good practices is essential.

### Research dimension

#### Full engagement of rangeland managers and users

While progress is being made, rangeland managers and users have not had the opportunity to contribute adequately to exploring and designing mitigation options. Climate change mitigation options in rangelands cannot be considered without considering their costs and benefits. Because of possible trade-offs between current land use and mitigation options, there is an urgent need to account for these trade-offs in any economic feasibility study on climate change mitigation in rangeland environments.

#### Limited data on potential for carbon storage

Thus far there is little insight in the total carbon storage that can be achieved in these systems, a lack in knowledge which limits our ability to assess how much carbon these systems could store, for how long they could play a role in sequestering carbon and when these systems, when managed to sequester carbon, would saturate. Proper assessment of the potential of rangeland systems to accumulate and store carbon requires more data on current carbon levels, the carbon gap and the maximum amount that the system could store. One issue plaguing this effort is definitional in nature as data related to drylands, rangelands and grasslands tend to be conflated. Further, there is limited data on the impacts of different management practices on carbon stocks in developing countries (FAO 2010) including standing stock rates, rotational grazing, Holistic Planned Grazing, exclusion of animals, integration of trees, farmer-managed natural regeneration and burning. While there are examples of large-scale reversal of degradation on grazing lands, there is still a lack of area-specific evidence. Further, there is limited data and knowledge on the potential for sequestration and avoidance of emissions in African smallholder rangelands (Tennigkeit and Wilkes 2008) and the trade-offs of this with existing land, livestock and crop-based livelihoods. Evidence is needed to support practices that regenerate ecosystem processes at a landscape level, including carbon accumulation and storage to advocate for national, regional and global policies that foster sustainable pro-carbon management of rangeland environments.

#### Short- versus long-term gains

Carbon accumulation takes time and changes are often not discernable in short-term projects. Long-term mitigation and adaptation programs need to be put in place that will allow biocarbon accumulation. For development efforts, project-level predictions related to changes in management practices such as the Ex-Act tool (FAO 2010) can be useful in anticipating gross accumulation rates over long-term implementation based on individual and bundles of practices.

*Variability*

While good practices can readily be initiated, it is likely that the potential for sequestration and avoidance of emissions in rangelands varies spatially much more so than in forested areas, as soil carbon varies significantly across landscapes. This may have cost implications for monitoring, reporting and verification.

*Albedo and geo-engineering*

Managing land to sequester carbon affects climate in other ways than through GHGs alone. Tree and shrub species differ in albedo (they typically reflect less of the incoming solar radiation) from grasses or bright, bare soil surfaces. Otterman (1977) suggested, based on observed impacts of grazing on rangeland albedo in the Middle East, that the introduction of grazing and associated deforestation since the mid-Holocene led to increased albedo (i.e., higher reflectance), and he postulated that this has led to a global cooling by 1–2 °C. A recent study on afforestation in the Negev revealed that during the first 80 years after afforestation, negative albedo effects outweighed positive effects of the sequestration in above- and below-ground carbon (Rotenberg and Yakir 2010; Schimel 2010).

Geo-engineering, the direct manipulation of the radiative balance of the earth, is increasingly promoted as options to store carbon in terrestrial ecosystems may not suffice (Lenton and Vaughan 2009). One geo-engineering option that has attracted recent attention is the planting of, or management for, plant species with increased albedo, thus promoting the reflection of a greater proportion of the incoming solar radiation out of the earth's system. Hamwey (2007) suggested that rangelands offer an opportunity for geo-engineering, by planting or managing for higher albedo rangeland species.

On the other hand low albedo land cover stimulates rainfall (Charney 1975; Samain *et al.* 2008), as the higher energy load captured by the land surface stimulates convection that triggers thunderstorms. The albedo effects of large- scale carbon management in rangelands on local and global climate have yet to be accounted for.

*Analogue site species*

Understanding the grassland species (e.g., grasses, legumes, shrubs, forage trees) that best contribute to carbon sequestration or accumulation and increase forage quality for reduced enteric fermentation will be important

341

for mitigation in grazing lands. For rangelands, it will be important to characterize analogue sites associated with expected climate change over the next 20–50 years to identify and anticipate the appropriate species for mitigation and adaptation.

### Life cycle analysis and landscape assessment

Life cycle analyses provide information on the carbon emissions per unit of product (e.g., milk, meat). These analyses must be carried out within the context of the system in which the animals are produced—either on-farm or at the landscape level to ensure an understanding of net GHG emissions of the whole system. Sustainable practices including other non-livestock elements represented in the farming system (agroforestry, conservation agriculture, etc.) may offset livestock-related emissions, constituting a potentially unanticipated *decrease* or *increase* in GHG benefits as associated with overall production activities (leakage). In the case of grazing lands, a landscape assessment should be carried out to characterize net mitigation as well as other ecosystem services.

## Recommendations and next steps

Sustainable grazing systems in drylands offer a high potential for climate change mitigation, both through sequestration in vegetation and soil and through avoidance of emissions. Mitigation measures should be compatible with sustaining livestock-based livelihoods, improving productivity, food security and nutrition, and enhancing biological diversity and water quality and quantity. Because of their extensive nature and the number of people these environments support, and the multiple co-benefits associated including those associated with adaptation and risk reduction, these systems warrant greater attention to scale up improved practices.

Mitigation efforts must take a long and multi-dimensional view and be framed in the broader context of a systems approach for sustainability. There is a need to design projects and programmes that will:

- result in greater recognition and technical and political support at local to global scale for sustainable pastoral and silvo-pastoral systems in view of their contributions to climate change mitigation, adaptation, disaster risk management (Neely *et al.* 2009) and agriculture in a green economy;
- provide evidence of the mitigation potential of different, emerging rangelands and degraded land management practices taking into account variability, time of soil accumulation, integration of different perennial species, albedo effects, water use efficiency and expected changes in climate;

- engage rangeland managers (pastoralists and others) in the exploration and design of mitigation and adaptation efforts;
- explore and develop with service providers innovative incentive structures that promote improved grazing lands management;
- build on successes and focus capacity development on managing ecosystem processes through improved grazing lands management;
- establish clear and appropriate tools and methods for accounting of carbon emissions and accumulation in rangelands.

## Notes

1 This paper reviews literature on drylands: lands with mean annual potential evapotranspiration (PET) at least 1.5 times greater than mean annual precipitation (Middleton and Thomas 1997), and rangelands—extensively used, mostly grazed lands with native vegetation. The two systems overlap significantly, rangelands being mostly located in drylands. We use the term rangelands when generalizing and are specific when referring to literature.
2 www.worldpress.org/Africa/2861.cfm.
3 Figures include above- and below-ground carbon unless specified otherwise.

## References

Andreae, M.O. (1991) 'Biomass burning: Its history, use and distribution and its impact on environmental quality and global climate. In Levine, J.S. (ed.), *Global Biomass Burning: Atmospheric, Climatic and Biospheric Implications*, pp. 3–21. Massachusetts Institute of Technology Press, Cambridge, MA.

Andreae, M.O. and Merlet, P. (2001), Emission of trace gases and aerosols from biomass burning, *Global Biogeochemical Cycles*, vol. 15, no. 4, pp. 955–966.

Andreae, M.O. and Warneck, P. (1994) Emission of trace gases and aerosols from biomass burning, *Global Biogeochemical Cycles*, vol. 15, pp. 955–966.

Batjes, N.H. (2004) Estimation of soil carbon gains upon improved management within croplands and grasslands in Africa, *Environ. Dev. Sustain.*, pp. 6: 133–143.

Bruins, H.J. and Lithwick, H. (1998) 'The arid frontier: Interactive management of environment and development'. Springer, pp. 381.

Campbell, A., Miles, L., Lysenko, I., Hughes, A. and Gibbs, H. (2008) 'Carbon storage in protected areas', Technical report. UNEP World Conservation Monitoring Centre.

Charney, J.G. (1975) Dynamics of deserts and droughts in the Sahel, *Quarterly Journal of the Royal Meteoreological Society*, vol. 101, no. 428, pp. 193–202.

Conant, R.T. and Paustian, K. (2002) Potential soil carbon sequestration in overgrazed grassland ecosystems, *Global Biogeochem. Cycles*, vol. 16, no. 4, pp. 1143.

Conant, R.T., Paustian, K. and Elliott, E.T. (2001) Grassland management and conversion into grassland: Effects on soil carbon, *Ecol. Appl.*, vol. 11, no. 2, pp. 343–355.

De Leeuw, J., Georgiadou, Y., Kerle, N., De Gier, A., Inoue, Y., Ferwerda, J. Smies, M. and Narantuya, D. (2010) The function of remote sensing in support of environmental policy, *Remote Sensing*, vol. 2, pp. 1731–1750.

FAO (1995) 'Sustainable dryland cropping in relation to soil productivity', *FAO Soils Bulletin 72*. Rome, Italy.

FAO (2006) *FAO statistical database* [Online]. Available at: http://faostat.fao.org/default.aspx/. Rome, Italy.

FAO (2009) 'State of food and agriculture 2009', *Livestock in the Balance*. Rome, Italy.

FAO (2010) 'Challenges and opportunities for carbon sequestration in grassland systems', *A Technical Report on Grassland Management and Climate Change Mitigation Integrated Crop Management*, vol. 9–2010, compiled by Rich Conant. Rome, Italy.

Gerber, P. and Steinfeld, H. (2008). 'Livestock's role in global climate change', Presentation to British Society of Animal Science Conference on Livestock and Global Climate Change, Hammamet, Tunisia, May 2008, Cambridge Press.

Guo, L. and Gifford, R. (2002) Soil carbon stocks and land use change: A meta analysis, *Global Change Biol.*, vol. 8 pp. 345–360.

Hamwey, R., (2007) Active amplification of the terrestrial albedo to mitigate climate change: An exploratory study, *Mitigation and Adaptation Strategies for Global Change*, vol. 12, pp. 419.

ILRI (International Livestock Research Institute). (2006) 'Pastoralist and poverty reduction in East Africa'. International Livestock Research Institute (ILRI) Conference, June, 2006, Nairobi, Kenya.

IPCC (2000) 'Land use, land use change, and forestry', *A special report of the IPCC*. Cambridge University Press, Cambridge, UK.

IPCC (2007) 'Climate Change 2007 Mitigation', *Contribution of Working Group III to the Fourth Assessment Report of the Intergovernmental Panel on Climate Change*. Cambridge University Press, Cambridge, UK and New York, NY.

IUCN (2006) *Global Review of the Economics of Pastoralism* [Online]. Available at: http://cmsdata.iucn.org/downloads/global_review_ofthe_economicsof_pastoralism_en.pdf. Hatfield, R. and Davies, J., The World Initiative for Sustainable Pastoralism.

Lal, R. (2004) Carbon sequestration in dryland ecosystems, *Environ. Manag.*, vol. 33, no. 4, pp. 528–544.

Lal, R., Kimble, J., Follet, R. and Cole, C.V. (1998) 'Potential of US cropland for carbon sequestration and greenhouse effect mitigation'. Sleeping Bear Press, Chelsea, MI.

Lenton, T.M. and Vaughan, N.E. (2009) The radiative forcing potential of different climate geoengineering options, *Atmos. Chem. Phys. Discuss*, vol. 9, pp. 2559–2608.

Levine, J., Bobbe, T., Ray, N., Witte, R. and Singh, A. (1999) 'Wildfires and the environment: A global synthesis', *Environmental Information and Assessment Technical Report 1*, UNEP/DEIA&EW/TR.99-1.

Little, P.D., McPeak, J., Barrett, C.B and Kristjanson, P. (2008) Challenging orthodoxies: Understanding poverty in pastoral areas of East Africa, *Dev. & Change*, vol. 39, pp. 587–611.

Middleton, N. and Thomas, D. (1997) *World Atlas of Desertification*. Arnold, London.

Milchunas, D.G. and Lauenroth, W.K. (1993) A quantitative assessment of the effects of grazing on vegetation and soils over a global range of environments, *Ecol. Monogr.*, vol. 63, pp. 327–366.

Millennium Ecosystem Assessment (2005) Available at: www.millenniumassessment.org/en/index.aspx.

Neely, C., Bunning, S. and Wilkes, A. (2009) 'Review of evidence on dryland pastoral systems and climate change: Implications and opportunities for mitigation and adaptation', *FAO Land and Water Discussion Paper 8*. Rome, Italy.

Nori, M., Switzer, J. and Crawford, A. (2005) *Herding on the brink: Towards a global survey of pastoral communities and conflict* [Online]. Available at: www.iisd.org/publications/pub.aspx?id=705. An occasional paper from the IUCN Commission on Environmental, Economic and Social Policy, Gland, Switzerland

Oldeman, L.R. (1994) 'The global extent of land degradation', in: *Land Resilience and Sustainable Land Use*, Greenland, D.J. and Szabolcs, I. (eds), pp. 99–118. Wallingford: CABI.

Otterman, J. (1977) Anthropogenic impact on the albedo of the earth, *Climatic Change*, vol. 1, pp. 137.

Pye-Smith, C. (2010) 'A rural revival in Tanzania: How agroforestry is helping farmers to restore the woodlands in Shinyanga Region World Agroforestry Centre', *Trees for Change, No. 7*.

Reeder, J.D. and Schuman, G.E. (2002) Influence of livestock grazing on C sequestration in semi-arid and mixed-grass and short-grass rangelands, *Environ. Pollut.*, vol. 116, pp. 457–463.

Reid, R.S., Thornton, P.K., McCrabb, G.J., Kruska, R.L., Atieno, F. and Jones, P.G. (2004) Is it possible to mitigate greenhouse gas emissions in pastoral ecosystems of the tropics? *Environ. Dev. Sustain.*, vol. 6, pp. 91–109.

Rotenberg, E. and Yakir, D. (2010) Contribution of semi-arid forests to the climate system, *Science*, 327: 451–454.

Rubanza, C.D.K., Otsyina, R. Chibwana, A. and Nshubekuki, L. (2009) 'Characterization of agroforestry interventions and their suitability of climate change adaptation and mitigation in semi-arid areas of northwestern Tanzania'. Poster presented at the Second World Agroforestry Congress, Nairobi, Kenya.

Samain, O., Kergoat, L., Hiernaux, P., Guichard, F., Mougin, E., Timouk, F., Lavenu, F. (2008) Analysis of the in situ and MODIS albedo variability at multiple timescales in the Sahel, *J. Geophysical Research*, 113: doi:10.1029/2007JD009174.

Savory, A. (1999) 'Holistic Management: A new framework for decision making'. Island Press, New York, NY.

Schimel, D.S., (2010) Drylands in the earth system: A study of one of the world's driest forests elucidates the climatic effects of drylands, *Science*, vol. 327, pp. 418–419.

Schuman, G.E., Janzen, H.H. and Herrick, J.E. (2002) Soil carbon dynamics and potential carbon sequestration by rangelands, *Environ. Pollut.*, vol. 116, no. 3, pp. 391–396.

Smith, P., Martino, D., Cai, Z., Gwary, D., Janzen, H.H., Kumar, P., McCarl, B., Ogle, S., O'Mara, F., Rice, C., Scholes, R.J. and Sirotenko, O. (2007) 'Agriculture', *In* Metz, B., Davidson, O.R., Bosch, P.R., Dave, R. and Meyer, L.A. (eds) *Climate Change 2007. Mitigation. Contribution of Working Group III to the Fourth Assessment Report of the Intergovernmental Panel on Climate Change*. Cambridge University Press, Cambridge, UK and New York, NY.

Smith, P., Martino, D., Cai, Z., Gwary, D., Janzen, H.H., Kumar, P., McCarl, B., Ogle S., O'Mara, F., Rice, C., Scholes, R.J., Sirotenko, O., Howden, M., McAllister, T., Pan, G., Romanenkov, V., Schneider, U., Towprayoon, S., Wattenbach, M. and Smith, J.U. (2008) Greenhouse gas mitigation in agriculture, *Phil. Trans. R. Soc. B.*, pp. 363: 789–813.

Steinfeld, H., Gerber, P., Wassenaar, T., Castel, V., Rosales, M. and de Haan, C. (2006) *Livestock's Long Shadow – Environmental Issues and Options*. FAO: Rome, Italy.

Swaminathan (2009) Presentation during Second World Congress of Agroforestry. Nairobi, Kenya.

Tennigkeit, T. and Wilkes, A. (2008) 'An assessment of the potential of carbon finance in rangelands', *World Agroforestry Centre Working Paper, No. 68.*

UNCCD (2010) *United Nations Decade for Deserts and the Fight against Desertification Fact Sheet* [Online]. Available at: http://unddd.unccd.int/docs/factsheet.pdf.

UNEP (2008) 'Carbon in drylands: desertification, climate change, and carbon finance', A UNEP/UNDP/UNCCD technical note for discussions at CRIC 7, 3–14 November, Istanbul, Turkey. Prepared by Trumper, K., Ravilious, C. and Dickson, B.

Vagen, T.G., Lal, R. and Singh, B.R. (2005) Soil carbon sequestration in sub-Saharan Africa: A review, *Land Degrad. Dev.*, pp. 16: 53–71.

White, R., Murray, S. and Rohweder, M. (2000) *Pilot analysis of global ecosystems: grassland ecosystems*. World Resources Institute, Washington, DC, pp. 112.

WOCAT (World Overview of Conservation Approaches and Technologies) (2009) 'Benefits of sustainable land management', *UNCCD World Overview of Conservation Approaches and Technologies*. Swiss Agency for Development and Cooperation, FAO, Centre for Development and Environment, pp. 15.

Woomer, P.L., Touré, A. and Sall, M. (2004) Carbon stocks in Senegal's Sahel transition zone, *J. Arid Environ.*, pp. 59: 499–510.

World Bank (2007) 'World Development Report 2008', *Agriculture for Development*, Washington, DC.

# 29

# INTEGRATED NUTRIENT MANAGEMENT AS A KEY CONTRIBUTOR TO CHINA'S LOW-CARBON AGRICULTURE

*David Norse, David Powlson, and Yuelai Lu*

China has drawn up plans to move to a low-carbon growth path as part of its contribution to global climate change mitigation. It aims to decrease national carbon intensity (the ratio of carbon dioxide ($CO_2$) emissions per unit of economic activity measured as gross domestic product (GDP)) in 2020 to 40–45% of its 2005 value largely by increasing energy efficiency in the industry and transport sectors. At present agriculture is not a formal part of the low-carbon plans in spite of the research findings presented in this chapter, which indicate that in 2007 agriculture and the agrochemical industry accounted for over 10% of China's total fossil energy use, and 19–22% of its greenhouse gas (GHG) emissions. Moreover, most of the measures to mitigate agricultural greenhouse gases (GHGs) have low or negative costs because of the high economic and environmental benefits they provide whereas carbon abatement costs in most other sectors of the economy are up to US$100 per tonne of carbon eliminated (McKinsey 2009). Some of the measures will also increase carbon sequestration but the net contribution is likely to be modest (see next section).

**Box 1**

---

**Improved nitrogen management (INM)** aims to maximize resource efficiency and agricultural sustainability by a combination of five main actions:

- preventing the overuse of synthetic nitrogen (N) fertilizers by ensuring that application rates allow for the N already in the soil, in manure and in irrigation water, and do not exceed the amount needed for high crop yields;

- ensuring that N fertilizer is applied at the right (optimum) time;
- using the most efficient technologies for the application of fertilizers;
- minimizing the non-point pollution, GHG emissions and other environmental impacts of N fertilizers using slow-release fertilizers and "inhibitors" where they are cost-effective, and by expanding the availability and use of organic fertilizers;
- getting the balance right between the amount of nitrogen, phosphate and potassium given to crops.

China has not produced an official national GHG inventory since 1994 when crop-related activities accounted for nearly 10% of total national GHG emissions and the livestock sector for less than 7% (NDRC 2004). Since then there have been major changes in the structure of the total economy and of agriculture—particularly the expansion of vegetable production and intensive livestock production, which have had a substantial impact on agricultural GHG emissions. Consequently, this chapter, which is based on the findings of an ongoing UK–China policy research project[1] (see Lu *et al.*, this volume), will first assess the state of current knowledge of the main sources of agricultural emissions and their biophysical and socio-economic drivers. It will then focus on integrated nutrient management (Box 1) as a package of GHG mitigation measures that have wide social, economic and environmental benefits, before drawing out some of the implications of the Chinese situation for policy planners in other developing countries. Finally, it will highlight research gaps and priorities for policy development.

## China's agricultural GHG emissions

The project has adopted a life cycle analysis approach (Brentrup *et al.* 2001) because the fossil energy inputs to synthetic nitrogen fertilizer production account for about 70% of the total fossil energy inputs to crop production (Table 29.1, column 2) and the methane emissions from the mining of the coal used are not insignificant (Table 29.1, column 4). In 2007 the GHG emissions related to crop production were <10% as in 1994, but this could be a significant underestimate if indirect emissions from synthetic nitrogen (N) fertilizer are much greater than the IPCC default value (IPCC 2006) as argued by Crutzen *et al.* (2007) and Davidson (2009). Emissions from livestock, on the other hand, increased by about three percentage points between 1994 and 2007 to account for ~10% of national emissions. Thus, agricultural GHG emissions now account for 19–22% of China's total emissions (see Table 29.1) and ~30% of the crop-related emissions are a consequence of the misuse or overuse of synthetic N fertilizers (Ma *et al.* 2009), and the underutilization and mismanagement of livestock wastes, which are also a major source of water pollution (Ma and

Table 29.1 Agriculture's contribution to China's energy use and GHG emissions in 2007 (Mt CO$_2$e) (SAIN 2011)

| Source | Fossil energy (Mt sce*) | CO$_2$ | Methane | Nitrous oxide | Total |
|---|---|---|---|---|---|
| N fertilizer production and transport (43Mt) | 117 | 235 | 26 | 13 | 274 |
| Phosphorous and potassium fertilizer production and transport | | 18 | | | 18 |
| N fertilizer use for crops (32Mt) | | 57 | (170 rice**) | 176*** | 233(403) |
| Other agricultural uses (3–5Mt) | | 15–25 | | 15–25 | 30–50 |
| Livestock – enteric and manure | | | 295–443 | 172–258 | 467–701 |
| Direct fossil energy inputs to agriculture | 82 | 190 | | | 190 |
| Total agricultural emissions | | 515–25 | 491–639 | 376–472 | 1382–1636 |
| Total economy emissions | 2656 | 6000 | | | 7230 |
| Agricultural emissions as % of total national emissions | | | | | 19–22 |

Notes

*sce = standard coal equivalent;

**methane emissions are primarily a consequence of the anaerobic conditions in rice fields and are often not closely N related;

***direct emissions plus conservative estimate for indirect N$_2$O emissions.

Zhang 2008). Both of these causes should be seen as an opportunity, rather than a problem, since they both open the way for GHG mitigation measures (Zhu and Chen 2002) and achieving a number of other local, national and regional environmental benefits (Norse 2005). However, the underlying socio-economic driving force has been the substantial improvement in per capita consumption of crop and livestock products and government policies to maintain close to 100% net self-sufficiency in grains. This has resulted since 1980 in a circa fivefold increase in GHG emissions from N fertilizer production, a sixfold increase in N fertilizer use and average N application rates to arable crops of ~230 kg N/ha, 2–5 times that in most other developed and developing countries (Brazil 50 kg, India 90 kg USA and UK ~100 kg). Table 29.1 does not allow for the increase in soil organic carbon arising from the use of fertilizer, which will make a modest contribution to offsetting the large GHG emissions arising from N fertilizer manufacture (Lu *et al.* 2009; Huang and Sun 2006).

## GHG mitigation, INM and misuse or overuse of synthetic N fertilizers

A number of issues need to be clarified before selecting the mitigation options to be given priority and how best to implement them. First, is it a spatially restricted problem needing local or regional action rather than a national strategy? Second, is it restricted to certain crops? Third, is it related to farm size or farm income? Fourth, does it stem from the lack of awareness of the problem or of the response options? Fifth, what are the main economic and institutional barriers to the adoption of mitigation measures?

*Spatial and cropping system dimension.* Misuse and overuse of synthetic N fertilizer is a problem across China. Most studies conclude that overuse is generally in the range 30–50% (Ju *et al.* 2009; Peng *et al.* 2010). Moreover, it has been widespread since the early 1980s (Zhang *et al.* 2006) including, for certain situations, in those provinces where average application rates are not excessive such as Shaanxi (Table 29.2). It is a problem for all of the main grain, fruit and particularly vegetable crops (Table 29.2; Zhang *et al.* 2006) although in some provinces, e.g., Jilin, 10% or more of farmers may be using too little synthetic N fertilizer.

*Effect of farm size and farm income.* Overuse is an issue on all farm sizes from smallholdings of <1 hectare to state farms of several hundred hectares. The greatest impact, as a proportion of income, is for low-income farm households dependent on agriculture and for farm households whose income comes largely from non-agricultural employment (Lu *et al.* 2006; Table 29.3). The costs of synthetic N can be 40–50% of the non-labour costs of production for low-income households and overuse can represent a 5–15% loss in their net farm incomes and substantially more

*Table 29.2* Examples of N overuse by province and crop

| Province | Crop | Farmers' rate kg N/ha | Recommended rate* kg N/ha | % overuse | Yield with recommended rate |
|---|---|---|---|---|---|
| Jiangsu | rice | 300 | 200 | 50 | +3% |
| 6 provinces** | rice | 195 | 133 | 47 | >+5% |
| N. China plain | wheat | 325 | 128 | 150 | +4% |
| N. China plain | maize | 263 | 158 | 66 | +5% |
| Shaanxi | wheat | 249 | 125–225 | ~100 | same |
| Shaanxi | maize | 405 | 165–255 | >60 | ~8% |
| Shandong | tomato | Up to 630 | 150–300 | 80–200 | +10% |

*Source:* Peng *et al.* 2010; Ju *et al.* 2009., Zhang *et al.* 2006., unpublished project data.

*Notes*

* Based on soil tests or field experiments.
** Guangdong, Heilongjiang, Hubei, Huan, Jiangsu and Zhejiang.

Table 29.3 Potential income gains from reducing synthetic N overuse

| Income level | Total household income (US$) | Savings from 30% fertilizer use reduction | | Savings from 50% fertilizer use reduction | |
|---|---|---|---|---|---|
| | | Savings (US$) | % of household income | Savings (US$) | % of household income |
| 1st Quartile | 252 | 23 | 9 | 39 | 15 |
| 2nd Quartile | 983 | 38 | 4 | 63 | 6 |
| 3rd Quartile | 1582 | 34 | 2 | 57 | 4 |
| 4th Quartile | 3070 | 34 | 1 | 56 | 2 |
| Average | 1474 | 32 | 2 | 53 | 4 |

*Note*
Unpublished project data for 2008.

where overuse actually lowers yields (Table 29.2, column 6). Yet the project has never come across this loss of income being used as part of the extension message to farmers.

*Awareness of the problem, its environmental impacts and of response options.* The lack of awareness goes deeper than the income loss issue raised above. Of particular importance is the lack of awareness by farmers and extension workers of the amount of N available from organic matter in the soil, unused fertilizer remaining from previous crops, and the N content of manure, irrigation water, dust and rain. These can total >3,000 kg N/ha/yr. These sources can often provide at least 50% of the N required to give maximum crop yield (Table 29.4) and may even remove the need for synthetic N fertilizer, at least in the short term. A major national programme for soil testing and another one for demonstrating the need for and benefits of balanced fertilization, together with a number of provincial or crop-based demonstration programmes on nutrient management (Ju *et al.* 2009; Peng *et al.* 2010, Fan *et al.* 2009) are progressively reaching many farmers, but there is still widespread unawareness of the benefits of INM or unwillingness to adopt it for a range of institutional and economic reasons as discussed in the next section.

Farmers and advisers are also generally unaware of the large emissions of ammonia and nitrous oxide ($N_2O$) that arise because of N overuse and misuse and the role these play in climate change, soil acidification, eutrophication of lakes and rivers, and harmful algal blooms in coastal waters (Lu *et al.* 2006; Guo *et al.* 2010; Fan *et al.* 2006; Li *et al.* 2009).

*Economic and institutional barriers to GHG mitigation.* It is clear from the foregoing that there are major constraints stemming from weaknesses in China's extension system (Hu *et al.* 2008). Extension workers are commonly poorly trained and have other public service duties such that they may spend only 10% of their time on agricultural extension. Moreover, there are a number of economic disincentives to optimal synthetic N fertilizer use. Two should be highlighted. First, subsidies to N fertilizer production (>US$1 billion/yr). Second, the low cost of overuse to many farmers,

*Table 29.4* N budget for "greenhouse" tomatoes, Shouguang, Shandong, 2003–2005

| Inputs kg N/ha/yr* | Losses and retentions % |
|---|---|
| N from previous crops 1000 | Leaching 30 (12–40) |
| Synthetic N 1360 | Losses to air 25** |
| Manure 1880 | Crop removals 15 (10–30) |
| Irrigation 402 | Retained in top 90 cm 30 |

Source of table data: Zhang *et al.* (2006) and Ji *et al.* (2006).

*Notes*
  * Covered production so no N from atmospheric deposition, but this can be 50–100 kg N/ha/yr.
** largely as nitrogen and ammonia.

because they are part-time farmers who earn most of their income (80–90%) by working in the industrial or service sector. For them the opportunity cost of farm labour is high, and consequently, although they know that it is better to apply fertilizer some time after planting and preferably in split applications, they apply it at sowing or transplanting to limit labour inputs. Moreover, they tend to apply more fertilizer than is needed as an insurance against unfavourable conditions. For them the cost of overuse is only 1–4% of their total household income (Table 29.3, columns 4 and 6).

## INM and GHG mitigation

INM in China has substantial potential to reduce emissions of GHGs, particularly both direct and indirect emissions of nitrous oxide ($N_2O$) from all crops, and methane from rice, without any risk to crop yields and food security (SAIN 2010). Although methane emissions are generally not closely related to N use, overuse of N can reduce crop yields by >5% (Table 29.2 and Peng *et al.* 2010) and increase methane emissions by ~ 5% for a given output. Moreover, where INM involves a switch away from flood irrigation, a further reduction in both $N_2O$ and methane ($Ch_4$) is possible though there can be complex trade-offs between $N_2O$ and $Ch_4$ emissions, with reduced flood irrigation lowering $Ch_4$ emissions but raising $N_2O$ emissions (Liu *et al.* 2010). Turning to $N_2O$ emissions and assuming that INM can lower synthetic N use by 30%, a decline in emissions of 38% is possible (Liu *et al.* 2010). These estimated gains have been made on the assumption that there is a linear relationship between N inputs and direct $N_2O$ emissions. In fact it is quite possible that the response is non-linear, so the emissions saving could be even greater (Crutzen *et al.* 2007).

INM also provide significant reductions in indirect GHG emissions and other environmental problems (Table 29.5). The N loss by leaching is a source of indirect $N_2O$ emissions and cause of eutrophication. Similarly a

*Table 29.5* INM and reduction of indirect GHG emissions and other environmental impacts

| N application | Jiangsu/Hubei | | Beijing/Hebei/Shandong | |
|---|---|---|---|---|
| | Farmers' rate | INM rate | Farmers' rate | INM rate |
| kg synthetic N fertilizer/ha/yr | 550 | 353 | 588 | 286 |
| **N loss kg/ha/yr** | | | | |
| by leaching | 12 | 8 | 56 | 23 |
| by ammonia emissions | 38 | 24 | 135 | 46 |
| by denitrification | 206 | 75 | 9 | 3 |

*Source:* Ju *et al.*, 2009

major proportion of the large emissions of ammonia are returned to the soil through dry and wet deposition (resulting from sedimentation of aerosols and rainfall respectively), where they add to soil acidification and a small fraction is re-emitted as $N_2O$.

Table 29.5 illustrates the importance of applying the optimal rate of synthetic N for the realistic target yield that takes account of the residual N in the soil and all of the other sources of N (Table 29.4) using a combination of soil tests, nutrient budgets and fertilizer trial results. However, the mitigation options do not end there. First, it is possible to use subsurface fertilizer placement to minimize losses (Cui *et al.* 2010). Second, the economics of using slow-release fertilizers (particularly resin-coated formulations) and urease or nitrification inhibitors has improved greatly in recent years and now account for 3–5% of the Chinese market. Collectively, these mitigations measures could reduce N fertilizer-related GHG emissions by ~50% by 2020 (Table 29.6). The gains from these measures are not simply additive, but they could represent a significant contribution to China's low-carbon agriculture and to the country's energy efficiency and GHG mitigation targets for 2020 without endangering food security.

## Implications for other developing countries

Few countries have average N application rates close to those in China or are major producers of N fertilizer, so INM will not play such an important role in their low-carbon agriculture. None the less, there are high N hotspots in many countries where the findings of this case study are relevant to the formulation of GHG mitigation policies. First, there is the contribution that Life Cycle Analysis (LCA) can make to the identification of where the largest fossil energy inputs and GHG emissions occur in the full crop production and processing chain. Second, there is the clear conclusion that the overuse of N can lower crop yields (by up to an average of 10%) as well as increasing GHG emissions, so there need not be a trade-off between reduced synthetic N inputs, GHG mitigation and food security. Third, GHG mitigation by INM can be a cost-effective solution where farmers are applying >150 kg N/ha/crop. Fourth, technological progress (e.g., improved resin-coated fertilizers, nitrification inhibitors) is increasing the cost-effectiveness of some mitigation options. Fifth, farm income analysis has an important role to play in identifying how farmers might respond to different policy approaches, for example, the low-income farmers (Quartile 1 in Table 29.3) may be strongly motivated by the cost savings from reducing overuse, whereas Quartile 4 farmers with large off-farm incomes may need different policy incentives.

## Research gaps and possible priorities for policy action

At the global level one of the most critical priorities is to reassess the IPCC default value for indirect $N_2O$ emissions from N fertilizer production and

*Table 29.6* INM-related GHG mitigation potential in China by 2020

| Improvement activity | Reduction in N use (%) | Reduction in N demand/ production* | Reduction in GHG emission Mt $CO_2e$ |
|---|---|---|---|
| Soil testing, better recommendations and focus on reducing <overuse | 20–30 | 7–10 | 84–120 |
| Improved application and timing | 10 | 3 | 36 |
| Improved fertilizers including slow-release formulations and inhibitors | 5–10 | 2–3 | 24–36 |
| Replacement of synthetic N by organic fertilizers | 30 | 10 | 60–120? |
| Implicit reduction in production** | | >24 | > 280 [641] |

*Notes*
Unpublished project results.
 * Assuming Business as Usual (BAU) demand of 35 Mt and GHG emissions of 12 t $CO_2e$.
** Arising from the reduced demand caused by the adoption of INM.

use, which may be far too low (Crutzen *et al.* 2007; Davidson, 2009), and to determine national default values for the major N uses, livestock production systems and aquaculture (Williams and Crutzen 2010). At the production system level two priorities stand out: (1) gaining a better understanding of the processes affecting GHG emissions arising from agricultural inputs of N to rivers, lakes and coastal waters (agriculture accounts for 25–80% of N exports to some major Chinese rivers) and how mitigation may increase the profitability of inland and coastal fisheries by improving water quality and reducing the frequency and extent of harmful algal blooms, and (2) improving the understanding of how N overuse and misuse affects agricultural sustainability, and particularly soil biological processes and crop vulnerability to pests and diseases. At the food and farming sector level more policy analysis is needed to integrate waste recycling along the whole of the crop and livestock production, food processing and household waste chain.

## Acknowledgements

Thanks to Profs Ju Xiaotang and Zhang Weifeng, China Agricultural University, Beijing, China, for access to unpublished data and helpful discussions.

## Note

1 The China–UK Project "Improved Nutrient Management in Agriculture—a Key Contribution to the Low Carbon Economy" is funded by the UK's Foreign and Commonwealth Office and by China's Ministry of Agriculture. It is led by Prof Zhang Fusuo, China Agricultural University, Beijing, and Prof David Powlson, Rothamsted Research, UK. The project forms part of the China–UK Sustainable Agriculture Network—SAIN (see www.sainonline.org/English.html).

## References

Brentrup, F., Kusters, J., Kuhlmann, H. and Lammel, J. (2001) Application of the life cycle assessment methodology to agricultural production: An example of sugar beet production with different forms of nitrogen fertilisers, *European J. Agron*, vol. 14, pp. 221–233.

Cui, Z., Chen, X. and Zhang, F.S. (2010) Current nitrogen management status and measures to improve the intensive wheat-maize system in China, *Ambio*, vol. 39, pp. 376–384.

Crutzen P.J., Mosier, A.R., Smith, K.A. and Winiwarter, W. (2007) $N_2O$ release from agro-biofuel production negates global warming reduction by replacing fossil fuels, *Atmos. Chem. Phys. Discuss.* vol. 7, pp. 11191–11205.

Davidson, E.A. (2009) The contribution of manure and fertilizer nitrogen to atmospheric nitrous oxide since 1860, *Nature Geoscience*, vol. 2, (September), pp. 659–662.

Fan, M., Zhang, F. and Jiang, R. (2009) *Integrated nutrient management for sustainable agriculture in China* [online]. Available at: http://escholarship.org/uc/

item/5hc5k91x. Proceedings of the International Plant Nutrition Colloquium XVI, UC Davis.

Fan, C., You, B. and Zhong, J. (2006) 'Non-point and point source pollution and the eutrophication of Chinese lakes' in Zhu, Z.L., Norse, D., Sun, B. (eds), *Policy for Reducing Non-point Pollution from Crop Production in China.* China Environmental Science Press, Beijing, China.

Guo, J.H., Liu, X.J., Zhang, Y., Shen, J.L., Han, W.X., Zhang, W.F., Christie, P., Goulding, K., Vitousek, P. and Zhang, F.S. (2010) 'Significant acidification in major Chinese croplands' *Science,* vol. 327, pp. 1008–1010 http://dx.doi.org/10.1126/science.1182570.

Hu, R., Yang, Z., Kelly, P. and Huang, J. (2008) Agricultural extension system reform and agent time allocation in China, *China Economic Review,* doi:10.1016/j.chieco.2008.10.009.

Huang, Y. and Sun, W.J. (2006) Change trend of topsoil organic carbon content of cropland in China for the recent 20 years, *Chinese Science Bulletin,* vol. 51, no. 7, pp. 750–763 (in Chinese).

IPCC (2006) Guidelines for National Greenhouse Gas Inventories Volume 4: Agriculture, Forestry, and Other Land Use. OECD Press, Paris, pp. 475–505.

Ji, L., Zhang, H.D., Gong, J.H., He, Y. and Norse, D. (2006) 'Agrochemical use and nitrate pollution of groundwater in typical crop production areas of China—case studies in Hubei, Hunan, Shandong and Hebei provinces' in Zhu, Z.L., Norse, D. and Sun, B. (eds), *Policy for Reducing Non-point Pollution from Crop Production in China.* China Environmental Science Press, Beijing, China.

Ju, X.T., Xing, G.X. and Chen, X.P. (2009) Reducing environmental risk by improving N management in intensive Chinese agricultural systems, *Proc. Natl. Acad. Sci. U.S.A.,* vol. 106, no. 9, pp. 3041–3046.

Li, J., Glibert, P.M., Zhou, M., Lu, S. and Lu, D. (2009) Relationship between nitrogen and phosphorus forms and ratios and the development of dinoflagellate blooms in the East China Sea, *Mar. Ecol. Prog.,* Ser 383:11.

Liu, S., Qin, Y., Zou, J. and Liu, Q. (2010) Effects of water regime during the rice-growing season on annual direct $N_2O$ emission in a paddy rice–winter wheat rotation system in southeast China, *Sci. total. Env,* vol. 408, pp. 906–913.

Lu, Y., Zhang, L. and Liu, H. (2006) 'Fertilizer use and rural livelihoods: Links and policy implications' in Zhu, Z.L., Norse, D. and Sun, B. (eds), '*Policy for Reducing Non-point Pollution from Crop Production in China'.* China Environmental Science Press, Beijing, China.

Lu, F., Wang, X.K, Han, B. *et al.* (2009) Soil carbon sequestrations by nitrogen fertilizer application, straw return and no-tillage in China's cropland, *Global Change Biology,* vol. 15, pp. 281–305.

Ma, W., Li, J., Ma, L., Wang, F., Sisak, I., Cushman, G. and Zhang, F. (2009) Nitrogen flow and use efficiency in production and utilization of wheat, rice and maize in China, *Agricultural Systems,* vol. 99, pp 53–63.

Ma, W. and Zhang, F. (2008) Nutrient management in the food chain: A challenge for sustainable development of China, *Sci. Technol. Rev.,* vol. 26, pp. 68–73 (in Chinese with English abstract).

McKinsey (2009) *China's green revolution: Prioritizing technologies to achieve energy and environmental sustainability.* McKinsey & Company.

NDRC (2004) 'The People's Republic of China initial national communication on climate change'. China Project Press, Beijing, China (in Chinese).

Norse, D. (2005) Non-point pollution from crop production: Global, regional and national issues, *Pedosphere*, vol. 15, no. 4, pp. 499–508.

Peng, S., Buresh, R.J., Huang, J., Zhong, X., Zou, Y., Yang, J., Wang, G., Liu, Y., Hu, R., Tang, Q., Cui, K., Zhang, F. and Dobermann, A. (2010) Improving nitrogen fertilization in rice by site-specific N management, *A Review. Agron. Sustain. Dev 30*, pp. 649–656. doi: 10.1051/agro/2010002.

SAIN (2010) *Greater food security and a better environment through improved nitrogen fertilizer management* [Online]. Available at: www.sainonline.org/SAIN-website(English)/download/PolicyBriefNo2FinalEn.pdf (Accessed: 30 January 2011).

SAIN (2011) *Improved nutrient management in agriculture – A neglected opportunity for China's low carbon growth path* [Online]. Available at: www.sainonline.org/SAIN-website(English)/download/PolicyBriefNo1final.pdf (Accessed: 15 August 2011).

Williams, J. and Crutzen, P.J. (2010) Nitrous oxide from aquaculture, *Nature Geoscience*, vol. 3, p. 143.

Zhang, F.S., Chen, Q., Zhang, X.S., He, F.F. and Ma, W.Q. (2006) 'Environmental assessment of nutrient management in vegetable production in China' in Zhu, Z.L., Norse, D. and Sun, B. (eds), *Policy for Reducing Non-point Pollution from Crop Production in China*. China Environmental Science Press, Beijing, China.

Zhu, Z.L. and Chen, D.L. (2002) Nitrogen fertilizer use in China—Contributions to food production, impacts on the environment and best management strategies, *Nutrient Cycling in Agroecosystem*, vol. 63, pp. 117–127.

# 30

# CLIMATE CHANGE MITIGATION IN AGROFORESTRY SYSTEMS

## Linking smallholders to forest carbon markets

*Bambi L. Semroc, Götz Schroth, Celia A. Harvey,*
*Yatziri Zepeda, Terry Hills, Saodah Lubis,*
*Candra Wirawan Arief, Olaf Zerbock, and Frederick Boltz*

## Introduction

Smallholder agricultural producers manage vast areas of land that could store or sequester important carbon stocks when managed under agroforestry systems. However, the potential for smallholder production systems to contribute to climate change abatement has not been adequately incorporated into the design of international or national policies addressing climate change (Montagnini and Nair 2004). Agroforestry systems intentionally integrate tree cultivation with agricultural crops, pastures or livestock production (Harvey *et al.* 2008). Smallholders with less than two hectares of land under production account for an estimated 85% of farms globally (Nagayets 2005).

Zomer *et al.* (2009) estimated that between 17% and 46% of all agricultural land contains some degree of tree cover and more than 1 billion hectares of land have at least 10% tree cover. The majority of these lands are located in South America, sub-Saharan Africa and Southeast Asia and are managed by smallholder producers (Zomer *et al.* 2009). These systems, managed appropriately, have the potential to maintain important stocks of carbon and may present opportunities for additional carbon sequestration (Trexler and Haugen 1994; Brown *et al.* 1996; Bloomfield and Pearson 2000; Cacho *et al.* 2003), which could deliver important climate benefits while simultaneously contributing to poverty reduction.

Forest carbon projects tend to fit into two categories: afforestation/reforestation (A/R) (the planting of trees in areas with no current forest cover);

and reducing emissions from deforestation or forest degradation, commonly referred to as REDD+.[1] Opportunities exist for smallholder agroforestry systems to contribute to both types of mitigation activities. Farmers may increase tree cover on their land through the introduction of agroforestry practices within productive areas, for example by introducing shade cover within sun coffee plantations and by increasing and diversifying tree cover within existing agroforestry systems (Smith and Scherr 2002). Under a REDD+ mechanism, agroforestry systems can serve as the carbon stock to be protected, or they can be promoted on existing cropland to reduce the need to deforest new lands to maintain or increase productivity, thereby reducing pressures within a broader landscape (Phelps *et al.* 2010). In the first case, the forested areas of the agroforestry system may be protected from conversion to competing, lower-carbon land uses if coupled with incentives commensurate to the opportunity cost of conserving them. In the second case, agroforestry could serve to enhance the productivity of existing farms and diversify livelihoods, thereby diminishing pressures for expansion of less productive practices into remaining forests. Farmers can also increase the density of shade cover within existing agroforestry systems as a means of carbon stock enhancement under REDD+ (Phelps *et al.* 2010; see also chapter by Méndez *et al.*). A third opportunity is avoiding the degradation of agroforestry systems (e.g., the gradual conversion of shade-grown coffee through timber extraction) and thereby reducing greenhouse gas (GHG) emissions.

Further opportunities for climate change mitigation through agroforestry systems include the adoption of agricultural practices that reduce GHG emissions, either by reducing the need for fertilizer application or maintaining or enhancing the amount of carbon stored in agricultural soils. Realizing the carbon mitigation potential of smallholder agroforestry systems requires a greater focus on putting into place the incentives necessary to enable smallholder participation in the developing carbon market.

The projects presented in this chapter focus exclusively on the forest carbon potential of agroforestry systems, recognizing that additional GHG emission reductions could be generated through the integration of agricultural carbon initiatives. The three projects featured in this chapter apply a variety of approaches and standards. Two focus on forest restoration while the third explores the feasibility of introducing REDD+ among smallholder coffee producers (see also global review of smallholder coffee agroforestry by Méndez *et al.*). Each project is applying different carbon accounting standards and using different mechanisms for registering credits generated (Table 30.1).

## Case 1: Reforestation with coffee communities in Chiapas, Mexico

This project in the Sierra Madre de Chiapas in southern Mexico explores opportunities to expand tree cover in coffee communities surrounding

Table 30.1 Overview of Conservation International smallholder carbon projects

| Name of project | Sierra Madre de Chiapas | Aceh Tengah | Tetik' Asa Mampody Savoka (TAMS) |
|---|---|---|---|
| Location | Chiapas, Mexico | Sumatra, Indonesia | Madagascar |
| Type of project | Reforestation | REDD+ | Reforestation |
| Project stage | Implementation | Feasibility | Design/implementation |
| Agroforestry activities | Tree planting for live fencing | Enhancing shade for coffee production | Alternative agriculture systems including forest gardens |
| Other emissions reduction activities | Restoration of degraded lands, introduction of cookstoves | Tree planting to delineate forest boundary, extension services, conservation agreements | Tree planting, land titling, fuelwood plantations |
| Standard applied | Plan Vivo | VCS + CCB (envisioned) | CDM |
| Size (ha) | 100 | TBD | 411 |
| No. of farmers engaged | 147 | 194 | 130 |
| No. communities engaged | 27 | 8 | NA |
| No. nurseries established | 3 | 0 | 32 |
| No. seedlings produced | 112,356 | 0 | around 500,000 |
| Est. carbon potential ($CO_2e$) | NA | 224,000 | 180,446 |
| Tons Carbon Sold ($CO_2e$) | 5,042 | 0 | 0 |

Note
NA = not available

protected areas. The region is an important area for coffee production in Mexico and shade-grown coffee production systems play a key role maintaining forest cover in the buffer zones of four protected areas, three of which are biosphere reserves. These protected areas represent 535,197 hectares and conserve nearly 40,000 hectares of forest habitat inside the core protection areas in a region with high levels of species endemism (Schroth *et al.* 2009). Within the Sierra Madre region, coffee is produced under complex agroforestry systems that may consist of between 60% and 70% shade cover and up to four strata of shade trees (Moguel and Toledo 1999).

In 2007, Conservation International (CI) began working with local non-governmental organization (NGO) and academic partners with support from the Critical Ecosystem Partnership Fund (CEPF) to develop a coffee and carbon project. The goal was to create opportunities for coffee producers to increase tree cover in agricultural production areas and generate carbon credits via reforestation to sell on the voluntary carbon market, thereby diversifying smallholder income. The project built on the success of the Plan Vivo system (Plan Vivo 2008) in the region, a voluntary carbon market standard designed specifically for smallholders that uses a participatory farm planning approach to identify opportunities to expand tree cover within farms and communities. Over the course of the initial 2-year project, 54 farmers in 8 communities established agroforestry practices on 57 hectares. The sale of $1,594\,tCO_2e$ under the Plan Vivo system generated nearly US$16,000 in revenue and on average each producer received US$295. Expansion of live fencing in pasture lands and coffee areas generated the majority (83%) of the carbon credits, with the planting of additional shade trees and improved fallows supplying the remainder.

Since 2008, with funding from Starbucks Coffee Company, the project has expanded to include 27 communities across the 4 protected areas in the region. Community selection was based on the results of consultations with government agencies and local stakeholders, complemented with an analysis of current deforestation patterns and drivers of forest loss. Over the past 2.5 years, project partners have established 3 tree nurseries producing over 112,000 native tree seedlings, which will be used by 147 farmers to reforest 100 hectares. The Plan Vivo system is one mechanism by which many discrete smallholder parcels can be cost-effectively grouped for the purposes of establishing a small-scale carbon offset project, as the transaction costs associated with other carbon standards such as the Clean Development Mechanism (CDM) or the Verified Carbon Standard (VCS) can be rather prohibitive for the small-scale producers typical of Chiapas.

A principal challenge to the project, faced by many reforestation carbon projects, was the need to demonstrate the economic benefits of participation to farmers early on, rather than waiting until the trees matured to pay farmers for the credits. The Plan Vivo system allows for agreements between farmers and project developers that compensate farmers in the early years of tree planting through forward payments based on projected future carbon storage. In this way the project developers assume some of the risk associated with the long-term generation of carbon credits, which was required to build the necessary trust with the local communities and achieve the equally important objective of improving buffer zone management for the protected areas.

## Case 2: Avoided deforestation in coffee landscapes in Northern Sumatra, Indonesia

The second case study explores opportunities to engage smallholder coffee producers in REDD+ activities in an area historically subject to deforestation pressure. The pilot focuses on a forest corridor spanning the provinces of North Sumatra and Acch to cxplore if REDD+ revenue, in combination with the demarcation of forest boundaries, clarification of land tenure and initiatives related to coffee marketing, offers sufficient incentive for farmers to conserve remnant forests. The North Sumatra corridor provides important habitat for tigers and other species and is an important area for coffee production. Initial surveys identified declining coffee production as a driver of deforestation. Rather than investing in practices to sustain production, farmers tend to clear adjacent forested areas and expand the area under coffee cultivation to compensate for lost productivity. This pressure is exacerbated by migration and heightened demand for land in post-conflict Aceh.

As with the work in Chiapas, the project objective is to improve coffee cultivation in the North Sumatra corridor and engage smallholders in carbon markets. Site selection was based on analyses of deforestation actors and drivers to establish a deforestation baseline for the region. This was combined with forest biomass inventories to identify districts with the most potential to generate REDD+ revenues based on emission reductions through interventions targeted towards reducing shifting coffee production. Aceh Tengah emerged as the district with the highest potential for reduced carbon emissions due to high rates of historical deforestation, high levels of carbon stocks in remaining forested lands and the strong relationship between coffee and deforestation due to the high suitability of currently forested lands for coffee production. The baseline analysis identified the potential to conserve $224,000\,tCO_2e$ through a REDD+ initiative and the generation of over US\$179 million in carbon revenues over a 30-year time frame.[2]

Initial work with coffee farmers in the Dairi district in North Sumatra has demonstrated that in return for technical assistance and training on coffee pruning, shade tree introduction, and composting, as well as assistance in gaining legally-recognized land tenure rights, smallholder coffee producers were willing to sign conservation agreements in which they agreed to stop clearing the forest. CI is currently consulting communities in Aceh Tengah to forge similar conservation agreements in return for technical assistance and other services. Identification of communities is based on willingness to participate, the importance of the site for biodiversity conservation and ecosystem service provision, and the rate of participation in coffee cooperatives (which indicates a history of working as a cohesive unit and could facilitate administration of a conservation or carbon project). To date, the project has identified 194 farmers in 8 communities in Aceh Tengah willing to participate and is exploring opportunities to engage additional communities. Studies are also under way to document biomass and carbon stocks per hectare in sun coffee, shade-coffee, secondary forest and primary forest systems to establish site-specific baselines for REDD+ feasibility. The project is exploring future integration under both the VCS and the Climate Community and Biodiversity (CCB) Standard.

Although this project is only at the feasibility stage of development, some challenges have already been identified. The design of this project is occurring in parallel to efforts to develop provincial and national level policies on REDD+, necessitating close coordination with government institutions regarding policies and regulatory standards. In addition, the legal tenure of coffee producers located within protected forest areas is unclear, which could affect the feasibility of the project because land and carbon rights currently underpin eligibility for compensation under a REDD+ scheme such as the VCS. The lack of tried and tested methodologies for aggregating large numbers of small farmers in a cost-effective manner for REDD+ implementation also presents some challenges. This effort is among the first attempts to do so in Indonesia.

### Case 3: Smallholder reforestation in Toamasina Province, Madagascar

The third case study looks at opportunities for smallholders to restore forest connectivity between two protected areas in Madagascar by establishing a combination of forest gardens and timber plantations. In the Ankeniheny-Zahamena Corridor in eastern Madagascar, CI is working with local communities to direct compensation from the World Bank Bio-Carbon Fund for carbon sequestration through reforestation. The project is supporting sustainable livelihood activities, including forest gardens, fallow lands gardens, fruit gardens and fuelwood plantations. These activities will diversify production and reduce pressure on remaining native forests to meet basic household needs. The reforestation of abandoned fallow lands with native tree species will generate carbon credits while restoring biodiversity corridors.

Initiated in 2003, the project seeks to restore 411 hectares of degraded land and sequester $180,446\,tCO_2e$ through native species reforestation while promoting alternative agriculture systems and fuelwood plantations, and working with farmers to secure more formal land tenure and titles to mitigate leakage and deliver community co-benefits (Harvey *et al.* 2010). CI has applied a CDM methodology to quantify carbon sequestration gains and create carbon credits. While the government of Madagascar will own and manage the sale of the carbon credits, local landowners will receive compensation for the planted areas through agreements with government.

This project is an excellent example of the complex multi-stakeholder efforts necessary for effective conservation, restoration and livelihood gains. The project governance structure involves 14 different stakeholder groups, a complexity that reflects the three different types of land tenure in the project area—public, private and informal (Harvey *et al.* 2010). The national government is leading the initiative to resolve the key constraint of land tenure uncertainty and pave the way for carbon revenue to flow to participating farmers (Harvey *et al.* 2010).

## Enabling conditions for successful smallholder agroforestry carbon projects

CI's experience suggests that successful smallholder agroforestry projects require a sound understanding of the local social, economic and political context to ensure that constraints from land tenure, institutional weaknesses, incentive mechanisms and lack of technical capacity can be resolved over time. While each of the projects has faced significant challenges, each has demonstrated that engaging local stakeholders can yield appropriate, innovative strategies that enable smallholders to capitalize on forest carbon through improved agroforestry practices.

Where the limiting constraint is the carbon capacity of a specific agroforestry system, such as coffee, a more holistic approach to carbon mitigation potential in a larger landscape may be necessary to generate sufficient carbon credits to make the project feasible. This factor arose in the case of coffee producers in Mexico as described above, and was similarly identified in studies of shade-cacao systems in the southern Bahia region of Brazil (Gonzales and Marques 2008). Further research and policy development is needed to establish that degradation through conversion from shade to sun production can be included in REDD+.

Delay in the establishment of a consistent, globally recognized REDD+ mechanism and crediting standards has rendered local REDD+ pilot development significantly more complex. In each of the case study countries, and in many others, national and state-level policy frameworks are being developed in parallel with site-level project design and implementation. To ensure the success of individual projects as REDD+ regulatory frameworks, institutions and efforts progress, project developers must coordinate closely with government, community and external stakeholders. Alignment with the emerging policy frameworks can require negotiation, delays, reorientation and incur significant additional project costs. Project-based experience is critical to informing policy development and decision-making at state and national levels. In Mexico, efforts to work with smallholder coffee farmers are providing important case studies to guide the development of a state-level climate change action program for Chiapas by demonstrating the potential of agroforestry systems to mitigate GHG emissions.

A third challenge is the need to facilitate the participation of dispersed smallholders in forest carbon activities and payments to achieve significant emissions reductions. Smallholder methodologies must promote the aggregation of small individual parcels, while simultaneously applying rigorous yet cost-efficient monitoring and verification in order to demonstrate credibility of emissions reductions (see chapter by Shames *et al.*). The Plan Vivo system (Plan Vivo 2008) has proven to work well with small-scale producers, and the forthcoming release of the project grouping specification under the VCS (VCS 2010) may facilitate the development of more smallholder forest carbon projects (see chapter by Swickard and Nihart). In addition, there is a need to further develop the market for these types of 'specialty' smallholder credits.

The level of local technical and institutional capacity can also significantly affect the success of projects (Smith and Scherr 2002). Regions with a history of farmer organization in cooperatives, producer unions or coffee forums will be best positioned to access forest carbon markets through agroforestry interventions, although there may still be a steep learning curve for these organizations regarding technical aspects related to carbon. Our experience demonstrates that a history of collective farmer action through cooperatives and/or producer unions provides a basis for more efficient and effective management of aggregated smallholder projects achieving an economically viable scale for carbon finance. Local NGO capacity and experience is a key enabling factor in the design and implementation of forest carbon initiatives (Harvey *et al.* 2010). The success of the Chiapas project is based in large part on the 10 years' experience of a local partner organization implementing forest carbon projects in the region. This expertise and experience has enabled the project to expand more quickly, in spite of capacity constraints for local monitoring, reporting and verification.

Given these challenges to developing forest carbon markets that enable smallholder participation within current policy and market frameworks, attention should be given to payment for ecosystem services schemes that efficiently compensate smallholders for climate mitigation (Kroeger and Casey 2007). These might need to rely on public finance and government demonstration programs, at least initially until local institutions mature and carbon markets become more robust. Such locally-appropriate innovations are especially important to establish a basis for bundling carbon with benefits of biodiversity, freshwater and other ecosystem services from agroforestry efforts, and providing greater economic incentives for low-carbon production in smallholder landscapes.

## Acknowledgements

The authors are grateful for the peer review and advice provided by Lucio Bede, Timothy J. Killeen, Jeannicq Randrianarisoa and Joanne Sonenshine.

## Notes

1 Reducing Emissions from Deforestation and forest Degradation 'plus' conservation, the sustainable management of forests and enhancement of forest carbon stocks (UNFCCC 2010).
2 Based on an average biomass of 280 tons $CO_2e$ of dry weight per hectare, a 10% discount rate for potential leakage and a price per ton of US$10.

## References

Bloomfield, J. and Pearson, H.L. (2000) Land use, land-use change, forestry and agricultural activities in the Clean Development Mechanism: estimates of greenhouse gas offset potential, *Mitigation and Adaptation Strategies for Global Change*, vol. 5, pp. 9–24.

Brown, S., Sathaye, J., Cannell, M. and Kauppi, P. (1996) 'Management of forests for migitation of greenhouse gas emissions', in Watson, R.T., Zinyowera, M.C. and Moss, R.H. (eds), *Climate Change 1995: Impacts, Adaptation and Mitigation of Climate Change: Scientific-Technical Analyses*. Cambridge University Press, New York, NY.

Cacho, O.J., Marshall, G.R. and Milne, M. (2003) 'Smallholder agroforestry projects: Potential for carbon sequestration and poverty alleviation', *UN FAO ESA Working Paper No. 03–06*.

Gonzales, P. and Marques, A. (2008) 'Forest carbon sequestration from avoided deforestation and reforestation in Mata Atlântica (Atlantic Forest), Sul da Bahia, Brazil', Topical Report, U.S. Department of Energy, National Energy Technology Laboratory, Morgantown, WV.

Harvey, C.A., Zerbock, O., Papageorgiou, S. and Parra, A. (2010) 'What is needed to make REDD+ work on the ground? Lessons learned from pilot forest carbon initiatives'. Conservation International, Arlington, VA.

Harvey, C.A., Schroth, G., Soto-Pinto, L., Zerbock, O. and Philipsborn, J. (2008) 'The Role of reforestation and agroforestry in mitigating climate change,' in Mittermeier, R.A., Mittermeier, C.A., Totten, M., Pennypacker, L.L. and Boltz, F. (eds), *A Climate for Life*, CEMEX.

Kroeger, T. and Casey, F. (2007) An assessment of market-based approaches to providing ecosystem services on agricultural lands, *Ecological Economics*, vol. 64, pp. 321–332.

Moguel, P. and Toledo, V.M. (1999) Biodiversity conservation in traditional coffee systems in Mexico, *Conservation Biology*, vol. 12, pp. 1–11.

Montagnini, F. and Nair, P.K.R. (2004) Carbon sequestration: an underexploited environmental benefit of agroforestry systems, *Agroforestry Systems*, vol. 61, pp. 281–295.

Nagayets, O. (2005) 'Small farms: Current status and key trends information brief', *Paper prepared for the Future of Small Farms Research Workshop*, Wye College, 26–29 June 2005.

Phelps, J., Guerrero, M.C., Dalabajan, D.A., Young, B. and Webb, E.L. (2010) 'What makes a "REDD" country?', *Global Environmental Change*, doi: 10.1016/ j. gloenvcha.2010.01.002.

Plan Vivo (2008) *The Plan Vivo Standards* [Online]. Available at: www.planvivo.org/ documents/standards.pdf (Accessed: 16 August 2010).

Schroth, G., Laderach, P., Dempewolf, J., Philpott, S., Haggar, J., Eakin, H., Castillejos, T., Garcia Moreno, J., Soto Pinto, L., Hernandez, R., Eitzinger, A. and Ramirez-Villegas, J. (2009) Towards a climate change adaptation strategy for coffee communities and ecosystems in the Sierra Madre de Chiapas, Mexico, *Mitigation and Adaptation Strategies for Global Change*, vol. 14, pp. 605–625.

Smith, J. and Scherr, S.J. (2002) 'Forest carbon and local livelihoods: Assessment of opportunities and policy recommendations', *CIFOR Occasional Paper No. 37*.

Trexler, M.C. and Haugen, C. (1994) *Keeping it Green: Tropical Forestry Opportunities for Mitigating Climate Change*. World Resources Institute, Washington, DC.

UNFCCC (2010) *UNFCCC Decision 1/CP.13, Bali Action Plan, FCCC/CP/2007/6/ Add.1* [Online]. Available at: www.unfccc.int/resource/docs/2007/cop13/ eng/06a01.pdf (Accessed: 3 December 2010).

Verified Carbon Standard (2010) *VCS Program Update July 2010* [Online]. Available at: www.v-c-s.org/sites/v-c-s.org/files/VCS%20Program%20Update,%20New%20 Program%20Requirements.pdf (Accessed: 17 August 2011).

Zomer, R.J., Trabucco, A., Coe, R. and Place, F. (2009) 'Trees on farm: analysis of global extent and geographical patterns of agroforestry', *ICRAF Working Paper no. 89*, World Agroforestry Centre, Nairobi, Kenya.

# 31

# LIVELIHOOD AND ENVIRONMENTAL TRADE-OFFS OF CLIMATE MITIGATION IN SMALLHOLDER COFFEE AGROFORESTRY SYSTEMS[1]

*V. Ernesto Méndez, Sebastian Castro-Tanzi,*
*Katherine Goodall, Katlyn S. Morris,*
*Christopher M. Bacon, Peter Läderach, William B. Morris,*
*and Maria Ursula Georgeoglou-Laxalde*

## Introduction

In this chapter we examine interactions and trade-offs between climate mitigation, farmer livelihoods and environmental conservation for smallholder coffee agroforestry. Smallholder coffee systems are relevant to climate mitigation because they cover a considerable land area worldwide, which frequently surrounds forests that also contribute to climate mitigation (Jha *et al.* 2011; Perfecto and Vandermeer 2008; Schroth *et al.* 2009).

## Importance of the global coffee commodity

In 2008, coffee plantations covered roughly 9.71 million ha of land worldwide (Figure 31.1), generating approximately US$5.8 billion in exports from the top 20 producing countries (FAO 2010). A recent review calculated there are approximately 4.3 million coffee smallholders (≤10 ha) in 14 countries (Jha *et al.* 2011). Others have calculated a global figure of 25 million small-scale producers (Gresser and Tickell 2002). Coffee also holds social and cultural value—from coffee drinkers that enjoy the 'café culture', to coffee farmers that find pride in their occupation (Méndez *et al.* 2010b; Pendergrast 1999). Recent research also shows that certain shade-coffee management approaches have potential to conserve wild

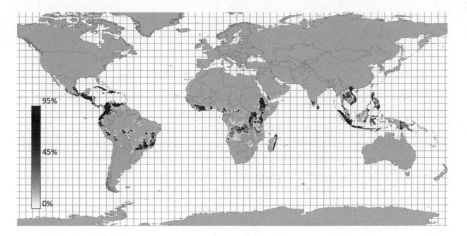

*Figure 31.1* Coffee-harvested area in percent cover per 10km grid cell. Data derived from Monfreda *et al.* (2008).

biodiversity and agrobiodiversity (Méndez *et al.* 2010b; Perfecto and Vandermeer 2008). Shade coffee's capacity to sequester carbon (C) has also gained increasing attention, and we review this body of work in the following sections.

## Climate mitigation in coffee agro-ecosystems

### *Coffee plantation management*

The type of shade and management used in coffee plantations is useful for evaluating climate mitigation potential because this affects above-ground biomass (AGB), which is where a large portion of carbon is stored. Based on shade and management characteristics, Moguel and Toledo (1999) classified coffee agro-ecosystems as follows: 1) *Rustic polycultures* use the existing forest canopy with coffee in the understory; they contain the highest levels of shade tree diversity, and use little or no synthetic inputs (i.e., fertilizers and pesticides); 2) *Traditional polycultures* have a shade canopy of remnant forest, naturally regenerated and planted trees; contain high species diversity, and use low or no levels of synthetic inputs; 3) *Commercial polycultures* contain a diversity of useful shade trees that replaced the original forest, and may use synthetic inputs; 4) *Shaded monocultures* have one species of shade tree, and usually use high levels of synthetic inputs; and 5) *Unshaded monocultures* have no shade trees (full-sun coffee), and there is heavy use of synthetic inputs and sometimes machinery.

Carbon is captured in coffee crops and soils across all systems, but we would expect C levels of AGB to increase as the density of shade trees increases. The more agrichemical-intensive systems would also be expected

371

to contribute more emissions from fertilization and mechanization (Albrecht and Kandji 2003).

## Carbon sequestration studies

A growing number of publications have documented C sequestration in coffee agro-ecosystems, including empirical studies (Table 31.1) and literature reviews on C mitigation in agroforestry (Albrecht and Kandji 2003; Montagnini and Nair 2004; Nair *et al.* 2009). Empirical research has estimated C sequestration levels in different types of coffee systems, and in different pools. However, a review of these findings reveals analytical challenges, which were summarized by Albrecht and Kandji (2003) and Nair *et al.* (2009), as follows: 1) Use of unstandardized methods limits comparisons; 2) Carbon stock measures are static, rather than as rates of accumulations per area over time; 3) Inaccuracies in AGB calculations through allometric equations; 4) Inaccuracies of C measured in root biomass; 5) Lack of consensus on vegetation age effects on C accumulation; 6) Failure to account for potentially important C pools (e.g., deep soil); and 7) Lack of analysis of other greenhouse gases (GHGs) (especially nitrous oxide ($N_2O$) and methane ($CH_4$)). These challenges were visible when examining the data presented in Table 31.1.

Data from Table 31.1 shows variations in sampling size, but standardized measures of C (t or Mg/ha) allowed for some comparison. High variation was also found in the number of C pools that were measured. Some studies analyzed up to six C pools, while others measured only one. However, in each shade category we found at least one study that measured C in the same pools, which allowed us to compare data across shade types (Table 31.2).

Traditional polycultures showed the highest mean C stocks (49 t C/ha), while unshaded monocultures showed the lowest (18.5 t C/ha). We found the highest C values in the soil organic pool, with a mean of 104 t C/ha, but with high variability. The next highest C pools were sequestered in the shade tree canopy with a mean of 15.3 t C/ha. All other C pools showed mean values less than or equal to 4.8 t C/ha.

Although we have restricted our discussion to C sequestration, climate mitigation research should also include the release of other GHGs from coffee plantations. Most important for coffee systems is probably $N_2O$, which originates from different nitrogen sources, including fertilizer and biologically fixed nitrogen (Hergoualc'h *et al.* 2008). One unit of this gas is equivalent to about 298 units of carbon dioxide ($CO_2$) (Scherr and Sthapit 2009).

Table 31.1 Summary findings of C sequestration studies on different types of coffee agro-ecosystems around the globe

| Region | Country | Mean total carbon (t C/ha) | | | | | C pools sampled[2] | Estimated plantation age (years)[3] | Mean C fixation rate (t/ha/year)[4] | Reference |
|---|---|---|---|---|---|---|---|---|---|---|
| | | R[1] | P | C | T | U | | | | |
| North America | Mexico | 55.2 | | | | | T,R,C,L,H | 30 | 1.85 | Soto-Pinto et al. 2010 |
| | | | | | 62.8 | | T,R,C,L,H | 30 | 2.1 | Avila 2000 |
| Central America | | | | | 10.8 | | T,C,L,H | 7 | 1.75 | de Miguel et al. 2004 |
| | | | | | | 10.4 | T,C,L,H | 9 | 1.2 | |
| | | | | | 14.6 | | T,C,L,H | 4.75 | 2.9 | |
| | | | | | | 11.4 | T,C,L,H | 14 | 0.8 | |
| | Costa Rica | | | | 37.5 | | T,C,L,H | 30 | 1.2 | Fournier 1995 |
| | | | | | 17.9 | | T,C,L,H | 7 | 2.55 | Harmand et al. 2007 |
| | | | | | 24.7 | | T,R,C,L,H | 14.6 | 1.58 | Mena 2008 |
| | | | | 31.6 | | | T,C,L,H | 40 | 0.8 | Polzot 2004 |
| | | | | | 17.2 | | T,C,L,H | 40 | 0.45 | |
| | El Salvador | | 16.5 | | | | T | 25 | 0.8 | Méndez et al. 2009 |
| | Guatemala | | | | 31.6 | | T,R,C,L,H | 4 | 7.88 | Alvarado et al. 1999 |
| | | 41.2 | | | | | T,C,L,H | 40 | 1 | |
| | Nicaragua | | | 14.9 | | | T,C,L,H | 20 | 2.85 | Suárez 2002 |
| | | | | | 20.0 | | T,C,L,H | 12.5 | 1.57 | |
| South America | Peru | | 80.2 | | | | T,L,H | 100 | 0.8 | Callo-Concha et al. 2002 |
| | | | | 33.7 | | | T,R,C,L,H | 3 | 11.2 | Rios et al. 2008 |
| | | | | | | 15 | T,C,L,H | 8 | 1.9 | Palm et al. 2005 |
| | Brazil | | | | 80.9 | | T,C,L,H | 12 | 6.75 | Rodriguez et al. 2000 |
| | | | | | | 16.6 | T,C,L,H | 12 | 1.4 | |
| Africa | Togo | | | | 82.3 | | T,R,C,L,H | 19 | 4.3 | Dossa et al. 2008 |
| | Togo | | | | | 23.4 | T,R,C,L,H | 19 | 1.2 | |
| Asia | East Java/Indonesia | | 53.13 | | | | T,R,C,L,H | 30 | 1.8 | Hairiah et al. 2010 |
| | | | | | 29.9 | | T,R,C,L,H | 30 | 1 | |
| | | | | | | 18 | T,C,L,H | – | | Lasco 2002 |
| | Sumatra/Indonesia | | 49.9 | | | | T,C,L,H | 27.4 | 1.87 | van Noordwijk et al. 2002 |
| | | | | 92 | | | T,C,L,H | 14 | 6.6 | |
| | | | | | | 22.6 | T,C,L,H | 15 | 1.62 | |

Notes
1  R = Rustic polyculture; P = Traditional polyculture; C = Commercial polyculture; T = Technified shade; U = Unshaded
2  T = Trees; R = Root biomass; C = Coffee shrubs; L = Leaf litter biomass, H = Herbaceous plants; S = Soil
3  Average ages in years
4  Calculated by dividing the reported carbon by the age of the system.

*Table 31.2* Comparison of C stocks between different types of shade-coffee systems[1]

| Shade type | No. of sampling units | C reported (t C/ha) | | | Mean C fixation rate (t C/ha)[2] |
|---|---|---|---|---|---|
| | | Min. | Mean | Max. | |
| Rustic polyculture | 1 | 41.2 | NA | 41.2 | 1.85 |
| Traditional polyculture | 10 | 25 | 49.9 | 74 | 1.9 |
| Commercial polyculture | 8 | 17.1 | 26.6 | 92 | 1.5 |
| Shaded monoculture | 21 | 14.2 | 23.1 | 97.2 | 2.2 |
| Unshaded monoculture | 9 | 15 | 18.5 | 44 | 1.5 |

Notes
1 References for data from Table 1.
2 Mean C fixation rate was calculated by dividing reported C sequestered by the age of the system. This represents an under-estimation, as it doesn't account for factors such as differences in C accumulation during different plantation stages, or for the removal of C through coffee and shade tree pruning, as well as weeding. A more accurate method that takes into account some of these issues has been developed by Palm *et al.* (2005) and termed 'time averaged carbon'.

# Livelihood and environmental trade-offs and climate mitigation governance

## Climate mitigation and environmental governance

Mechanisms for climate mitigation can be perceived as a form of environmental governance, which incorporates the regulatory processes and organizations driving environmental actions and outcomes (Lemos and Agrawal 2006). For smallholder farmers, the outcomes of climate mitigation governance can best be assessed by how these affect their livelihoods (Nelson and de Jong 2003), as issues related to land-use management and conservation.

### Market-based mechanisms for climate mitigation compensation

#### Direct payments

Externally funded climate mitigation projects can improve the livelihoods of smallholder coffee farmers in several ways. The most common is through direct monetary payments for carbon credits through the Clean Development Mechanism (CDM) or voluntary markets, such as the Chicago Climate Exchange (CCX) (https://www.theice.com/ccx.jhtml) and Plan Vivo (www.planvivo.org/), among others. Through these arrangements, farmers can receive payments for carbon stored per unit area over time (t C/ha/yr). This requires documenting a baseline of C stocks, and of expected C additionality through afforestation or reforestation (A/R) over time. A well-known example of this type of arrangement is the Scolel'te project, which was started in the 1990s with coffee growers in Chiapas, Mexico (Nelson and de Jong 2003; www.planvivo.org/?page_id=49). Farmers have planted timber trees in their coffee plantations for C credits. In addition, they expect to gain income from timber sales after 25 years and price premiums for their eco-friendly coffee (see also chapter by Semroc *et al.*).

In order to access monetary compensation (through C payments and market premiums), smallholders need to organize and develop capacities to negotiate with various national and international actors (Rosa *et al.* 2004). This is usually done through local associations or cooperatives, which partner with non-governmental organizations (NGOs) and/or government institutions. For example, The Scolel'te project included Ambio, a Mexican NGO; El Colegio de la Frontera Sur (a university in Chiapas); researchers from the University of Edinburgh, Scotland; Plan Vivo and the carbon buyers. The establishment of such networks has also been reported as an important benefit derived from coffee certifications (Bacon *et al.* 2008; Méndez *et al.* 2010a), and which could be replicated by mitigation initiatives.

Direct payments to farmers from climate mitigation face similar challenges as payment for ecosystem services (PES) mechanisms. These include: 1) the need for strong organizational and administrative capacities; 2) high implementation and monitoring costs (human, financial and administrative); 3) long establishment and investment recuperation time periods; and 4) exclusion of marginalized farmers due to insecure land tenure, limited information, and small land holdings (Corbera *et al.* 2007a; Rosa *et al.* 2004; Tschakert 2007). While payments for climate mitigation can benefit smallholders, they are also risky because they require upfront investments, C prices are volatile and expected changes in farming practices could threaten food security (Bumpus and Liverman 2011; Henry *et al.* 2009).

*Incorporation of climate modules in certification and verification schemes*

Several certifications and verification schemes are evaluating the potential of incorporating climate modules. The Common Code for The Coffee Community (4C) verification system is exploring a standard climate component for coffee producers to use synergies between climate adaptation and mitigation. The Sustainable Agricultural Network (SAN), which works through Rainforest Alliance, is also developing a SAN Climate Module for coffee and other tropical crops (http://clima.sanstandards.org/webroot/userfiles/file/SAN%20Climate%20Module-%20DRAFT%20version-%20July%202010.pdf). The challenge for these climate modules is determining and quantifying their added value, since many of the agricultural practices promoted in current standards already contribute to climate adaptation and mitigation.

On the other hand, recent studies show that coffee price premiums provided through certifications have limited effects on smallholder livelihoods (Barham *et al.* 2011; Méndez *et al.* 2010a). This puts into question the utilization of similar schemes to implement climate mitigation initiatives. However, one positive finding from these studies is related to the importance of farmer cooperatives as vehicles to support smallholder livelihoods (Bacon 2010; Raynolds *et al.* 2004), albeit with some pervasive challenges related to organizational governance (Bacon 2010; Méndez *et al.* 2010a).

### *Non-market mechanisms for climate mitigation compensation*

Non-monetary alternatives can deliver different types of relevant support to smallholders. These include providing technical assistance related to mitigation, supporting agro-ecological production and expanding livelihood options. These types of support may not require new and complex mechanisms that are difficult for farmers, and could be run through established institutions or NGOs. Challenges related to these initiatives are well known, and include overly 'top-down' and paternalistic models; lack of smallholders' participation in project design and objectives; and short project timelines, among others (Bebbington 2000; Rosa *et al.* 2004). Alternatives to these models include participatory approaches in research and development implementation, as well as researcher involvement in actions for change (Bacon *et al.* 2005; Kindon *et al.* 2007). Examples of participatory experiences with coffee farmers can be found in Mexico (Scolel'te; Nelson and de Jong 2003; Soto-Pinto *et al.* 2010) and Central America (Bacon *et al.* 2005). These approaches may be ideal for climate mitigation, which is research and capacity-intensive, but they may require integration with broader development processes in order to scale up their impact (Méndez *et al.* 2010b).

### *Climate mitigation, tropical habitat and ecosystem services*

Ecosystem services can be defined as the benefits that humans obtain from ecosystems, which can be categorized into 'provisioning', 'regulating', 'cultural', and 'supporting' services (Millennium Ecosystem Assessment 2005). Shaded coffee plantations have great potential to contribute to biodiversity and ecosystem services conservation at local, regional and global scales (Jha *et al.* 2011; Perfecto and Vandermeer 2008). However, these ecosystem services interact with one another in complex ways (Bennett *et al.* 2009), and it is necessary to examine how enhancing one ecosystem service may affect others (Bennett *et al.* 2009). Analyzing the trade-offs between provisioning (i.e., food and coffee production) and regulating services (i.e., climate mitigation) is of critical importance. Mitigation might be enhanced by increasing shade-coffee areas, but this could compromise food production (Henry *et al.* 2009; Méndez *et al.* 2009). It is also important to examine how potential increases in new coffee areas may threaten protected areas in coffee regions (Jha *et al.* 2011).

## Discussion and conclusions

Diversified shade-coffee systems show great potential for C sequestration. Smallholdings are especially well suited given that they tend to manage farms with a high density and diversity of shade trees. Mitigation in small-holder coffee systems could have significant global impact, since there are between 4 and 25 million small-scale farmers worldwide. However, mitigation strategies will face a number of challenges when working with small-holder coffee farmers.

First, it is unclear what type(s) of mechanisms would yield the best synergies in terms of livelihood support, climate mitigation and other environmental conservation. Market-based, direct payments to farmers, such as PES and certifications, have shown limited livelihood effects on smallholders, due to a need for high levels of financial, administrative and monitoring capacities. The low coffee volumes that smallholders produce also limit the impacts of premiums based on production. Participatory research alternatives with shade-coffee farmers have been able to raise capacities and incorporate livelihood and conservation concerns, but they have not been able to scale up these successes. Climate mitigation may provide an opportunity for innovative initiatives, which could integrate market-based schemes with direct support for research, capacity building and livelihood improvements through participatory approaches.

A complex environmental trade-off related to climate change is the predicted shift in the suitability of landscapes to produce coffee, including areas that will become unsuitable and others that will become suitable. Shifting from coffee to other crops, or from other land uses to coffee, could have severe effects on protected areas and local livelihoods.

Climate mitigation initiatives may be at odds with the conservation of other ecosystem services and biodiversity. Specifically, mitigation schemes that require increased levels of tree cover may compete for space and resources with food and coffee production. Overcoming this conflict will require assessing all ecosystem services that are conserved in coffee landscapes and prioritizing initiatives that support the most critical services. For smallholders, provisioning services are necessary for survival, so these will have to be prioritized. This calls for mitigation projects to consider directly supporting provisioning services such as food and coffee production as part of their interventions.

## Note

1 This chapter is a summarized, modified version of Méndez *et al.* (2011).

# References

Albrecht, A. and Kandji, S.T. (2003) Carbon sequestration in tropical agroforestry systems, *Agriculture, Ecosystems & Environment*, vol. 99, no. 1–3, pp. 15–27.

Alvarado, J., Leon, E.L.d. and Medina, B. (1999) 'Cuantificación estimada del dióxido de carbono fijado por el agrosistema café en Guatemala', *Boletin PROMECAFE*, pp. 1–13.

Avila, G. (2000) 'Fijación y almacenamiento de carbono en sistemas de café de sombra, café a pleno sol, sistemas silvopastoriles y pasturas a pleno sol', M.Sc. thesis, CATIE, Turrialba.

Bacon, C.M. (2010) Who decides what is fair in fair trade? The agri-environmental governance of standards, access, and price, *Journal of Peasant Studies*, vol. 37, no. 1, pp. 111–147.

Bacon, C., Méndez, V.E. and Brown, M. (2005) 'Participatory action research and support for community development and conservation: Examples from shade coffee landscapes of El Salvador and Nicaragua'. Center for Agroecology and Sustainable Food Systems (CASFS), University of California, Santa Cruz.

Bacon, C.M., Méndez, V.E., Gliessman, S.R., Goodman, D. and Fox, J.A. (eds). (2008) *Confronting the coffee crisis: Fair Trade, sustainable livelihoods and ecosystems in Mexico and Central America.* MIT Press, Cambridge, MA.

Barham, B.L., Callenes, M., Gitter, S., Lewis, J. and Weber, J. (2011) Fair trade/organic coffee, rural livelihoods, and the 'agrarian question': Southern Mexican coffee families in transition, *World Development*, vol. 39, no. 1, pp. 134–145.

Bebbington, A. (2000) Reencountering development: Livelihood transitions and place transformations in the Andes, *Annals of the Association of American Geographers*, vol. 90, no. 3, pp. 495–520.

Bennett, E.M., Peterson, G.D. and Gordon, L.J. (2009) Understanding relationships among multiple ecosystem services, *Ecology Letters*, vol. 12, no. 12, pp. 1394–1404.

Bumpus, A.G. and Liverman, D.M. (2011) 'Carbon colonialism? Offsets, greenhouse gas reductions, and sustainable development', in Peet, R., Robbins, P. and Watts, M.J. (eds), *Global Political Ecology*. Routledge, London and New York, NY.

Callo-Concha, D., Krishnamurthy, L. and Alegre, J. (2002) Secuestro de carbono por sistemas agroforestales amazónicos, *Revista Chapingo Serie Ciencias Forestales y del Ambiente*, vol. 8, no. 2, pp. 101–106.

Corbera, E., Brown, K. and Adger, W.N. (2007a) The equity and legitimacy of markets for ecosystem services, *Development and Change*, vol. 38, no. 4, pp. 587–613.

de Miguel Magana, S., Harmand, J.M. and Hergoualc'h, K. (2004) Cuantificación del carbono almacenado en la biomasa aérea y el mantillo en sistemas agroforestales de café en el suroeste de Costa Rica, *Agroforestería en las Américas*, vol. 41–42, pp. 98–104.

Dossa, E.L., Fernandes, E.C.M., Reid, W.S. and Ezui, K. (2008) Above- and belowground biomass, nutrient and carbon stocks contrasting an open-grown and a shaded coffee plantation, *Agroforestry Systems*, vol. 72, no. 2, pp. 103–117.

FAO (2010) *FAOSTAT: Crops* [Online]. Available at: http://faostat.fao.org/site/567/default.aspx#ancor (Accessed: 3 June 2010).

Fournier, L. (1995) 'Fijación de carbono y diversidad biológica en el agroecosistema cafetero'. Paper presented at the XVII Simposio sobre caficultura Latinoamericana, San Salvador, El Salvador.

Gresser, C. and Tickell, S. (2002) *Mugged: poverty in your coffee cup*. Oxfam International, London, UK.

Hairiah, K., Kurniwana, S., Aini, F., Lestari, N.D., Lestariningsih, I.D., Widianto, Zulkarnaen, T. and Van Noordwijk, M. (2010) 'Carbon stock assessment for a forest-to-coffee conversion landscape in Kalikonto Watershed'. Paper presented at the International Conference on Coffee Science (ASIC), Denpasar, Bali.

Harmand, J.M., Hergoualc'h, K., De Miguel, S., Dzib, B., Siles, P., Vaast, P. and Locatelli, B. (2007) 'Carbon sequestration in aerial biomass and derived products from coffee agroforestry systems in Central America'. Paper presented at the Second International Symposium on Multi-Strata Agroforestry Systems With Perennial Crops: Making Ecosystem Services Count for Farmers, Consumers and the Environment, oral and poster presentations, Turrialba, Costa Rica.

Henry, M., Tittonell, P., Manlay, R.J., Bernoux, M., Albrecht, A. and Vanlauwe, B. (2009) Biodiversity, carbon stocks and sequestration potential in aboveground biomass in smallholder farming systems of western Kenya, *Agriculture, Ecosystems & Environment*, vol. 129, no. 1–3, pp. 238–252.

Hergoualc'h, K., Skiba, U., Harmand, J.M. and Henault, C. (2008) Fluxes of greenhouse gases from Andosols under coffee in monoculture or shaded by Inga densiflora in Costa Rica, *Biogeochemistry*, vol. 89, no. 3, pp. 329–345.

Jha, S., Bacon, C.M., Philpott, S.M., Rice, R.A., Méndez, V.E. and Laderach, P. (2011) 'A review of ecosystem services, farmer livelihoods, and value chains in shade coffee agroecosystems', in Campbell, B.W. and Lopez-Ortiz, S. (eds), *Integrating Agriculture, Conservation, and Ecotourism: Examples From the Field*. Springer Academic Publishers, New York, NY and Berlin.

Kindon, S., Pain, R. and Kesby, M. (eds) (2007) *Participatory Action Research Approaches and Methods*. Routledge, Oxon, England.

Lasco, R. (2002) Forest carbon budgets in Southeast Asia following harvesting and land cover change, *Science in China Series C-Life Sciences*, vol. 45, pp. 55–64.

Lemos, M.C. and Agrawal, A. (2006) Environmental governance, *Annual Review of Environment and Resources*, vol. 31, no. 1, pp. 297–325.

Mena, V. (2008) 'Relación entre el carbono almacenado en la biomasa total y la composición fisionómica de la vegetación en los sistemas agroforestales con café y en bosques secundarios del Corredor Biológico Volcánica Central-Talamanca, Costa Rica'. M.Sc. thesis, Tropical Agriculture Research and Higher Education Center (CATIE), Turrialba, Costa Rica.

Méndez, V.E., Bacon, C., Olson, M., Petchers, S., Herrador, D., Carranza, C., Trujillo, L., Guadarrama-Zugasti, C., Cordón, A. and Mendoza, A. (2010a) Effects of fair trade and organic certifications on small-scale coffee farmer households in Central America and Mexico, *Renewable Agriculture and Food Systems*, vol. 25, no. 3, pp. 236–251.

Méndez, V.E., Bacon, C.M., Olson, M., Morris, K.S. and Shattuck, A.K. (2010b) Agrobiodiversity and shade coffee smallholder livelihoods: A review and synthesis of ten years of research in Central America, *Professional Geographer*, vol. 62, no. 3, pp. 357–376.

Méndez, V.E., Castro-Tanzi, S., Goodall, K., Morris, K.S., Bacon, C.M., Läderach, P., Morris, W.B. and Georgeoglou-Laxalde, M.U. (2011). *Climate mitigation and smallholder livelihoods in coffee landscapes: synergies and tradeoffs*. Unpublished manuscript.

Méndez, V.E., Shapiro, E.N. and Gilbert, G.S. (2009) Cooperative management and its effects on shade tree diversity, soil properties and ecosystem services of coffee plantations in western El Salvador, *Agroforestry Systems*, vol. 76, no. 1, pp. 111–126.

Millennium Ecosystem Assessment (2005) *Ecosystems and Human Well-Being: Synthesis*. Island Press, Washington, DC.

Moguel, P., Toledo, V.M. (1999) Biodiversity conservation in traditional coffee systems of Mexico. *Conservation Biology*, vol. 13, no. 1, pp. 11–21.

Monfreda, C., Ramankutty, N. and Foley, J.A. (2008) Farming the planet: 2. Geographic distribution of crop areas, yields, physiological types, and net primary production in the year 2000, *Global Biogeochemical Cycles*, vol. 22, no. 1, GB1022, doi:10.1029/2007GB002947.

Montagnini, F. and Nair, P.K.R. (2004) Carbon sequestration: An underexploited environmental benefit of agroforestry systems, *Agroforestry Systems*, vol. 61, no. 1, pp. 281–295.

Nair, P.K.R., Nair, V.D., Kumar, B.M. and Haile, S.G. (2009) Soil carbon sequestration in tropical agroforestry systems: A feasibility appraisal, *Environmental Science & Policy*, vol. 12, no. 8, pp. 1099–1111.

Nelson, K.C. and de Jong, B.H.J. (2003) Making global initiatives local realities: Carbon mitigation projects in Chiapas, Mexico, *Global Environmental Change*, vol. 13, no. 1, pp. 19–30.

Palm, C.A., van Noordwijk, M., Woomer, P.L., Alegre, J.C., Arevalo, L., Castilla, C.E., Cordeiro, D.G., Hairiah, K., Kotto-Same, J., Moukam, A., Parton, W.J., Ricse, A., Rodrigues, V. and Sitompul, S.M. (2005) 'Carbon losses and sequestration after land use changes in the humid tropics', in Palm, C.A., Vosti, S.A., Sanchez, P.A. and Ericksen, P.J. (eds), *Slash and Burn Agriculture: The Search for Alternatives*. Columbia University Press, New York, NY.

Pendergrast, M. (1999) *Uncommon Grounds: The History of Coffee and How it Transformed the World*. Basic Books, New York, NY.

Perfecto, I. and Vandermeer, J. (2008) Biodiversity conservation in tropical agroecosystems—A new conservation paradigm, *Annals of the New York Academy of Sciences*, vol. 1134, no. 1, pp. 173–200.

Polzot, C.L. (2004) 'Carbon storage in coffee agroecosystems of southern Costa Rica: Potential applications for the clean development mechanism'. Master in Envioronmental Studies thesis, York University, Toronto.

Raynolds, L.T., Murray, D. and Taylor, P.L. (2004) Fair trade coffee: building producer capacity via global networks, *Journal of International Development*, vol. 16, pp. 1109–1121.

Rios, J., Bastos da Veiga, J. and Cordeiro de Santana, A. (2008) Quantification of carbon storage in land-use systems of José Crespo e Castillo district, Peru, *Archivos Lationamericanos de Producción Animal*, vol. 16, no. 3, pp. 139–152.

Rodriguez, V.C.S., Castilla, C., Correa da Costa, R.S. and Palm, C. (2000) 'Estoque de carbono em sistemas agroforestales com café Rondonia – Brasil', Paper presented at the Simposio de Pesquisa dos Cafés do Brasil, Minasplan, Brasilia.

Rosa, H., Kandel, S. and Dimas, L. (2004) Compensation for environmental services and rural communities: Lessons from the Americas, *International Forestry Review*, vol. 6, no. 2, pp. 187–194.

Scherr, S.J. and Sthapit, S. (2009) 'Mitigating climate change through food and land use'. Worldwatch Institute, Washington, DC.

Schroth, G., Läderach, P., Dempewolf, J., Philpott, S., Haggar, J., Eakin, H., Castillejos, T., Garcia Moreno, J., Soto Pinto, L., Hernandez, R., Eitzinger, A. and Ramirez-Villegas, J. (2009) Towards a climate change adaptation strategy for coffee communities and ecosystems in the Sierra Madre de Chiapas, Mexico, *Mitigation and Adaptation Strategies for Global Change*, vol. 14, no. 7, pp. 605–625.

Soto-Pinto, L., Anzueto, M., Mendoza, J., Ferrer, G.J. and de Jong, B. (2010) Carbon sequestration through agroforestry in indigenous communities of Chiapas, Mexico, *Agroforestry Systems*, vol. 78, no. 1, pp. 39–51.

Suárez, D. (2002) 'Cuantificación y valoración económica del servicio ambiental almacenamiento de carbono en sistemas agroforestales de café en la comarca Yassica Sur, Matagalpa, Nicaragua'. Magister Scientiae thesis, CATIE, Turrialba, Costa Rica.

Tschakert, P. (2007) Environmental services and poverty reduction: Options for smallholders in the Sahel, *Agricultural Systems*, vol. 94, no. 1, pp. 75–86.

van Noordwijk, M., Rahayu, S., Hairiah, K., Wulan, Y.C., Farida, A. and Verbist, B. (2002) Carbon stock assessment for a forest-to-coffee conversion landscape in Sumber-Jaya (Lampung, Indonesia): From allometric equations to land use change analysis, *Science in China Series C-Life Sciences*, vol. 45, Supp. Oct. 2002, pp. 75–86.

# 32

# AGRICULTURAL INTENSIFICATION AS A CLIMATE MITIGATION AND FOOD SECURITY STRATEGY FOR SUB-SAHARAN AFRICA

*James Gockowski and Piet Van Asten*

In 40 years, the population of sub-Saharan Africa is expected to double in size to 1,680 million people (UN 2009). The daunting challenge of meeting the food demands of an additional 840 million consumers in 2050 will be further complicated by stresses induced by climate change. Low productivity, the high incidence of poverty, and missing markets for inputs, seeds, irrigation, credit, land, labor, and crop insurance all limit the capacity of African farmers to deal with the food demands of the future in the face of a changing climate. To compound matters further, land clearing for agriculture has almost completely eliminated the West African Guinea Rainforest and is on track to do the same in the Congo basin. Deforestation and forest degradation associated with land preparation for agriculture are the largest sources of greenhouse gas (GHG) emissions and species extinction in Africa (Canadell *et al.* 2009).

This chapter considers the role that the sustainable intensification of agriculture may play in jointly addressing the multiple concerns of food security, economic growth, biodiversity conservation and GHG emissions. We dismiss the notion that the productivity gains needed to transform agriculture in sub-Saharan Africa can be achieved through "low-input agriculture," arguing instead that extensive low-input slash and burn agriculture and laissez-faire land-use policies have resulted in the unnecessary deforestation of millions of hectares of tropical rainforests in West and Central Africa.

The first part of the chapter examines the mitigation potential of sustainable intensification on GHG emissions in the humid and sub-humid tropics of West and Central Africa drawing on research conducted by the Alternatives to Slash-and-Burn (ASB) program in Cameroon (see also the chapter by Minang *et al.*). The second half explores a potential development pathway for sustainable intensification based on experiences with cocoa-based farming systems of West Africa and coffee–banana based farming systems of the Great Lakes region in East Africa. The chapter concludes with the recommendation to support a paradigm shift from the current consumptive model of natural resource exploitation to an equilibrium model of resource management that is coherent with national and regional development objectives.

## Agriculture as a source of global climate emissions

There are two major sources of agricultural GHG emissions—those directly associated with agricultural production and those associated with deforestation due to land clearing for agriculture. The former are mainly due to nitrous oxide ($N_2O$) emissions from the mineralization of nitrogenous fertilizers and plant residues and methane ($CH_4$) gas emissions from irrigated rice and livestock production. Canadell *et al.* (2009) report that global emissions from land-use change are the dominant source of emissions for 8 of the 10 largest emitters in sub-Saharan African nations. Achard *et al.* (2004) estimate that 850,000 ha/yr of deforestation and 390,000 ha/yr of forest degradation occurred in the 1990s in Africa. The 0.157 Gt/yr of carbon emitted from the deforestation and degradation of African forests represents 17% of the annual global land-use change emission (as compared to 44% for Latin America and 39% for Asia). Curbing these emissions is the objective of the Reducing Emissions from Deforestation and forest Degradation (REDD+) initiative of the United Nations Framework Convention on Climate Change (UNFCCC).

GHG emissions increase with the use of nitrogen fertilizers that can accompany agricultural intensification. On the other hand, use of fertilizer can result in higher crops yields and reduce the need to clear "wild or natural" lands. Green *et al.* (2005) were among the first to note this relationship of "land sparing" in the context of endangered species conservation. Burney *et al.* (2010) applied the principle of "land sparing" to a global analysis of agricultural intensification and carbon emissions. They showed that the emissions from avoided land-use change as a result of intensification was much larger than the increase in emissions from the intensified use of inputs. Depending on the scenario, net avoided emissions amounted to 18% to 34% of the estimated total global anthropogenic emissions since 1850. For "land conversion" emissions, they calculated a global average of 105 ± 21 t C/ha lost in conversion of uncultivated land to cropland.

In a similar analysis focused on the 7 countries comprising the West African Guinea Rainforest, Gockowski and Sonwa (2011) estimated that over 21,000 km² of deforestation and forest degradation, which occurred since 1988, could have been avoided, along with the emission of nearly 1.4 billion t carbon dioxide ($CO_2$), had intensified cocoa production systems been adopted. Land conversion emissions consisted of both deforestation and degradation with an average emission of 177 ± 26.0 t C/ha. This higher value (as compared to the Burney *et al.* estimate) reflects the abundant biomass of tropical forests.

In tropical forests the largest portion of plant nutrients are usually tied up in the above-ground biomass with only a relatively minor proportion retained in the acid soils common to the tropics. When the forest is converted to agricultural uses, the most common soil fertility practice is to burn the forest biomass to generate ash fertilizer and in order to allow for photosynthesis.[1] With annual crop systems in West and Central Africa (WCA), fallow periods determine the quantity of ash fertilizer produced. Over time, as the population increases, the biomass and associated stock of nutrients in the landscape decline as fallow periods shorten (Figure 32.1). At some point fallow fields disappear and depleted nutrient stocks have to be replaced by imports of organic and mineral fertilizer as well as other innovations such as improved varieties, new crops, irrigation, and pesticides.

Boserup (1965) was among the first to describe how these population pressures induced technical change in the humid tropics. Research conducted in southern Cameroon showed that by the time fallow lengths were less than 3 years, 98% of the original carbon of the forest had been emitted into the atmosphere (Gockowski *et al.* 2005). Ultimately, the farmer is either forced to abandon the degraded land or invest in fertilizers. Hill (1963) described how the quest for the capital gains from converting new forest led to frequent relocation by the cocoa migrants of Ghana from east to west.

The opportunity costs of land-use change in the tropics and the opportunity costs of avoided deforestation from the perspective of the smallholder has been a recent climate mitigation research topic in southern Cameroon. Remote sensing analysis is used to generate land-use transition matrices for the major land uses identified in southern Cameroon (Robiglio 2007). These transition matrices are then combined with estimates of the time-averaged carbon stocks to generate $CO_2$ emissions due to land-use change over some period of time. Under the assumption that the farmer would be indifferent between a cash payment equal to the net discounted present value per ha of land use $i$ and an equivalent payment not to undertake land use $i$, a supply curve of avoided emissions priced at their opportunity cost was generated using a spreadsheet model (Swallow *et al.* 2007).

The key finding in the analysis was that much of the carbon emitted due to land-use change had low opportunity costs (i.e., more than 50% of the carbon emitted was priced at less than US$1 per ton). This implies that paying farmers not to undertake land-use conversions that emit carbon could provide a relatively low-cost means of achieving significant emission reductions.

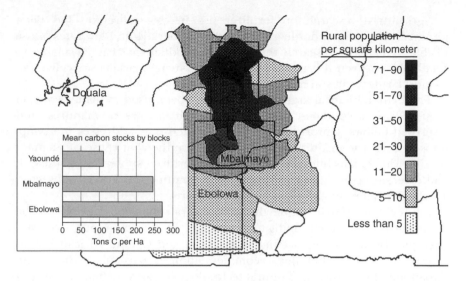

*Figure 32.1* Changes in mean carbon stocks/ha over a population density gradient by intensification blocks in the Forest Margins Benchmark of Southern Cameroon in 2004 (source: author's estimations based on ASB field data).

However institutional failures raise serious doubts about the feasibility of a project-based approach where farmers would be paid the "opportunity cost" of the avoided emissions associated with their extant cropping systems. First of all, the widespread poverty along the forest margins of the tropics underlies the low opportunity costs and thus the relatively meager payments run the risk of maintaining the households in chronic poverty. If farmers were paid not to produce, how would rural development proceed? Where would innovation take place? Another issue is the lack of well-developed food markets to substitute for forgone subsistence food production efforts. Finally, the technical capacity for measuring, reporting, and verifying avoided emissions does not exist and will have to be developed.

Meeting the food demands of a rapidly growing population will not be achieved by REDD schemes that pay farmers not to produce. Governments through the Comprehensive African Agricultural Development (CAADP) framework are seeking annual agricultural production growth of 6% by 2015. The gap between potential yield and actual yield in West and Central Africa is among the widest in the developing world. In essence, forest biomass is the farmers' source of fertilizer. The system operated well in the 1960s and 1970s, as long as there were more forests to cut down and convert to cocoa, oil palm, or food staples systems. Today in West Africa the rainforest has largely disappeared and that in Central Africa is on essentially the same nutrient mining pathway of deforestation.

J. GOCKOWSKI AND P. VAN ASTEN

Agricultural research on fertilizer use in this subregion has been minimal, reflecting underdeveloped fertilizer markets and low rates of use (Table 32.1). Why research the use of a product that isn't available to smallholders? Even if it is available, rural financial markets scarcely exist, so most farmers will still not have access.

These are all logical arguments that have been used to justify work on intermediate second-best technologies such as green manures and improved fallows. Unfortunately, this approach only addresses the symptoms and not the institutional failures that are the root cause of stagnant productivity. Meanwhile, the neglect of the fertilizer subsector by research and extension continues and instead of promoting this crucial subsector, governments impose tariffs on fertilizer imports.

Burney *et al.* (2010) argue that agricultural intensification should be encouraged as a climate mitigation strategy. In the low input, extensive slash and burn farming systems of the West and Central African humid forests, we believe that this recommendation is particularly pertinent. Expansion of extensive agricultural technology remains the major driver of deforestation and GHG emissions and, perhaps most importantly, slash and burn technology underlies chronic poverty among rural households. The primary role of extensive agricultural expansion in the deforestation of the rainforests of West and Central Africa suggests that the promotion of its converse, that is agricultural intensification, could be a strategy to mitigate emissions while tackling directly the strategic issue of food security.

## Agronomic research to support sustainable intensification

In Cameroon and Nigeria, the International Institute for Tropical Agriculture (IITA) is conducting diagnostic surveys on cocoa-based cropping systems in the major cocoa-producing areas. This work is conducted in partnership with the local research, cocoa authority partners, CIRAD (Centre de Coopération Internationale en Recherche Agronomique pour le Développement, the French international agricultural research center) and the private sector. The aim of these ongoing surveys is to determine current yields and constraints of cocoa and the main food crops (i.e., plantains and cassava) in these smallholder farming systems. Reliable quantitative data on yields, yield gaps, and the underlying factors is grossly missing in these systems. IITA has successfully developed and published a series of survey and analytical approaches to identify, quantify, and rank crop constraints for these production systems. These include tools such as boundary line analysis and balanced nutrient diagnostics. The surveys will not only help identify the crop growth constraints, but will also help to reveal which of the existing crop and soil management practices adopted by the smallholders lead to a reduction in the yield gaps.

After the analysis of the yield gaps and the underlying factors, a series of participatory feedback workshops will be organized to agree on options

Table 32.1 Mean annual fertilizer use, and ratio of arable land to forests in West and Central Africa, Brazil, Indonesia and USA, 2002 to 2008

| Country/region | Total fertilizer nutrients applied (tons) | Arable land permanent crops (000 ha) | Fertilizer application per ha (kg/ha) | Forest land (000 ha) | Ratio of arable to forest (%) |
|---|---|---|---|---|---|
| Cameroon | 51,156 | 7,152 | 7.1 | 21,016 | 34 |
| Congo | 1,709 | 541 | 3.2 | 22,475 | 2 |
| Democratic Republic of the Congo | 3,090 | 7,450 | 0.4 | 155,692 | 5 |
| Gabon | 2,326 | 431 | 4.8 | 22,000 | 2 |
| Eq. Guinea | 0 | 215 | 0.0 | 1,685 | 13 |
| Central African Republic | 0 | 2,011 | 0.0 | 22,755 | 9 |
| Congo basin countries | 58,280 | 17,860 | 3.3 | 245,623 | 7 |
| Côte d'Ivoire | 117,409 | 6,850 | 17.1 | 10,391 | 66 |
| Ghana | 39,895 | 6,852 | 5.8 | 5,517 | 124 |
| Guinea | 2,922 | 2,594 | 1.1 | 6,724 | 39 |
| Nigeria | 200,374 | 37,814 | 5.3 | 11,089 | 341 |
| Sierra Leone | 0 | 1,523 | 0.0 | 2,824 | 54 |
| Liberia | 0 | 598 | 0.0 | 4,479 | 13 |
| West Africa (GRF) | 360,600 | 56,231 | 6.4 | 41,024 | 137 |
| Brazil | 9,420,054 | 67,986 | 139 | 531,261 | 13 |
| Indonesia | 3,317,669 | 36,614 | 91 | 97,536 | 38 |
| USA | 19,690,900 | 175,689 | 112 | 302,108 | 58 |

Source: Compiled from FAOSTAT (2011).

to reduce the primary yield constraints. Ex-ante evaluation of investment trade-offs through partial budget analysis and crop response of proposed technologies will help identify the best-bet interventions for cocoa, plantain, and cassava. Identified technologies should be tested on-farm and are likely to include use of improved seed-fertilizer systems, including the involvement of local private-sector input suppliers. The on-farm trials and demonstrations should start in the second half of 2011. The proper identification of crop constraints can lead to targeted eco-efficient and cost-effective technology interventions. The aim is to develop, adapt, and demonstrate that agricultural intensification in the humid forest margins is profitable, and that this will reduce the pressure on farmers to exploit and degrade the forest areas in order to provide their desired food and income needs.

## A development pathway for agricultural intensification

In Ghana, recent results with intensification in the cocoa sector are positive. Baseline data from over 5,000 cocoa farmers trained by the Sustainable Tree Crop Program (STCP) in 2010 revealed that those who were able to access and apply the recommended quantity of fertilizer had an average yield (672 kg per ha) that was 160% greater than farmers who did not use fertilizers. This translated into a 140% increase in their gross margin per ha (IITA unpublished data). This result was achieved by producers applying a blanket recommendation of 370 kg per ha of blended nitrogen-phosphorous-potassium (NPK) fertilizer (i.e., NPK 0-18-22 plus calcium, sulfur, and magnesium).

The encouraging results with fertilizer use on cash crops can open the door for the development of the agrochemical input sector in the humid forest zone. The intensification of commodities with well-established markets such as cocoa, coffee, and oil palm is less risky and can lead to the rapid development of the input market and supporting services such as insurance and credit markets when the technology is profitable.

The IITA is actively seeking partnerships with the input suppliers and national agricultural research and extension organizations of West, Central and East Africa to sustainably intensify smallholder production. The strategy is to identify a leading cash crop and develop recommendations for soil fertility management for this crop as well as the other principal crops in the farming system, which are then demonstrated and promoted through public–private partnerships. Thus, in Cameroon and Nigeria, IITA, through the STCP, is focusing on cocoa, cassava and plantains, with cocoa as the anchor commodity. The approach was also piloted in Uganda with coffee and cooking bananas through the Consortium for improving agriculture-based livelihoods in Central Africa (CIALCA) project.

## Concluding remarks

From the standpoint of climate mitigation, the warm humid agro-ecological zone offers the highest returns to intensification because of the high carbon stock in its natural ecosystems. That does not necessarily mean that intensification should take place there. The three most common crops in this agroecology in West Africa are cocoa, oil palm, and cassava. Cocoa and oil palm are primarily confined to this agro-ecological zone, but cassava does very well in the warm subhumid savanna lands just to the north of the Guinea Forest and to the south of the Congo basin rainforest. With population expected to double in the next 40 years, food security is at the top of the development priority list. Intensification of cassava systems in the subhumid savannas of sub-Saharan Africa will help to meet a rapidly growing urban demand and relieve pressure on the forest. At the same time, intensification of cocoa, coffee, and oil palm in the forest zone will directly impact on the land-use trajectory in the humid tropics.

The REDD+ decision in the Cancún agreement recognizes implementation through a phased approach beginning with (1) the development of national strategies, policies, and capacity building, followed by (2) the implementation of national policies and strategies. Preparations are being made in anticipation of significant resource flows pledged by developed countries. One of the positive outcomes of COP-16 in Cancún was the creation of a Green Fund to administer these resource flows (Climate Focus 2011).

There is an important need for the design of institutional frameworks at the national level to finance and implement sustainable intensification mitigation strategies. It is our opinion that project-based approaches will be too complex to administer, at least in the short run, given local capacity. A national or regional framework for agricultural intensification supported by REDD finance would be most feasible. An important element in such a framework will be the issue of measuring, reporting, and verifying avoided emissions from the sustained intensification of production systems. Without information on baseline yields and the progress of those yields through time, developed country investors will not continue to invest.

In addition to the institutional framework, location-specific agronomic and socio-economic recommendations will be needed to support a comprehensive "fertilizers for forest" mitigation strategy. In conjunction with this, a better understanding of the impacts of a warmer, $CO_2$-enriched environment on the sustainable intensification mitigation strategies will help in adapting these mitigation strategies to a changing climate.

## Note

1 In theory, the 400 to 500 tons/ha of forest biomass that characterizes secondary forest should generate between 28 to 50 tons/ha of ash fertilizer and liming amendment.

## References

Achard, F., Eva, H.D., Mayaux, P., Stibig, H.J. and Belward, A. (2004) Improved estimates of net carbon emissions from land cover change in the tropics for the 1990s, *Global Biogeochemistry Cycles*, vol. 18.

Boserup, E. (1965) *The Conditions of Agricultural Growth: The Economics of Agrarian Change under Population Pressure.* Aldine, Chicago, Il.

Burney, J.A., Davis, S.J. and Lobell, D.B. (2010) Greenhouse gas mitigation by agricultural intensification, *PNAS*, vol. 107, no. 26, pp. 12052–12057.

Canadell, J.G., Raupach, M.R., and Houghton, R.A. (2009) Anthropogenic $CO_2$ emissions in Africa, *Biogeosciences*, vol. 6, pp. 463–468.

Climate Focus (2011) CP16/CMP 6: *The Cancún Agreements Summary and Analysis* [Online]. Available at: www.CLIMATEFOCUS.com/documents/files/Cancun Briefing Jan 2011 v.1.0.pdf (Accessed: 27 February 2011).

FAOSTAT (2011) *FAO Statistics Division* [Online]. Available at: http://faostat.fao. org (Accessed: 13 March 2011).

Green R.E., Cornell, S.J., Scharlemann, J.P.W. and Bamford, A. (2005) Farming and the fate of wild nature, *Science*, vol. 307, pp. 550–555.

Gockowski, J. and Sonwa, D. (2011) 'Cocoa intensification scenarios and their predicted impact on $CO_2$ emissions, biodiversity conservation, and rural livelihoods in the Guinea rain forest of West Africa', *Environmental Management*, vol. 48, no. 2, pp. 307–321.

Gockowski, J., Tonye, J., Diaw, C., Hauser, S., Kotto-Same, J., Moukam, A., Nomgang, R., Nwaga, D., Tiki-Manga, T., Tondoh, J., Tchoundjeu, Z., Weise, S., Zapfack, L. (2005) 'The forest margins of Cameroon', in: Palm, C.A., Vosti, S.A., Sanchez, P.A. and Ericksen, P.J. (eds) *Slash and Burn: The Search for Alternatives.* Columbia University Press, New York, NY, 463 pp.

Hill, P. (1963) *The Migrant Cocoa Farmers of Southern Ghana: A Study in Rural Capitalism*, Cambridge University Press, Cambridge, UK.

Robiglio, V. (2007) 'Impact of shifting cultivation on forest landscapes and secondary vegetation dynamics in the humid forest of the South of Cameroon'. Ph.D. dissertation, School of Environment and Natural Resources, Bangor University, UK.

Swallow, B., van Noordwijk, M., Dewi, S., Murdiyarso, D., White, D., Gockowski, J., Hyman, G., Budidarsono, S., Robiglio, V., Meadu, V., Ekadinata, A., Agus, F., Hairiah, K., Mbile, P.N., Sonwa, D.J. and Weise, S. (2007) 'Opportunities for avoided deforestation with sustainable benefits', *An Interim Report by the ASB Partnership for the Tropical Forest Margins.* ASB Partnership for the Tropical Forest Margins, Nairobi, Kenya.

UN Department of Economic and Social Affairs, Population Division, Population Estimates and Projections Section (2009) World Population Prospects, 2008 Revision. United Nations, New York.

# 33

# CARBON TRADE-OFFS ALONG TROPICAL FOREST MARGINS

## Lessons from ASB Partnership work in Cameroon

*Peter A. Minang, Meine Van Noordwijk, and*
*James Gockowski*

## Introduction

Changes in forest and tree cover can lead to rapid and substantial carbon emissions. Currently, about 7.3 million ha of forests are lost annually, releasing an estimated 5.8 gigatons of carbon dioxide (Gt $CO_2$) per year into the atmosphere. This represents 12–17% of human-generated greenhouse gas (GHG) emissions. On the other hand, forests have the capacity to act as sinks (the ability to absorb and hold $CO_2$ for long periods). The Intergovernmental Panel on Climate Change (IPCC) Third Assessment report puts the total potential for avoiding carbon emissions through aggressive forestry practice changes on 700 million ha of forest at about 60–80 billion tons, or about 12–25% of current fossil fuel emissions over a period of 50 years.

As a corollary, forests and trees have been at the heart of strategies and negotiations for mitigating GHG emissions in the last few years. The Clean Development Mechanism (CDM) and Reducing Emissions from Deforestation and forest Degradation (REDD+) are two key policy instruments that have emerged in the last few years to enable climate change mitigation through forests and trees. The CDM is a mechanism employed within the context of the Kyoto protocol to enable carbon sequestration through afforestation and reforestation (A/R), but has had little success so far in developing countries (Boyd *et al.* 2009).

Meanwhile agreement was reached on the key principles for a REDD+ mechanism at the 16th Conference of the Parties of the United Nations Framework Convention on Climate Change (UNFCCC) in Cancún. The REDD+ idea suggests a mechanism in which countries that reduce national level deforestation to below an agreed level would receive compensation or rewards. These countries may also make commitments to stabilize or further reduce deforestation in the future. Such reductions can be obtained through "reducing emissions from deforestation," "reducing emissions from forest degradation," "conservation of forest carbon stocks," "sustainable management of forests," and "enhancement of carbon stocks." The reductions and rewards must contribute to sustainable development.

The level of sustainable development benefits, costs and trade-offs to be expected from a REDD+ type of scheme remains a fundamental question, as very little evidence currently exists, but expectations of win–win situations are high. At what point, or when, will carbon represent a good proposition to land users seeking to put land into alternative livelihoods and biodiversity conservation uses? This chapter examines the trade-offs between carbon and alternative land uses in the southern forest zone of Cameroon. It uses evidence from long-term research by the Alternatives to Slash-and-Burn (ASB) Partnership for the Tropical Forest Margins (see also chapter by Gockowski and Van Asten).

In line with current global forest and tree climate mitigation strategies, the ASB Partnership is conducting research on how to reduce deforestation along tropical forest margins without compromising agricultural productivity and the provision of environmental services. ASB was created in 1994 as a partnership between international and national research organizations and universities, and is currently working in a network of sites across the humid and sub-humid tropics. ASB research primarily focuses on the sustainability of impacts in terms of livelihoods, agricultural productivity, environmental services, and the trade-offs therein. Uniform indicators, methods and approaches have been applied to understand land use, carbon, biodiversity, agricultural productivity and livelihood impacts across more than 12 sites (Palm *et al.* 2005). This chapter focuses on trade-offs between carbon, profitability and biodiversity in order to provide insights into the potential for REDD+ and other mechanisms to meet the triple objectives of climate, development and conservation using Cameroon as an example.

## The ASB benchmark site and context in Cameroon

The Cameroon ASB site was selected to represent the land use and population dynamics of humid tropical rainforest conditions of West and Central Africa. The 1.54 million ha stretch spans several different population characteristics and land uses with different market access conditions (Kotto-Same *et al.* 2002).

Population densities range between 4 to 100 persons/km$^2$ with a gradient from the northeast (around the capital city Yaounde) to the southeast (south of Ebolowa/Ambam towns). Precipitation peaks twice a year and ranges from 1,350 to 1,900 mm annually. In terms of plant diversity, over

200 plant species have been recorded within a 1,000 m$^2$ survey of natural forest in the area (Garland 1989 in Kotto-Same *et al.* 2002). Forests in the area vary from dense semi-deciduous forests in the north to dense humid Congo basin forests in the southeast.

Land-use analysis constitutes the basis for trade-off analysis. Table 33.1 presents a summary of major land uses in the humid forest zone of Cameroon. Agricultural land use occupied about 24% of the area, with cocoa being the dominant productive agricultural land use covering about 3.8%. Horticulture has been growing in the vicinity of the Yaounde urban area and is characterized by intense cultivation of vegetables and maize for the growing urban market. Smallholder oil palm has been growing but at a slower rate due to poor quality planting material and poor infrastructure for processing.

### Carbon stocks

In trade-off analysis within the ASB benchmark sites, all alternative land uses were evaluated in terms of carbon stocks. Time-averaged carbon stocks (estimates of carbon content over the rotation or life cycle of a land-use system) were calculated for 6 land uses namely primary forest, 2-year-old cropland, a cocoa plantation, 4-year-old bush fallow, 9-year-old tree fallow and 17-year-old secondary forest in the benchmark site (Kotto-Same *et al.* 1997; Palm *et al.* 2005). Estimates were obtained from measurements of trees, litter, roots and soil. Forests were used as the basis for comparison between land-use systems.

The carbon stocks of 6 selectively logged forests in the area averaged about 228 tons of carbon per hectare (t C/ha), ranging from 193 to 252 t C/ha. A mature jungle cocoa stand contains about 43% of the carbon of the forest, ranging from 54 to 141 t C/ha with an average of 89 t C/ha. Traditional long fallows recorded the maximum carbon stocks for crop-fallow systems, at about 167 t C/ha. Carbon accumulation rates varied with fallow age, ranging from 2.8 t C/ha during the first two years with *Chromolaena* domination and increasing to 8.5 t C/ha for the next 6–10 years (Kotto-Same *et al.* 2002).

### Profitability of land uses

Profitability was evaluated as a key determinant of adoption and hence land-use change in the area and represented in terms of net present value (NPV). Shaded intensive cocoa systems with fruit trees and short, fallow intercropped food systems posted the highest and lowest social profitability respectively. However, it should be noted, that since per-hectare profitability is measured on an annual basis and includes any non-productive fallow period, annual profitability of shifting cultivation is significantly reduced.

*Table 33.1* Main land uses at the forest margins of southern Cameroon

| Land use | Description |
|---|---|
| Natural forests | Undisturbed dense semi-deciduous or humid Congo basin forests. This is the reference point for all land uses. |
| *Community forests | Forest system of no more than 5,000 ha (ranging from natural to degraded) that is the subject of a management agreement between "a community" and government as defined in the 1994 forest law. A community forest could be subsequently exploited as a sale of standing volume. |
| *Commercial logging | Either a concession (up to 200,000 ha) or a sale of standing volume (2,500 ha maximum). |
| Extensive cocoa with fruit tree shade–long fallow | Complex agroforestry cocoa system with fruit trees (mango, avocado, African prune–*dacryodes edulis*, oranges). Normally established on forested land or long fallows and intercropped with plantain, melon seed and cocoyam in the first three years. Fungicide use is about 50% of intensive cocoa systems and yields average 265 kg per hectare (kg/ha). Farm sizes average 1.3 ha. |
| Intensive cocoa with fruit tree shade–short fallow | Complex agroforestry cocoa system with fruit trees (mango, avocado, African prune–*dacryodes edulis*, oranges) with high input practices. Normally established on 4-year fallow. Yields average 500 kg/ha. Average farm size is 1.3 ha. |
| Oil palm–long fallow | Small holder monoculture of oil palm at a density of 143 trees/ha. Established on forests and intercropped with plantain, cocoyam and melon seed during the first two years. Yields average 8,000 kg/ ha at maturity. Average farm size is 1 ha. |
| Intercropped food crop field–short fallow | Crop/fallow system planted into a 15-year fallow of *Chromollaena ordorata* fallow consisting of melon seed (*cucumeropsismannii*), plantain, maize and cocoyam. Mainly for subsistence. Average farm size is 0.25 ha. |
| Intercropped food crop field–long fallow | Crop/fallow system planted into a 4-year fallow of *Chromollaena ordorata* consisting of groundnuts, cassava, maize, cocoyams and plantain. Mainly for subsistence. Average farm size is 0.25 ha. |

*Note*
* denotes a land use that has not been subject to evaluation within ASB work in Cameroon though present in the landscape.

*Biodiversity*

ASB explored three indicators of plant biodiversity including species counts, Plant Functional Types (PFTs) and the Vegetation index ("V" index). The number of plant species was counted for each standard plot (40 m × 5 m) and for each land use in Cameroon. Plants were further classified through coding into functional groups (what they do and how they do it) and adaptive characteristics as described in (Gillison 2000b; Gillison and Carpenter 1997). The "V" index represents the relative position of each plot in terms of increasing structural complexity and richness in both species and PFTs. (Gillison 2000b; Williams *et al.* 2001). The index is standardized between 0.1 and 1.0, with 1 being the value of the forest. The PFT and "V" index are biodiversity measures developed within ASB to address mosaic land use/plot-level biodiversity dynamics and were found to highly correlate with land cover type, plant and animal richness and soil nutrient availability in Indonesia, Cameroon and Brazil, hence a potentially useful index (Gillison 2000a).

# Trade-offs

The assessment of trade-offs in ASB studies was given by a set of multiple indicators across the humid tropics. We have addressed one indicator each for time-averaged carbon stocks and NPV for profitability of land uses, and three indicators for biodiversity (species count, PFT/modi and the "V" index). Figure 33.1 shows the indicators for each of the land uses and

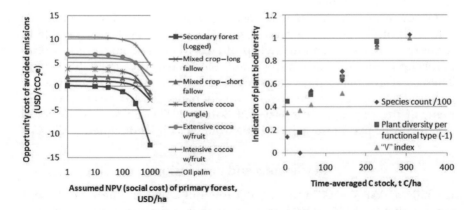

*Figure 33.1* A. Opportunity cost of avoiding emissions by conversion of natural forest to a range of land uses, expressed as the difference in Net Present Value (NPV) divided by the difference in time-averaged C stock (multiplied by 44/12 as $CO_2$/C ratio); values are given for a range of NPV estimates of natural forests, as this was not assessed by the ASB Cameroon report; 1B. Correlation of above-ground C stock with indicators of plant biodiversity for land-use systems in Cameroon.

their trade-offs. Figure 1B shows that scores for the three indicators are related to above-ground C stocks for the range of land uses found in Cameroon.

In a typical cycle in Cameroon, forests are first cleared for mixed-crop systems in short fallow rotation followed by long fallows. In cases where cocoa plantations are established, mixed crops do not go into fallows but are planted with cocoa and crops phased out slowly. Land that is left to fallow long enough (over 20 years) eventually reverts to forest (secondary forest). This transformative cycle entails various trade-offs.

Transformation from natural forest to mixed-crop systems constitutes the greatest loss in terms of carbon and biodiversity. More than 200 t C/ha and great amounts of biodiversity are lost. While returns on land increase, it remains the least profitable of all land-use conversions and continues to grow because of low input requirements. Intensification occurs in home garden systems near urban areas where market conditions and infrastructure are favorable. However, this intensification does not bring any carbon or biodiversity gains.

The second most significant system change with best benefits for carbon, biodiversity and profitability is the conversion of short fallows into intensive cocoa with fruit tree systems. With such conversions, carbon stocks increase from 6 to over 100 t C/ha. However, there are questions about the sustainability of such systems owing to susceptibility to pest attacks. Such systems have not been growing significantly in the area due to high input requirements, notably labor and pesticides. Labor is scarce in the region while capital and associated infrastructure to support the supply of inputs is limited.

On the other hand, extensive cocoa systems planted in long fallows or remnants of forests do not require high inputs, are fairly manageable in terms of pests, and are moderately profitable. They also come with considerably lower losses in carbon stocks and biodiversity during establishment. These features make the extensive cocoa system a widespread practice and, with slight improvements, potentially beneficial for carbon management in the region (Gockowski and Sonwa 2011). These systems could potentially be part of workable REDD strategies if well designed.

## Key lessons and conclusions

Looking at ASB's work in the last decade and a half, a number of lessons emerge that could be of relevance to current policy discussions and practice on emission reductions in developing countries.

First, current prices in the compliance carbon market are unlikely to ensure emission reductions because they will not meet the opportunity costs of land conversions in the forest landscape in southern Cameroon. Conversion of forests or long fallow to extensive cocoa gardens provides benefits above 10 USD/t $CO_2$e of emissions, while avoiding conversion to more intensive cocoa systems would have even higher opportunity costs (Figure 33.1a). The fact that opportunity costs represent only part of the carbon project costs makes it even more unlikely.

Second, conversions within non-forest land-use classes hold some of the best opportunities for emission reductions and co-benefits. REDD alone is unlikely to enable the warranted emission reductions. Studies in both Indonesia and Cameroon show that trade-offs are more beneficial in terms of carbon, profitability and biodiversity gains in areas that do not currently fall under the definition of a forest within the UNFCCC REDD+ framework, for example, conversions from mixed crops to intensive cocoa agroforestry (van Noordwijk and Minang 2009; van Noordwijk et al. 2009; Ekadinata et al. 2010). It is therefore imperative to move beyond the current REDD+ framework within the UNFCCC (which is mainly about forests) to a broader framework for reduced emissions from all land-use change.

Furthermore, ASB studies in Cameroon have highlighted the importance of institutional and policy issues as crucial factors for success in the design of emission reduction projects. We have learnt from Cameroon that a key strategy for reducing emissions, such as the intensification and diversification of cropping systems with trees—resulting in agroforestry with huge potential carbon, economic and biodiversity benefits—can be hampered by labor, market, infrastructure and land tenure challenges. Strategies for addressing drivers of deforestation as well as emission reduction strategies outside REDD will need serious analytical work and investment in institutions and policies to ensure success.

Lastly, the trade-off analysis in Cameroon demonstrates the need to understand whole landscape systems in order to develop a complete picture of how REDD+ and other emission reduction mechanisms might work. Economic analysis was not done for community forest and logging systems, hence, judging opportunity costs against a natural forest with little or no economic benefit presents only part of the picture. ASB work in Indonesia and Peru showed that with conservative values for logging, opportunity costs for REDD are potentially very high and therefore require a high carbon price to offset any potential conversion (Tomich et al. 2002; Swallow et al. 2007). More work is required in understanding trade-offs within logged forests (including reduced impact logging) likely to be applicable for emission reductions in the context of "degradation" within the REDD+ framework.

# References

Boyd, E., Hultman, N., Roberts, J.T., Corbera, E., Cole, J., Bozmoski, A., Ebeling, J., Tippman, R., Mann, P., Brown, K., Liverman, D.M. (2009) Reforming the CDM for sustainable development: lessons learned and policy futures, *Environmental Science and Policy*, vol. 12, pp. 820–831.

Ekadinata, A., van Noordwijk, M., Dewi, S. and Minang, P.A. (2010) 'Reducing emissions from deforestation, inside and outside the "forest"', ASB PolicyBrief 16. ASB Partnership for the Tropical Forest Margins, Nairobi, Kenya.

Gillison, A.N. and Carpenter, G. (1997) A plant functional attribute set and grammar for dynamic vegetation description and analysis, *Functional Ecology*, vol. 11, pp. 775–783.

Gillison, A.N. (2000a) 'Alternatives to Slash and Burn Project: Phase II Above-ground biodiversity assessment working group summary report'. ICRAF, Nairobi, Kenya.

Gillison, A.N. (2000b) 'VegClass Manual: A Field Manual for Rapid Vegetation Classification and Survey for General Purposes'. CIFOR, Bogor, Indonesia.

Gockowski, J. and Sonwa, D. (2011) Cocoa intensification scenarios and their predicted impact on $CO_2$ emissions, biodiversity conservation and livelihoods in the Guinea Rainforest of West Africa, *Environmental Management* (vol. 48, no. 2, pp. 307–321).

Kotto-Same, J., Woomer, P.L., Appolinaire, M. and Zefack, L. (1997) Carbon dynamics in slash-and-burn agriculture and land use alternatives of the humid forest zone in Cameroon, *Agriculture, Ecosystems & Environment*, vol. 65, pp. 245–256.

Kotto-Same, J., Moukam, A., Njomgang, R., Tiki-Manga, T., Tonye, J., Diaw, C., Gockowski, J., Hauser, S., Weise, S., Nwaga, D., Zafack, L., Palm, C., Woomer, P, Gillison, A., Bignell, D., Tondoh, J. (2002) 'Summary Report and Synthesis of Phase II in Cameroon', Alternatives to Slash and Burn. ICRAF, Nairobi, Kenya.

Palm, C.A., Vosti, S.A., Sanchez, P.A. and Ericksen, P.J. (eds) (2005) *Slash and Burn: The Search for Alternatives.* Columbia University Press, New York, pp. 387–414.

Swallow, B., van Noordwijk, M., Dewi, S., Murdiyarso, D., White, D., Gockowski, J., Hyman, G., Budidarsono, S., Robiglio, V., Meadu, V., Ekadinata, A., Agus, F., Hairiah, K., Mbile, P.N., Sonwa, D.J. and Weise, S. (2007) 'Opportunities for Avoided Deforestation with Sustainable Benefits', An Interim Report by the ASB Partnership for the Tropical Forest Margins. ASB Partnership for the Tropical Forest Margins, Nairobi, Kenya.

Tomich, T.P., Foresta, H.d., Dennis, R., Ketterings, Q., Murdiyaso, D., Palm, C., Stolle, F., v. Noordwijk, S. and van Noordwijk, M. (2002) Carbon offsets for conservation and development in Indonesia, *American Journal of Alternative Agriculture*, vol. 17, pp. 125–137.

Van Noordwijk, M. and Minang, P.A. (2009) 'If we cannot define it, we cannot save it: fuzzy forest definition as a major bottleneck in reaching REDD+ agreements at and beyond Copenhagen COP15', in Bodegom, A.J., Savenijeand, H. and Wit, M. (eds) *Forests and Climate Change: adaptation and mitigation* [Online]. Available at: www.etfrn.org/ETFRN/newsletter/news50/ETFRN_50_Forests_and_Climate_Change.pdf. Tropenbos International, Wageningen, The Netherlands, pp. 5–10.

van Noordwijk, M., Minang, P.A., Dewi, S., Hall, J. and Rantala, S. (2009) 'Reducing Emissions from All Land Uses (REALU): The Case for a whole landscape approach', *ASB PolicyBrief 13*, ASB Partnership for the Tropical Forest Margins, Nairobi, Kenya.

Williams, S., Gillison, A., and van Noordwijk, M. (2001) 'Biodiversity: issues relevant to integrated natural resource management in the humid tropics', ASB Lecture Note 5, in van Noordwijk, M. Williams, S. and Verbist B. (eds) *Towards integrated natural resource management in the humid tropics: local action and global concerns to stabilize forest margins*, Alternatives to Slash-and-Burn. Nairobi, Kenya.

# 34

# MITIGATION OPTIONS FROM FORESTRY AND AGRICULTURE IN THE AMAZON

*Jan Börner and Sven Wunder*

## Introduction

This chapter seeks to identify "low-hanging fruits" for agricultural and forestry-based climate change mitigation in the Amazon region and derive implications for mitigation initiatives. Mitigation potential, especially in the forestry sector, looms large. Niles *et al.* (2002) estimate for Latin America that forest restoration and avoided deforestation together could result in up to 128 Mt of carbon emission reductions annually (68% of the world total) as opposed to 9.3 Mt through sustainable agricultural change (see chapters by Minang *et al.* and Gockowski and Van Asten for examples from Africa).

The Amazon forest is the world's largest continuous tropical rainforest covering over 4% of the earth's land surface. Between 1977 and 2010, 618,000 km² of forest area was converted to pasture and agricultural crops in the Brazilian Amazon alone.[1] Carbon emissions from Amazon deforestation are estimated to account for 24% of global carbon emissions from land-use change (Aragão and Shimabukuro 2010). In the countries with large shares of their territory in the Amazon region (Bolivia, Brazil, Colombia, Ecuador, and Peru), combined AFOLU (Agriculture, Forestry and other Land Use) emissions account for over 83% of total greenhouse gas (GHG) emissions[2] thus representing the most important sector for climate change mitigation.

AFOLU-based climate change mitigation has been ridden with obstacles. Afforestation and reforestation (A/R) initiatives, for example, the only AFOLU measure eligible under the Clean Development Mechanism (CDM) of the Kyoto Protocol, account for only 17 out of a total of 2,970 registered projects[3] approved since the CDM's inception—complicated rules and high transaction costs have been binding constraints, and pilot experiences were often confined to the voluntary market (Michaelowa and Jotzo 2005).

The slow take-off of AFOLU-based mitigation suggests that absolute mitigation potential is only one among several indicators of scope for climate change mitigation in the sector. Ideally, AFOLU mitigation options would:

1    come with **high potential emission reductions or removals**;
2    be adoptable at **low opportunity costs and risk** of economic failure;
3    be promoted with **low transaction costs**; and
4    be disseminated with **low risk of economic and environmental spillover effects**.

Mitigation potential thus becomes more than a matter of the biophysical, technological and economic characteristics (criteria 1 and 2) of mitigation options and is inherently related to intervention design and the local context (criteria 3 and 4).

The chapter is organized according to the four criteria above and concludes with a discussion of implications. The prospects for AFOLU-based mitigation in the Amazon region are discussed based on case studies and lessons learned from past and ongoing mitigation initiatives. Sections 2 and 3 draw heavily on two case studies in the eastern and western Brazilian Amazon, in which consistent biophysical and socio-economic data have been collected over several years (see Figure 34.1).

Between early 1992 and 2003, the Alternatives to Slash-and-Burn Program (ASB)[4] and the Studies on Human Impact on Forests and Flood-

*Figure 34.1* Case study sites in the Brazilian Amazon.

plains in the Tropics Program (SHIFT)[5] conducted extensive research in the western and eastern Brazilian Amazon, respectively. The western Amazon sites represented typical new agricultural frontier settings with dynamic land-use patterns and large tracts of primary forests on smallholdings. Research activities focused on the districts of Acrelândia and Theobroma in the states of Acre and Rondônia (Vosti *et al.* 2002). In contrast, the eastern Amazon sites lie in a consolidated agricultural zone close to the urban center Belém, where colonization had already started by the 19th century. Practically no primary forest is left in this region and a relatively stable landscape mosaic of agricultural crops, pastures, fallows and secondary forests has evolved (Metzger 2002). The three studied districts—Castanhal, Igarapé-Açu, and Bragança, in the state of Pará—exhibit a gradient of market to subsistence-oriented farming systems, due to their increasing distance to Belém (Börner *et al.* 2007).

## 2 Biophysical and technological mitigation potential

Agriculture and forestry can contribute to climate change mitigation by (1) enhancing or maintaining above-ground biomass or soil carbon storage through land cover change, soil management or avoided vegetation loss, and/or (2) adopting land-use technologies with low-carbon footprints.

Carbon storage in Amazonian vegetation is highly heterogeneous. Above-ground live biomass is on average 276 t/ha (approximately 138 t C/ha) in dense forest, but below 50 t/ha in most secondary forests and savanna-type ecosystems (Saatchi *et al.* 2007). Closed natural forests dominate the Amazonian landscape in the Central and Western Amazon, whereas agriculture–forest mosaics and grassland pastures are typical landscape features at its southern and eastern borders, particularly in Brazil.

After 100 years of agricultural activities in the Eastern Amazon, the current landscape exhibits a comparatively low biomass density. Avoiding the conversion of secondary forest fallows to annual crops or pastures could nonetheless result in average carbon gains of over 50 t C/ha. Switching between agricultural land-use options is less effective, unless trees are introduced, for example in agroforestry systems. Moving from permanent mechanized cassava production to a fallow-based mulch cropping system represents the most promising agricultural change mitigation option in terms of above-ground biomass carbon gains (6.4 t C/ha) (Börner *et al.* 2007).

In the Western Amazon sites, avoided deforestation could preserve up to 145 t C/ha in forest biomass. Agricultural change options, including pasture improvement, exhibit approximately the same low biomass carbon gains as in the Eastern Amazon, but agroforestry options analyzed by Vosti *et al.* (2001) could yield up to 51 t C/ha through above-ground biomass accumulation in tree components.

401

The presence of trees in land-use systems thus generally boosts carbon gains, which is why agroforestry and tree plantations are often praised for their mitigation potential (Albrecht and Kandji 2003). Yet, biomass density in Amazonian secondary forests after natural regeneration can be higher than in a range of agroforestry and monoculture plantations (Schroth *et al.* 2002). A/R could nonetheless come with substantial mitigation benefits if end uses included the substitution of fossil fuels, e.g., charcoal instead of coke for pig iron production (Fearnside 1995).

A highly effective agricultural mitigation option is the avoidance of fire for land preparation (Aragão and Shimabukuro 2010). Switching from slash-and-burn based annual cropping to a mechanized mulch system could curb global warming potential of traditional farming practices by a factor of five, primarily through a reduction in methane ($CH_4$) emissions from the use of fire during land preparation (Davidson *et al.* 2008). Multiple mitigation options also exist in the livestock sector ranging from increasing stocking rates, thus reducing $CH_4$ emissions from soil biological processes after forest-to-pasture conversion, to improved feeding and manure management, reducing emissions per unit of output (Cerri *et al.* 2010). The attractiveness of mitigation options, however, depends crucially on their adoption costs.

## 3 Opportunity costs and adoption risks

Climate change mitigation measures must be financially additional, i.e., if mitigation alternatives are more profitable than business-as-usual (BAU) land uses they are generally not eligible for financial support. While technology adoption is a complex process involving more than purely financial factors, opportunity costs (i.e., the forgone profits of abandoning the first best land-use alternative or technology) represent a fundamental barrier to the adoption of innovative land-use systems and technologies (Rogers 2003).

In both Brazilian case-study areas, mitigation through forest conservation clearly delivers the most "bang for the buck." Per-hectare opportunity costs can be high for some options, e.g., avoided logging of high-value timber species, but the sheer volume of carbon benefits reduces the opportunity costs of avoided forest conversion to values between less than 1 and 8 US$/t C. Similar ranges were found in regional-level studies for the Brazilian and Peruvian Amazon (Armas *et al.* 2009; Börner *et al.* 2010). Mitigation at more recent colonization areas, like the Western Amazon case study sites, tends to be more cost-effective than in the Eastern Amazon, where farmers are better connected to markets and capture higher returns to agricultural activities.

Some agricultural change-based mitigation options appear financially attractive on their own, while others exhibit higher per-hectare opportunity costs than forest conservation options. For example, switching from mechanized continuous to traditional fallow-based annual crop production in the Eastern Amazon comes with opportunity costs of over US$/t C 300. For the few existing low opportunity cost options, e.g., switching from slash-and-burn to mechanized mulching, biomass gains are minimal. The attractiveness of agricultural change mitigation options thus also crucially depends on their potential to reduce GHG emissions independent from land cover changes, e.g., through fire-free land preparation (Davidson *et al.* 2008).

Mitigation options, however, also often face adoption barriers beyond opportunity costs. In both study areas, researchers had spotted potential mitigation options that seemed to provide per-hectare net returns far above those from traditional practices, such as high-yielding agroforestry systems. Why has adoption for most of these alternatives remained negligible?

While limited information about the performance of alternative land-use practices may only subjectively increase adoption risk, some technological alternatives are indeed more risky. Eastern Amazonian farmers are well-acquainted with chemical fertilizers, but only apply them to certain crops. Though on-farm trials demonstrate that fertilizer could almost double expected cassava yields (Kato *et al.* 1999), only 3% of cassava producers have adopted fertilization. These farmers were almost exclusively large-scale producers with secure commercialization channels, as opposed to smallholders who depended on highly fluctuating local wholesale prices. Economic analyses confirmed that low fertilizer use among smallholders can indeed be explained by aversion to price and production risks (Börner *et al.* 2007). Moreover, accidental fire risk was identified as a potentially important adoption barrier for agroforestry and reforestation in areas where slash-and-burn farming dominated (Hoch *et al.* 2009; Börner 2009). For the western Amazon, Vosti *et al.* (1997) identified high upfront investments and long time spans between investment and positive cash flow as additional adoption barriers for otherwise profitable agroforestry systems.

In sum, cost-effectiveness, risks, and the temporal distribution of cash flows often work against technological change-based mitigation, including plantation forestry, in our two case study areas. Setting aside degraded areas for natural regeneration or threatened natural forests for conservation is operationally simpler and almost risk free, at least from the perspective of land users whose food security remains unaffected by conservation measures.

## 4 Transaction costs

Several past and ongoing environmental service initiatives have tried to promote agricultural and forestry-based mitigation options in the Amazon (Table 34.1).

*Table 34.1* Forest and agricultural mitigation schemes in the Amazon region

| Scheme | Start year | Scale | Objectives | Strategy | Mechanism |
|---|---|---|---|---|---|
| Noel Kempff Mercado Climate Action Project–Bolivia | 1997–ongoing | Local (642,458 ha in a protected area) | Emission reductions and biodiversity conservation | Forest conservation | Buying out logging concessions, local community support |
| Proambiente–Brazil | 2004–discontinued due to lack of funding and appropriate national legislation | Regional (~250,000 ha)–12 pilot areas in the Brazilian Amazon with 250 families each) | Emission reduction, biodiversity and water conservation, sustainable agricultural production | Agricultural change and forest conservation | Subsidized credits, conditional payments, technical assistance |
| Socio Bosque–Ecuador | 2008–ongoing | National (539,703 ha in 2010,* target: 4 million ha) | Emission reductions, other forest environmental services | Forest conservation | Conditional payments to communities and individual landholders |

*Note*
*Presentation by German Mosquera, Socio Bosque legal specialist, 12 November 2010, Santa Cruz, Bolivia.

Table 34.1 presents initiatives that differ in at least three fundamental aspects with implications for implementation and/or transaction costs: (1) scale (e.g., local versus regional or national): larger-scale projects tend to benefit from lower transaction costs per mitigation unit; (2) objective (carbon versus environmental services bundle): multiple objectives are expected to increase monitoring and verification needs and costs; and (3) choice of on-the-ground mechanisms to induce mitigation: intervention strategies that require frequent interaction with land users suffer from higher transaction costs.

In the Noel Kempff initiative, transaction costs[6] related to monitoring and administration amounted to about 29% of the project budget, but costs per unit of avoided carbon emissions over a 30-year horizon were still rather low (US$0.45–0.76/t $CO_2$[7]).

For Proambiente, initial estimates of costs for technical assistance and other administrative outlays represented over half of the total program budget (Mattos *et al.* 2001). Based on estimated mitigation potential over 15 years, total program costs per unit of avoided emissions and seques-tered carbon would have been US$/$CO_2$ 1.09 (see also Stella *et al.* this volume).

Since 2008, Socio Bosque's transaction costs went down from 51% to a projected 31% in 2010 (Bustamante and Albán 2010). During the first year, start-up investments such as dissemination campaigns represented between 25% (2008) and projected 7% (2010) of total program costs. Recurrent transaction costs may thus reduce to well below 30% in coming years. Based on 2010 performance and costs of the Socio Bosque program, and assuming a conservative average carbon content of 80 t C/ha of pro-tected forest land, the total cost of retaining the forest carbon of the enrolled area over a 15-year period (to allow for comparison with the Proambiente estimate) amounts to US$0.41/t $CO_2$.

The two forest conservation schemes, Noel Kempff and Socio Bosque, cover slightly larger areas than the Proambiente proposal, which may result in economies of scale in program execution. The Proambiente program was never fully implemented, but its proposed intensive technical assistance component clearly featured as a major additional transaction cost item compared to the two other initiatives, whose mitigation com-ponents merely require monetary transfers and field presence to establish and enforce contractual relationships.[8]

Although transaction costs represent a considerable budget share in all three mitigation initiatives, per-unit implementation costs appear rather low if compared to carbon offset prices on existing market places (Kossoy and Ambrosi 2010). The estimates are nonetheless rather sensitive to assumptions about additionality. Leakage or exaggerated assumptions about deforestation pressure would result in higher actual per-unit costs of carbon benefits.

## 5 Economic and environmental spillover risks

Mitigation incentives can create environmental spillover effects in three ways: first, mitigation measures can produce unintended environmental externalities, e.g., fast-growing monoculture plantations can reduce biodiversity and groundwater reserves (Jackson *et al.* 2005) (i.e., *losing other services*); second, unless conservation-based, mitigation options could become so profitable that they are expanded on previously unused land, such as natural forests (*overshooting adoption scale*); and third, mitigation incentives can trigger unintended land-use change outside the mitigation scheme (*leakage*).

In both Brazilian study sites, economic models were used to assess the likely impacts of mitigation incentives (Carpentier *et al.* 2000; Carpentier *et al.* 2002; Vosti *et al.* 2002; Börner *et al.* 2006; Börner *et al.* 2007). The models suggest that incentives for both forest conservation and land retirement could be competitive strategies for carbon mitigation at offset prices above US\$5/t $CO_2$. The Eastern Amazon model, however, also showed that such liquidity-providing payments and increased land scarcity would increase farmers' input-intensive cash-crop production at the expense of annual staple crops. The resulting additional use of pesticides and fertilizers could increase nitrous oxide emissions and water pollution, thus providing an example of the first spillover mechanism, i.e., losing other environmental services.

Intensification of land use, e.g., through mechanization and fertilizers, is often proposed as a means to increase productivity on existing cropland and thus reduce the need of continuous agricultural expansion. Model simulations, however, suggest that introducing labor-saving mechanization would allow farmers to raise per-hectare productivity and expand the new high-yield cropping system (overshooting adoption scale). A global compilation of case studies shows that this is not an exception (Angelsen and Kaimowitz 2001): it is quite common that agricultural innovations cause locally higher deforestation, even when the innovative technology allegedly is 'land-sparing'.

The third spillover effect (leakage) is particular relevant for use-restricting conservation schemes, since they can 'push' economic activities into non-target areas. In the Noel Kempff REDD pilot project in Bolivia (see previous section), the logging-ban component was estimated in different models to provoke leakage in the 2–42% range—i.e., demand for timber would cause additional timber extraction elsewhere, including abroad (Sohngen and Brown 2004). These large ranges illustrate the uncertainties involved in quantifying leakage effects.

## 6 Discussion and conclusions

We have reviewed the potential for climate change mitigation in the AFOLU sector for the Amazon according to four criteria: (1) bio-physical and technological potential; (2) opportunity costs and adoption risks; (3) transaction costs; and (4) spillover effects (see Table 34.2 for a summary). Like previous studies (e.g., Fearnside 1999), our analysis confirmed that use-restricting conservation options—especially REDD and agricultural land retirement—represent low-hanging fruits not only in terms of technological mitigation potential, but also with respect to farm-level opportunity costs and adoption risk. Reducing or eliminating the widespread use of fire in fallow-based annual crop production and pastures, probably one of the most effective agricultural mitigation options, often comes with higher adoption cost, production and market risks, or long time spans between investment and returns, e.g., under plantation forestry or agroforestry-based mitigation strategies.

The low-cost mitigation potential, as well as perceived operational simplicity in combination with a promising outlook for an international REDD mechanism may thus have helped forest conservation-based mitigation initiatives to come off the ground relatively easily in the region. One large-scale attempt to promote mitigation through agricultural change has remained non-operational and relied on a transaction cost-intensive rural extension component. Transaction costs are, nonetheless, also highly context-dependent, and expanding forest conservation-based mitigation towards forest margins with high deforestation pressure and weak institutional environments may considerably increase their implementation costs vis-à-vis interventions in well-established agricultural zones.

Spillover effects from mitigation initiatives, too, are context-dependent. Nonetheless, the risk of losing other environmental services, e.g., water quality, is generally higher when mitigation initiatives promote input-intensive technological change in agriculture or forestry. Depending on

*Table 34.2* Comparison of key findings on agricultural and forestry mitigation options in the Amazon

|  | Agricultural options + plantation forestry (use modification) | Forest conservation and set-aside (use restriction) |
| --- | --- | --- |
| Mitigation potential (per ha) | Low–medium for agricultural options, medium–high for plantation forestry | Generally high |
| Adoption costs/risks | Costs vary, but adoption risks are often high | Low risk, many low-cost opportunities |
| Transaction costs | Context dependent, tend to increase with population density of intervention area (agricultural zones) and poor infrastructure or weak institutional environments (forest margins) | |
| Spillover effects | Prone to environmental externalities and overshooting adoption scale | Prone to leakage |

the economic setting, increasing the profitability of agricultural land-use alternatives can also often result in the expansion of agricultural land, and thus more pressure on forestland. Meanwhile, forest conservation-based mitigation is prone to leakage through the spatial relocation of demand for timber or agricultural expansion.

In a land-abundant humid forest frontier, such as the Amazon, forest conservation-based mitigation appears more cost-effective and operationally feasible than agricultural change-based mitigation options. Incentives to maintain forests and take marginal lands out of production, e.g., through payments for environmental services, may ultimately also encourage the adoption of more land-intensive production technologies in the agricultural and forestry sectors, where mitigation benefits could be particularly high in cattle production (Cerri *et al.* 2010). In other words, only by taking the carbon-wise biggest bull (deforestation) by the horns and shifting the rules of the land-use game to close forest frontiers (incentives and disincentives vis-à-vis land grabbing), will mitigation through agricultural intensification or plantation forestry become ultimately interesting to land users.

## Notes

1 INPE-PRODES accessed in 2011: www.obt.inpe.br/prodes/.
2 World Resource Institute—CAIT (accessed 2010), combining Land Use Change & Forestry with emissions from Agriculture: http://cait.wri.org.
3 http://cdm.unfccc.int/Statistics/index.html.
4 www.asb.cgiar.org.
5 www.shift-capoeira.uni-bonn.de.
6 All estimates exclude project development costs.
7 Estimate range based on leakage risk reported in Robertson and Wunder (2005).
8 The Noel Kempff project also provided community infrastructure support, but primarily as a means to compensate the employees of the timber concessionaries, who were bought out to avoid emissions from forest degradation.

## References

Albrecht, A., and Kandji, S.T. (2003) Carbon sequestration in tropical agroforestry systems, *Agriculture, Ecosystems & Environment*, vol. 99, no. 1–3, pp. 15–27.
Angelsen, A. and Kaimowitz D. (2001) 'Agricultural technologies and tropical deforestation'. CIFOR, CABI Publishing New York, NY and Oxon, England.
Aragão, L.E.O.C. and Shimabukuro, Y.E. (2010) The incidence of fire in Amazonian forests with implications for REDD, *Science*, vol. 328, no. 5983 pp. 1275–1278.
Armas, A., Börner, J., Tito, M.R., Cubas, L.D., Coral, S.T., Wunder, S., Reymond, L. and Nascimento, N. (2009) 'Pagos por Servicios Ambientales para la conservación de bosques en la Amazonía peruana: Un análisis de viabilidad'. SERNANP Lima, Perú.
Börner, J. (2006) 'A bio-economic model of small-scale farmers' land use decisions

and technology choice in the eastern Brazilian Amazon in Faculty of Agriculture'. University of Bonn: Bonn, Germany pp. 193.

Börner, J., Mendoza, A. and Vosti, S.A. (2007) Ecosystem services, agriculture, and rural poverty in the Eastern Brazilian Amazon: Interrelationships and policy prescriptions, *Ecological Economics*, vol. 64, pp. 356–373.

Börner, J., Denich, M., Mendoza-Escalante, A., Hedden-Dunkhorst, B. and Abreu, Sá T. (2007) 'Alternatives to slash-and-burn in forest-based fallow systems of the eastern Brazilian Amazon region: Technology and policy options to halt ecological degradation and improve rural welfare', in Zeller, M. *et al.*, (eds) *Stability of Tropical Rainforest Margins*. Springer: Berlin-Heidelberg, pp. 333–361.

Börner, J. (2009) 'Serviços ambientais e adoção de sistemas agroflorestais na Amazônia: Elementos metodológicos para análises econômicas integradas', in Porro, R. (ed.) *Alternativa agroflorestal na Amazônia em Transformação*. Embrapa Informação Tecnológica: Brasilia, Brasil, pp. 411–434.

Börner, J., Wunder, S., Wertz-Kanounnikoff, S., Tito, M.R., Pereira, L. and Nascimento, N. (2010) Direct conservation payments in the Brazilian Amazon: Scope and equity implications, *Ecological Economics*, vol. 69, pp. 1272–1282.

Bustamante, M. and Albán, M. (2010) 'Evaluación del Programa Socio Bosque 2008–2009'. Secretaría Nacional de Planificación y Desarrollo, Quito, Ecuador.

Carpentier, C.L., Vosti, S.A. and Witcover, J. (2000) Intensified production systems on western Brazilian Amazon settlement farms: Could they save the forest? *Agriculture, Ecosystems & Environment*, vol. 82 no. 1–3 pp. 73–88.

Carpentier, C.L., Vosti, S. and Witcover, J. (2002) 'Small-scale farms in the western Brazilian Amazon: Can they benefit from carbon trade?', in EPTD discussion papers 2002. International Food Policy Research Institute (IFPRI): Washington, DC.

Cerri, C.C., Bernoux, M., Maia, S.M.F., Cerri, C.E.P., Costa Junior, C., Feigl, B.J., Frazao, L.A., Mello, F.F.d.C., Galdos, M.V., Moreira, C.S. and Carvalho, J.L.N. (2010) Greenhouse gas mitigation options in Brazil for land-use change, livestock and agriculture, *Scientia Agricola*, vol. 67, no. 1, pp. 102–116.

Davidson, E.A., Abreu Sá, T.D.d., Carvalho, C.J.R., Oliveira Figueiredo, R.D., Kato, M.d.S.A., Kato O.R. and Ishida, F.Y. (2008) 'An integrated greenhouse gas assessment of an alternative to slash-and-burn agriculture in eastern Amazonia', pp. 998–1007.

Fearnside, P.M. (1999) Forests and global warming mitigation in Brazil: Opportunities in the Brazilian forest sector for responses to global warming under the 'clean development mechanism', *Biomass and Bioenergy*, vol. 16. no. 3, pp. 171–189.

Fearnside, P.M. (1995) Global warming response options in Brazil's forest sector: Comparison of project-level costs and benefits, *Biomass and Bioenergy*, vol. 8, no. 5, pp. 309–322.

Hoch, L., Pokorny, B. and Jong, W.D. (2009) How successful is tree growing for smallholders in the Amazon? *International Forestry Review*, vol. 11, pp. 299–310.

Jackson, R.B., Jobbagy, E.G., Avissar, R., Roy, S.B., Barrett, D.J., Cook, C.W., Farley, K.A., le Maitre, D.C., McCarl, B.A. and Murray, B.C. (2005) Trading water for carbon with biological carbon sequestration, *Science*, vol. 310, no. 5756 pp. 1944–1947.

Kato, M.S.A., Kato, O.R., Denich, M. and Vlek, P.L.G. (1999) Fire-free alternatives to slash-and-burn for shifting cultivation in the eastern Amazon region: The role of fertilizers, *Field Crops Research*, vol. 62, no. 2–3, pp. 225–237.

Kossoy, A. and Ambrosi, P. (2010) 'State and trends of the carbon market 2010'. The World Bank, Washington, DC.

Mattos, L., Faleiro, A. and Perreira, C., (2001) 'Uma Proposta Alternativa para o Desenvolvimento da Produção Familiar Rural da Amazônia: O Caso do Proambiente', in IV Encontro Nacional da ECOECO – 2001. Belém (PA), Brazil.

Metzger, J.P. (2002) Landscape dynamics and equilibrium in areas of slash-and-burn agriculture with short and long fallow period (Bragantina region, NE Brazilian Amazon), *Landscape Ecology*, vol. 17, no. 5, pp. 419–431.

Michaelowa, A. and Jotzo, F. (2005) Transaction costs, institutional rigidities and the size of the clean development mechanism, *Energy Policy*, vol. 33, no. 4, pp. 511–523.

Niles, J.O., Brown, S., Pretty, J., Ball, A.S. and Fay, J. (2002) Potential carbon mitigation and income in developing countries from changes in use and management of agricultural and forest lands, *Philosophical Transactions of the Royal Society of London. Series A: Mathematical, Physical and Engineering Sciences*, vol. 360 (1797), pp. 1621–1639.

Robertson, N. and Wunder, S. (2005) 'Fresh Tracks in the Forest: Assessing Incipient Payments for Environmental Services Initiatives in Bolivia', CIFOR, Bogor, Indonesia.

Rogers, E.M. (2003) 'Diffusion of innovations', 5 ed. Free Press, New York, NY.

Saatchi, S.S., Houghton, R.A., Dos Santos, A.R.C., Soares, J.V. and Yu, Y. (2007) Distribution of aboveground live biomass in the Amazon basin, *Global Change Biology*, vol. 13, no. 4, pp. 816–837.

Schroth, G., D'Angelo, S.A., Teixeira, W.G., Haag, D. and Lieberei, R. (2002) Conversion of secondary forest into agroforestry and monoculture plantations in Amazonia: Consequences for biomass, litter and soil carbon stocks after 7 years, *Forest Ecology and Management*, vol. 163(1–3), pp. 131–150.

Sohngen, B. and Brown, S. (2004) Measuring leakage from carbon projects in open economies: A stop timber harvesting project in Bolivia as a case study, *Canadian Journal of Forest Research*, vol. 34, no. 4, pp. 829–839.

Vosti, S., Witcover, J., Oliveira, S. and Faminow, M. (1997) Policy issues in agroforestry: Technology adoption and regional integration in the western Brazilian Amazon, *Agroforestry Systems*, vol. 38: 195–222.

Vosti, S.A., Witcover, J., Carpenter, C.L., Oliveira, S.J.M. and de Santos, J.C. dos (2001) 'Intensifying small-scale agriculture in the western Brazilian Amazon: Issues, implications, and implementation', in *Tradeoffs or synergies?: Agricultural intensification, economic development and the environment*, Lee, D.R. and Barret, C. B. (eds), pp. 245–266. Oxon, England.

Vosti, S.A., Witcover, J., and Carpentier, C.L. (2002) 'Agricultural intensification by smallholders in the Western Amazon: From deforestation to sustainable land use', in *Research Report 130*. IFPRI, (ed.), Washington, DC.

# INDEX

Page numbers in *italics* denote tables, those in **bold** denote figures.